T0315243

TIME-TRIGGERED COMMUNICATION

Embedded Systems

Series Editor

Richard Zurawski

SA Corporation, San Francisco, California, USA

TIME-TRIGGERED COMMUNICATION

Edited by
ROMAN OBERMAISSER

CRC Press
Taylor & Francis Group
Boca Raton London New York

CRC Press is an imprint of the
Taylor & Francis Group, an **informa** business

CRC Press
Taylor & Francis Group
6000 Broken Sound Parkway NW, Suite 300
Boca Raton, FL 33487-2742

© 2012 by Taylor & Francis Group, LLC
CRC Press is an imprint of Taylor & Francis Group, an Informa business

No claim to original U.S. Government works

ISBN-13: 978-1-4398-4661-2 (hbk)

Visit the Taylor & Francis Web site at
http://www.taylorandfrancis.com

and the CRC Press Web site at
http://www.crcpress.com

Contents

4   Core Algorithms   53

M. Paulitsch, W. Steiner, R. Obermaisser and C. El Salloum viii

8 Time-Triggered Ethernet   **181**
W. Steiner, G. Bauer, B. Hall and M. Paulitsch

List of Figures

List of Tables

Editor

Roman Obermaisser is a full professor for embedded systems at the Department of Electrical Engineering and Computer Science of the University of Siegen in Germany. He studied computer sciences at Vienna University of Technology and received his master's degree in 2001. In February 2004, Professor Obermaisser finished his doctoral studies in computer science at Vienna University of Technology with Professor Hermann Kopetz as research advisor. In July 2009, he received the habilitation ("Venia docendi") certificate for Technical Computer Science. He is the author of numerous journal papers, books and conference publications.

Professor Obermaisser has participated in European research projects (e.g., universAAL, DECOS, NextTTA, INDEXYS) and was the technical coordinator of the FP7 research projects GENESYS (GENeric Embedded SYStem Platform) and ACROSS (Artemis Cross-Domain Architecture). He was also a member of the working groups "reference designs/architectures" and "middleware/seamless connectivity" in the European technology platform ARTEMIS, where a roadmap for European research in the area of embedded systems was defined. His leading role in the scientific community is shown through his chairing of and participation in many program committees (e.g., chair of the program committee of the IEEE Symposiums for Object and Component-Oriented Real-Time Distributed Computing, chair of the IEEE Workshop on Architectures and Applications for Mixed-Criticality Systems, chair of the IFIP Workshops on Software Technologies for Future Embedded and Ubiquitous Systems).

Professor Obermaisser's research focuses on system architectures, which provide the scientific and engineering foundation for the construction of embedded systems. The goals of his research are to discover design principles and to develop platform services that enable a component-based development of embedded systems in such a way that the ensuing systems can be built cost-effectively and exhibit key nonfunctional properties (e.g., dependability, timeliness, composability, maintainability). His investigations have resulted in contributions ranging from conceptual models of component-based system architectures to model-based development solutions and distributed algorithms for fault-tolerance and embedded operating system technologies for safety-relevant applications.

Contributors

Günther Bauer
TTTech Computertechnik AG
Vienna, Austria

Kenan Bilic
Vienna University of Technology
Vienna, Austria

Kevin Driscoll
Honeywell International Inc.
Maple Grove, MN

Christian El Salloum
Vienna University of Technology
Vienna, Austria

Petru Eles
Linkoping University
Linköping, Sweden

Wilfried Elmenreich
Alpen-Adria-Universität Klagenfurt
Klagenfurt, Austria

Alois Goller
TTTech Computertechnik AG
Vienna, Austria

Brendan Hall
Honeywell International Inc.
Eden Prairie, MN

Roland Kammerer
Vienna University of Technology
Vienna, Austria

Heinz Kantz
Thales Austria GmbH
Vienna, Austria

Hermann Kopetz
Vienna University of Technology
Vienna, Austria

Roman Obermaisser
University of Siegen
Siegen, Germany

Michael Paulitsch
EADS Innovation Works
Munich, Germany

Paul Pop
Technical University of Denmark
Kongens Lyngby, Denmark

Traian Pop
Ericsson AB
Linköping, Sweden

Christoph Scherrer
Thales Austria GmbH
Vienna, Austria

Eric Schmidt
TTTech Automotive GmbH
Vienna, Austria

Wilfried Steiner
TTTech Computertechnik AG
Vienna, Austria

Contributors

1

Introduction

R. Obermaisser

University of Siegen

CONTENTS

Embedded computers are by far the most common type of computer in use to-day. Ninety-eight percent of all computing devices are embedded in different kinds of electronic equipment such as automotive, industrial automation, telecommunications, consumer electronics and health/medical systems. Due to the many different and, partially, contradicting requirements, there exists no single model for building embedded systems. Well-known tradeoffs are predictability versus flexibility or resource adequacy versus best-effort strategies. Therefore, the chosen system model depends strongly on the requirements of the application.

For example, in safety-critical control applications such as by-wire systems in the avionic and automotive industries, a system's inability to provide its specified services can lead to a catastrophe endangering human lives. Failure rates in the order of 10^{-9} failures/hour are demanded in these systems, which are called ultra-dependable systems [349]. Since today's technology does not support the manufacturing of electronic components with failure rates low enough to meet these reliability requirements, ultra-dependable systems can only be built by utilizing fault-tolerant strategies that enable the continued operation of the system in the presence of component failures [48]. Ultra-dependable real-time systems must be designed according to the resource adequacy policy by providing sufficient communication and computational resources to handle the worst-case load and fault scenarios.

At present, two types of communication networks can be distinguished. The communication activities in *event-triggered networks* are triggered by the occurrence of significant events in the environment or the computer system. For example, a node computer requests the transmission of a message whenever an interrupt arrives from a sensor. In *time-triggered networks*, on the other hand, communication activities are controlled by the progression of a global time base. Each correct node sends messages with predefined periods and phases regardless of the events occurring within the node and in the environment. The difference between event-triggered and time-

1

triggered networks is the location of control. While event-triggered networks react to stimuli as they occur, time-triggered networks provide autonomous temporal control based on a statically computed communication schedule.

While event-triggered and time-triggered models exhibit duality in the sense that each of them is sufficiently expressive to model a system [331], the two communication approaches result in significant differences of non functional properties [165, 293]. Time-triggered networks are beneficial in safety-critical systems, because they help in managing the complexity of fault-tolerance and analytical dependability models. The static schedule of a time-triggered system maximizes predictability, while the schedule in an event-triggered network unfolds dynamically at runtime depending on the occurrence of events. In a time-triggered network, the predetermined instants of the periodic message exchanges enable rigorous error detection and fault isolation. Redundancy can be provided transparently to applications without modifications in the function and timing of the application software [25]. Time-triggered systems also support replica determinism, which is essential in establishing fault tolerance through active redundancy [258]. Furthermore, time-triggered systems support temporal composability via a precise specification of the interfaces between subsystems.

1.1 Scope of the Book

The *scope* of the book includes the conceptual foundations and fundamental principles of time-triggered communication, prevalent members of time-triggered communication protocols, industrial applications and development tools.

The *conceptual foundation* covers key concepts, properties and algorithms of time-triggered communication. This knowledge allows us to understand the different time-triggered communication protocols, as well as their differences and commonalities.

The implementation of the time-triggered concepts is discussed in-depth for specific *communication protocols*. The described protocols range from low-cost time-triggered field-bus networks to time-triggered networks for safety-critical applications.

Time-triggered communication protocols are widely used in today's embedded systems. Therefore, the book explains typical *industrial applications* along with the rationale and benefits of the time-triggered paradigm in these systems. The book provides information about the use of FlexRay in cars, TTP in railway and avionic systems and TTEthernet in aerospace applications.

An important aspect in deploying a time-triggered communication system is the generation, optimization and verification of time-triggered communication schedules. Therefore, the book presents requirements and algorithms of *development tools* for time-triggered communication networks. These concepts are also illustrated based on commercially available tool chains.

1.2   Structure of the Book

The chapters of the book are as follows:

Chapter 2 presents the underlying concepts and principles of time-triggered communication. Key concepts such as the global time base, the autonomous control of the communication system and the temporal firewall interface are introduced.

Chapter 3 is devoted to the properties of a time-triggered communication system. The strengths of the time-triggered paradigm (e.g., composability, determinism, diagnosability, fault containment) are contrasted with weak points (e.g., flexibility, average performance). This discussion enables an application developer to decide when a time-triggered communication protocol is most suitable.

The core algorithms, which can be found in many time-triggered communication protocols, are the focus of Chapter 4. In particular, algorithms for clock synchronization, startup, membership and fault isolation are addressed.

Chapters 5 to 13 describe widely used time-triggered communication protocols, which incorporate these algorithms. The protocols TTP, FlexRay, SAFEbus, TTEthernet, ROBUS, TTCAN, LIN, TTP/A, BRAIN and ASCB realize the concepts of time-triggered communication for specific domains and corresponding requirements (e.g., backward compatibility to legacy systems, reliability requirements).

Industrial applications of time-triggered communication are the focus of Chapter 14. Applications and platforms for the aerospace, automotive and railway domains are detailed.

Chapter 15 describes tooling requirements and system integration aspects and potential solutions and approaches. One important integration aspect is the generation of a schedule for a time-triggered system.

2

Basic Concepts and Principles of Time-Triggered Communication

R. Obermaisser

University of Siegen

H. Kopetz

Vienna University of Technology

CONTENTS

2.1   Introduction

This introductory chapter describes the basic concepts used throughout the book. The first section introduces the *structural elements* of an embedded computer system with a time-triggered communication network. Thereby, a terminology is established for describing time-triggered protocols and ensuing systems.

The following section addresses principles of *dependability* with a focus on fault containment and the role of a fault hypothesis. The fault-tolerance mechanisms of time-triggered communication protocols build on these concepts in order to mask different types of node and network failures.

The following section describes the notions of *state and time*. A concise model of time is required for a time-triggered communication protocol, which is autonomous in the sense that all activities are controlled by the progression of a global time base.

This *autonomous temporal control* is the topic of the final section, which also elaborates on the implications for flow control, supported information semantics and synchronization.

2.2   System Structure

A real-time application can be decomposed into an *embedded computer system* and a *controlled object*. The embedded computer system interacts with the controlled object using sensors and actuators, each of which performs a transformation between a physical variable in the controlled object (e.g., temperature, acceleration, speed) and digital information in the embedded computer system. Such a real-time application is also called a cyber-physical system [193] in order to reflect the integration of computation and physical processes.

In any given application, the purpose of the embedded computer system is defined by the requirements at the interface to the controlled object. For example, in a control application the embedded computer system is required to monitor physical variables using sensors, compute set-points and perform outputs via actuators in order to keep the actual value of the physical variable close to an intended value. In an alarm monitoring application, abnormal conditions are detected by observing significant variables via sensors.

The scope of this book is distributed embedded computer systems, which are realized using time-triggered communication networks. Such a distributed embedded system contains a set of node computers (*nodes* for short). Each node is a self-contained composite hardware/software subsystem, which interacts with the other nodes using the time-triggered communication network.

A node is internally structured into a *host computer* and a *time-triggered communication controller* (see Figure 2.1). The time-triggered communication controller runs the communication protocol, while the host computer executes the operating

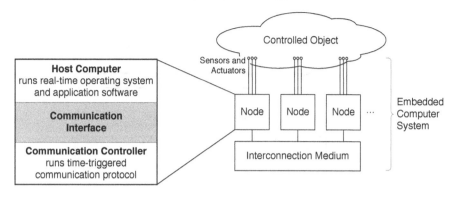

FIGURE 2.1
Embedded Computer System and Internal Structure of a Node

system and the application software. The communication controllers of all nodes and the interconnection medium comprise the *time-triggered communication network*. The set of nodes in conjunction with the time-triggered communication network is called a *cluster*. Clusters can be interconnected by nodes that are part of more than one cluster. Such a node is denoted as a *gateway* and serves for the construction of multi-cluster systems.

The borderline between the host and the communication controller is called the *communication interface*. This interface is of central importance for developing applications based on a time-triggered communication network. At this interface the services of the time-triggered communication network are provided, such as the exchange of messages and the access to the global time base (cf. Chapter 4).

The services of the time-triggered communication network provide the basis for the implementation of a node's *application services*. The application services are provided to other nodes or to the controlled object. Examples in an automotive context would be a braking actuation service provided by a node to the controlled object or a diagnostic node providing a persistent memory service for the storage of break-down logs from other nodes.

Conceptually, an application service is the intended behavior of a node according to the specification. The application service of a node at the time-triggered communication network is the sequence of intended messages that is produced by a node in response to the progression of time, input and state [109, page 28].

For the time-triggered communication network, different topologies can be distinguished such as bus, star and ring topologies. Independently of the topology, different redundancy degrees of the communication network are possible. A single communication channel is typically used in non safety-critical applications. In safety-critical systems, redundant communication channels support the masking of channel failures.

Figure 2.2 depicts examples of specific topologies with two redundant communication channels. The simplest topology is a bus, where for each communication

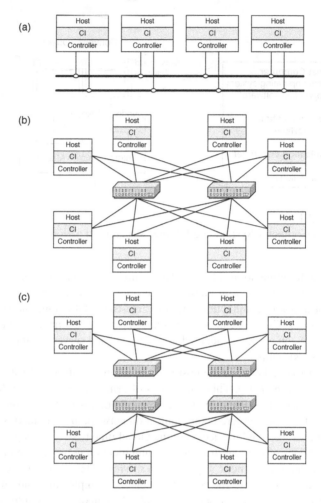

FIGURE 2.2
Bus Topology (a), Star Topology (b) and Cascaded Star (c)

channel every node is connected to a single cable. A terminator is required at each end of the bus to prevent signal reflections. Since the bus topology consists of only one cable for a channel, it is inexpensive compared to other topologies. This topology inherently supports broadcasting, because all nodes can receive a message transmitted on the bus within the propagation delay. The propagation delay is the time for a bit to travel from one end of the bus to its other end. It can be computed as the ratio between the wire length and the propagation speed (e.g., 2/3 of the speed of light in copper wires).

In a star topology, every node is connected to a central device with a point-to-point connection. A message sent by a node passes through this central device, which redirects the message to a selected set of receiver nodes. The possibility for multicasting a message only to a subset of the nodes enables a more efficient use of communication bandwidth, because different messages can be exchanged between subgroups of nodes at the same time. The broadcast communication of a bus topology is supported as a special use-case of multicasting.

Another benefit of a star topology is improved fault isolation for node failures (e.g., for slightly-off-specification failures or during start-up [104]). The main disadvantages are the higher cost of installation and wiring. Also, since the central device represents a single-point-of-failure, redundant stars are essential in safety-critical systems.

In order to improve scalability, the cascading of star topologies can be performed. Thereby, the physical dimensions of the distributed system can be extended and it is not necessary to connect all nodes to a single device.

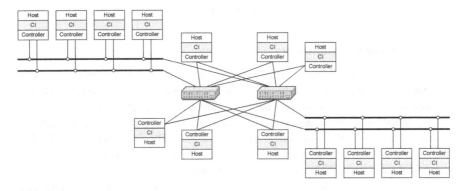

FIGURE 2.3
Combination of Star Topology and Bus Topology

Topologies can also be combined such as the integration of bus and star topologies in Figure 2.3. Such a hybrid topology has the benefit of improved fault isolation between subsystems arranged in the star topology, while reducing cost of the wiring harness for the network within such subsystems.

2.3 Concepts of Dependability

Dependability is the ability of a computing system to deliver services that can justifiably be trusted [52]. The service delivered by a system is its behavior as it is perceptible by another system (human or physical) interacting with the former [21].

2.3.1   Dependability Threats – Failure, Error, Fault

A *failure* occurs when the delivered service deviates from fulfilling the functional specification. An *error* is that part of the system state which is liable to lead to a subsequent failure. A failure occurs when the error reaches the service interface. A *fault* is the adjudged or hypothesized cause of an error. As stated in [21], the concept of fault is introduced to stop recursion.

It is important to discriminate faults based on the fault boundaries. An internal fault of a node can be a physical fault (e.g., a short circuit, wire break), a design fault in the software (a program error) or a design fault in the hardware (an erratum). An external fault can be a physical disturbance (e.g., electro magnetic interference) or the provision of incorrect input data.

Due to the recursive definition of systems, a failure at a particular level of decomposition can be interpreted as a fault at the next upper level of decomposition, thereby leading to a hierarchical causal chain.

2.3.2   Fault Containment

A *Fault Containment Region (FCR)* is defined as a subsystem that operates correctly regardless of any arbitrary logical or electrical fault outside the region [187]. The justification for building ultra-reliable systems from replicated resources rests on an assumption of failure independence among redundant units. For this reason the independence of FCRs is of critical importance [48]. The independence of FCRs can be compromised by shared physical resources (e.g., power supply, timing source), external faults (e.g., EMI, spatial proximity) and design.

A node of an embedded computer system or a communication channel can be considered as a FCR in a properly designed system where aspects such as physical separation between nodes, separate timing sources, computing hardware and power supplies have been considered.

2.3.3   Failure Modes

Failure modes of FCRs are defined through the effects as perceived by the service user, i.e., independently of the actual cause or rate of failures. A formal definition in terms of assertions on the sequences of value-time tuples can be found in [268]. Failure modes determine the degree of redundancy required to ensure correct error processing. Based on the rigidity of assumptions, the following hierarchy of failure modes can be established [69]:

- **Fail-Stop Failures:** A fail-stop failure is defined as a node behavior, where the node does not produce any outputs. The node omits to produce output to subsequent inputs until it restarts. It is additionally assumed that all correct nodes detect the fail-stop failure.

- **Crash Failures:** A node with a crash failure does not produce any outputs. In contrast to fail-stop failures, a crash failure can remain undetected for correct nodes.

- **Omission Failures:** An omission failure occurs if the sender node fails to send a message or the receiver fails to receive a sent message. As a consequence, the receiver does not respond to an input. The detection of an omission failure is not guaranteed.

- **Timing Failures:** The node does not meet its temporal specification. Outputs of a node are delivered too early or too late.

- **Byzantine or Arbitrary Failures:** There is no restriction on the effects a service user may perceive. Arbitrary failures include the forging of messages and "two-faced" node behaviors [190].

A different classification of failure modes can be found in [172] and distinguishes the following additional types of failure modes:

- **Babbling Idiot:** In case of a babbling idiot failure, the FCR does not obey its temporal specification by sending untimely messages. For example, a node constantly sends messages and would monopolize the network without fault-containment mechanisms.

- **Slightly-off-Specification (SoS):** Such a failure is a special type of Byzantine failure. One can distinguish temporal and value SoS failures. An example for a value SoS failure is an intermediate electrical voltage that is close to the threshold between logical 0 and logical 1 and can be perceived with different logical values by different observers. An example for a temporal SoS failure is a message with a receive instant that is slightly outside the boundary of the interval of correct receive instants. In such a case, due to the inability to perfectly synchronize clocks one node can classify the message as timely, whereas another node may detect a message timing failure.

- **Masquerading:** Masquerading is defined as the sending or receiving of messages using the identity of another principal without authority [66, p. 480].

2.3.4 Fault Hypothesis

The fault hypothesis specifies assumptions about the types of faults, the rate at which nodes fail and how nodes may fail [237]. The fault hypothesis is a central part in any safety-relevant system and provides the foundation for the design, implementation and test of the fault-tolerance mechanisms.

For a time-triggered communication protocol, the fault hypothesis partitions the fault space into two sets: a set of faults that is tolerated by the fault-tolerance mechanisms of the communication protocol and a set of faults outside the fault hypothesis. The latter faults, which must be rare events, can be addressed (without guaranteed masking) by a never-give-up strategy [174]. In addition, the application can be informed about a scenario outside the fault hypothesis in order to perform an appropriate reaction at the application-level (e.g., emergency shutdown in a fail-stop application, algorithmic redundancy [110]).

The assumption coverage is the probability that these assumptions hold in reality. Since fault-tolerance mechanisms of a system are based on these assumptions, the complete system may fail in case the assumptions concerning faults, failure rates, and failure modes are violated.

The fault hypothesis defines the fault containment regions and the failure mode assumptions. In addition, the fault hypothesis contains failure rate assumptions. These assumptions typically perform a differentiation of failure rates with respect to different failure modes and the failure persistence. For example, fault injection experiments [160] have shown that restrictive failure modes, such as omission failures, are more frequent by a factor of 50 compared to arbitrary failures. Also, transient failures are more likely than permanent failures by at least a factor of 1000.

Another part of the fault hypothesis are assumptions concerning the maximum number of faults. This parameter denotes the maximum number of FCR failures, which must be handled by the system. The maximum number of failures depends on the failure rate and the recovery interval of FCRs.

2.4 Global Time and State

In many engineering models (e.g., Newtonian mechanics) of physical phenomena, time is introduced as an independent variable that determines the sequence of states of a system. The basic constants of physics are defined in relation to the standard of time, the physical second. This is why the time base in a cyber-physical real-time system should be based on the metric of the physical second.

In a typical real-time application, the distributed computer system performs a multitude of different functions concurrently, e.g., the monitoring state variables in the environment (both their value and rate of change), the detection of alarm conditions, the display of the observations to the operator and the execution of control algorithms to find new set-points. These different functions are normally executed at different nodes. In addition, replicated nodes are introduced to provide fault tolerance by active redundancy. To guarantee a consistent behavior of the entire distributed system, it must be ensured that all nodes process all events in the same consistent order, preferably in the same temporal order in which the events occurred in the controlled object. A proper global time base helps to establish such a consistent temporal order on the basis of the timestamps of the events.

2.4.1 Time and Clocks

In ancient history, the measurement of durations between events was mainly based on subjective judgment. With the advent of modern science, objective methods for measuring the progression of time by using *physical clocks* have been devised.

A *(digital physical) clock* is a device for measuring time. It contains a *counter* and a *physical oscillation mechanism* that periodically generates an event to increase the counter. The periodic event is called the *microtick* of the clock. (The term *tick* is introduced later to denote the events generated by the global time.)

The duration between two consecutive microticks of a digital physical clock is called a *granule* of the clock. The granularity of a given clock can be measured only if there is a clock with a finer granularity available. The granularity of any digital clock leads to a digitalization error in time measurement. There exist also analog physical clocks, e.g., the *sundial*, that do not have a granularity. In the following, we only consider digital physical clocks.

In subsequent definitions, we use the following notation: clocks are identified by natural numbers $1, 2, \ldots, n$. If we express properties of clocks, the property is identified by the clock number as a superscript with the microtick or tick number as a subscript. For example, microtick i of clock k is denoted by $microtick_i^k$.

Let us assume an *omniscient external observer* who can observe all events that are of interest in a given context. This observer possesses a *unique reference clock z* with frequency f^z which is in perfect agreement with the international standard of time. The counter of the reference clock is always the same as that of the international time standard. We call $1/f^z$ the *granularity* g^z of clock z. Let us assume that f^z is very large, say 10^{15} microticks/second, so that the granularity g^z is 1 femtosecond (10^{-15} seconds). Since the granularity of the reference clock is so small, the digitalization error of the reference clock is considered a second order effect and disregarded in the following analysis.

Whenever the omniscient observer perceives the occurrence of an event e, she/he will instantaneously record the current state of the reference clock as the time of occurrence of this event e, and, will generate a *timestamp* for e. *Clock(event)* denotes the timestamp generated by the use of a given *clock* to timestamp an *event*. Because z is the single reference clock in the system, $z(e)$ is called the absolute timestamp of the event e.

The *duration* between two events is measured by counting the microticks of the reference clock that occur in the interval between these two events. The *granularity* g^k of a given clock k can now be measured and is given by the nominal number n^k of microticks of the reference clock z between two microticks of this clock k.

The temporal order of events that occur between any two consecutive microticks of the reference clock, i.e., within the granularity g^z, cannot be reestablished from their absolute timestamps. This is a fundamental limit in time measurement.

The *drift* of a physical clock k between microtick i and microtick $i+1$ is the frequency ratio between this clock k and the reference clock, at the instant of microtick i. The drift is determined by measuring the duration of a granule of clock k with the reference clock z and dividing it by the nominal number n^k of reference clock microticks in a granule:

$$\text{drift}_i^k = \frac{z(microtick_{i+1}^k - microtick_i^k)}{n^k} \qquad (2.1)$$

Because a good clock has a drift that is very close to 1, for notational convenience the notion of a *drift rate* ρ_i^k is introduced as

$$\rho_i^k = \left| \frac{z(microtick_{i+1}^k - microtick_i^k)}{n^k} - 1 \right| \qquad (2.2)$$

A perfect clock will have a drift rate of 0. Real clocks have a varying drift rate that is influenced by environmental conditions, e.g., a change in the ambient temperature, a change in the voltage level that is applied to a crystal oscillator, or aging of the crystal. Within specified environmental parameters, the drift rate of an oscillator is bounded by the *maximum drift rate* ρ_{max}^k which is documented in the data sheet of the resonator. Typical maximum drift rates ρ_{max}^k are in the range of 10^{-2} to 10^{-7} sec/sec, or better, depending on the quality (and price) of the oscillator. Because every clock has a non-zero drift rate, *free-running* clocks, i.e., clocks that are never resynchronized, leave any bounded relative time interval after a finite time, even if they are fully synchronized at startup.

A physical digital clock can exhibit two types of failures. The counter could be mutilated by a fault so that the counter value becomes erroneous, or the drift rate of the clock could depart from the specified drift rate (the shaded area of Figure 2.4) because the clock starts ticking faster (or slower) than specified.

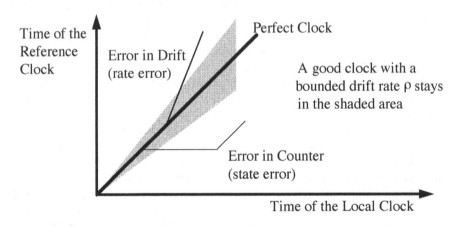

FIGURE 2.4
Failure Modes of a Physical Clock

2.4.2 Precision and Accuracy

The *offset* at microtick i between two clocks j and k of an ensemble of clocks with the same granularity is defined as

$$\text{offset}_i^{jk} = \left| z(microtick_i^j) - z(microtick_i^k) \right| \tag{2.3}$$

The offset denotes the time difference between the respective microticks of the two clocks, measured in the number of microticks of the reference clock.

Given an ensemble of n clocks $1, 2, \ldots, n$, the maximum offset between any two clocks of the ensemble

$$\Pi_i = \max_{\forall 1 \leqslant j, k \leqslant n,} \{offset_i^{jk}\} \tag{2.4}$$

is called the *precision* Π_i of the ensemble at microtick i. The maximum of Π_i over an *interval of interest* is called the precision Π of the ensemble. The precision denotes the *maximum offset* of respective microticks of any two clocks of the ensemble during the period of interest. The precision is expressed in the number of microticks of the reference clock.

Because of the drift rate of any physical clock, the clocks of an ensemble will drift apart if they are not resynchronized periodically (i.e., brought closer together). The process of mutual resynchronization of an ensemble of clocks to maintain a bounded precision is called *internal synchronization*.

The offset of clock k with respect to the reference clock z at microtick i is called the *accuracy*$_i^k$. The maximum offset over all microticks i that are of interest is called the *accuracy*^{k} of clock k. The accuracy denotes the maximum offset of a given clock from an external time reference during the time interval of interest.

To keep a clock within a bounded interval of the reference clock, it must be periodically resynchronized with an external time reference. This process of resynchronization of a clock with an external time reference is called *external synchronization*.

If all clocks of an ensemble are externally synchronized with an accuracy A, then the ensemble is also internally synchronized with a precision of at most $2A$. The converse is not true. An ensemble of internally synchronized clocks will drift from the external time if the clocks are never resynchronized with the external time base.

In the last decades a number of different time standards have been proposed to measure the time difference between any two events and to establish the position of an event relative to some commonly agreed origin of a time base, the *epoch*. Two of these time bases are relevant for the designer of a distributed real-time computer system, the International Atomic Time (TAI) and the Universal Time Coordinated (UTC).

International Atomic Time (TAI-Temps Atomique International): The need for a time standard that can be generated in a laboratory gave birth to the International Atomic Time (TAI). TAI defines the second as the duration of 9 192 631 770 periods of the radiation of a specified transition of the cesium atom 133. The intention was to define the duration of the TAI second so that it agrees with the second derived from astronomical observations. TAI is a *chronoscopic* timescale, i.e., a timescale without any discontinuities (e.g., leap seconds). The epoch of TAI starts on January 1, 1958

00:00 hours Greenwich Mean Time (GMT). The time base of the global positioning system (GPS) is based on TAI with the epoch starting on January 6, 1980 at 00:00 hours.

Universal Time Coordinated (UTC): UTC is a time standard that has been derived from astronomical observations of the rotation of the earth relative to the sun. It is the basis for the time on the "wall-clock." However, there is a known offset between the local wall-clock time and UTC determined by the timezone and by the political decisions about when daylight savings time must be used. The UTC time standard was introduced in 1972, replacing the Greenwich Mean Time (GMT) as an international time standard. Because the rotation of the earth is not smooth, but slightly irregular, the duration of the GMT second changes slightly over time. In 1972, it was internationally agreed that the duration of the second should conform to the TAI standard, and that the number of seconds in an hour would have to be modified occasionally by inserting a *leap second* into the UTC to maintain synchrony between the UTC (wall-clock time) and astronomical phenomena, like day and night. Because of this leap second, the UTC is not a chronoscopic time scale, i.e., it is not free of discontinuities. It was agreed that on January 1, 1958 at midnight, both the UTC and the TAI had the same value. Since then the UTC has deviated from TAI by about 30 seconds. The point in time when a leap second is inserted into the UTC is determined by the Bureau International de l'Heure and publicly announced, so that the current offset between the UTC and the TAI is always known.

2.4.3   Global Time

If the real-time clocks of all nodes of a distributed system were perfectly synchronized with the reference clock z, and all events were timestamped with this reference time, then it would be easy to measure the interval between any two events or to reconstruct the temporal order of events, even if variable communication delays generated differing delivery orders. In a loosely coupled distributed system where every node has its own local oscillator, such a tight synchronization of clocks is not possible. A weaker notion of a universal time reference, the concept of *global time*, is therefore introduced into a distributed system. Suppose a set of nodes exists, each one with its own local physical clock c^k that ticks with granularity g^k. Assume that all of the clocks are internally synchronized with a precision Π, i.e., for any two clocks j, k and all microticks i

$$\left| z(microtick_i^j) - z(microtick_i^k) \right| < \Pi \qquad (2.5)$$

It is then possible to select a *subset of the microticks* of each local clock k for the generation of the local implementation of a global notion of time. We call such a selected local microtick i a *macrotick* (or a *tick*) of the global time. For example, every tenth microtick of a local clock k may be interpreted as the global tick, the macrotick i_i^k, of this clock. If it does not matter at which clock k the (macro)tick occurs, we denote the tick t_i without a superscript. A global time is thus an *abstract notion* that is *approximated* by properly selected microticks from the synchronized local physical clocks of an ensemble.

The global time t is called *reasonable*, if all local implementations of the global time satisfy the condition

$$g > \Pi \tag{2.6}$$

the *reasonableness condition* for the global granularity g. This reasonableness condition ensures that the synchronization error is *bounded* to less than one *macrogranule*, i.e., the duration between two (macro) ticks. If this reasonableness condition is satisfied, then for a single event e, that is observed by any two different clocks of the ensemble,

$$\left| t^j(e) - t^k(e) \right| \leqslant 1 \tag{2.7}$$

i.e., the global timestamps for a single event can differ by at most one tick. *This is the best we can achieve.* Because of the impossibility of synchronizing the clocks perfectly, and the granularity of any digital time, there is always the possibility of the following sequence of events: clock j ticks, event e occurs, clock k ticks. In such a situation, the single event e is timestamped by the two clocks j and k with a difference of one tick.

Provided we have established a *reasonable global time*, the following four conditions represent fundamental limits of time measurement in a distributed system:

1. If a single event is observed by two different nodes, there is always the possibility that the timestamps differ by one tick. A one-tick difference in the timestamps of two events is not sufficient to reestablish the temporal order of the events from their timestamps.

2. If the observed duration of an interval is d_{obs}, then the true duration d_{true} is bounded by

$$d_{obs} - 2g < d_{true} < d_{obs} + 2g \tag{2.8}$$

3. The temporal order of events can be recovered from their timestamps, if the difference between the measured timestamps is equal to or greater than 2 ticks.

4. The temporal order of events can *always* be recovered from their timestamps, if the events are at least 3 ticks apart.

These fundamental limits of time measurement are also the *fundamental limits to the faithfulness* of a digital model of a physical system.

2.4.4 Sparse Time

Assume a set E of events that are of interest in a particular context. This set E could be the ticks of all clocks, or the events of sending and receiving messages. If these events are allowed to occur at any instant of the timeline, we call the time base *dense*. If the occurrence of these events is restricted to some *active intervals* of duration ε with an interval of silence of duration Δ between any two active intervals, then we call the time base ε/Δ-sparse, or simply *sparse* for short (Figure 2.5). If a system is based on a sparse time base, there are time intervals during which no significant

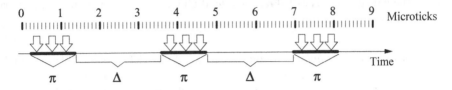

Events ⇩ are only allowed to occur within the intervals   π.

FIGURE 2.5
Sparse Time-Base

event is allowed to occur. Events that occur only in the active intervals are called *sparse events*.

It is evident that the occurrences of events can only be restricted if the given system has the authority to control these events, i.e., these events are in the sphere of control of the computer system [73]. The occurrence of events outside the sphere of control of the computer system cannot be restricted. These external events are based on a dense time base and cannot be forced to be sparse events.

Consider a distributed system that consists of two clusters: cluster A generates events, and cluster B observes these generated events. Each one of the clusters has its own cluster-wide synchronized time with a granularity g, but these two cluster-wide time bases are not synchronized with each other. Under what circumstances is it possible for the nodes in the observing cluster to reestablish the *intended temporal order* of the generated events without the need to execute an agreement protocol?

If two nodes, nodes j and k of cluster A, generate two events at the same cluster-wide tick t_i, i.e., at tick t_i^j and at tick t_i^k, then these two events can be a distance Π apart from each other, where $g > \Pi$, the granularity of the cluster-wide time. Because there is no intended temporal order among the events that are generated at the same cluster-wide tick of cluster A, the observing cluster B should *never* establish a temporal order among the events that have been sent at about the same time. On the other hand, the observing cluster B should *always* reestablish the temporal order of the events that have been sent at different cluster-wide ticks. Is it sufficient if cluster A generates a $1g/3g$ precedent event set, i.e., after every cluster-wide tick at which events are allowed to be generated there will be silence for at least three granules?

If cluster A generates a $1g/3g$ precedent event set, then it is possible that two events that are generated at the same cluster-wide granule at cluster A will be time-stamped by cluster B with timestamps that differ by 2 ticks. The observing cluster B should not order these events (although it could), because they have been generated at the same cluster-wide granule. Events that are generated by cluster A at different cluster-wide granules ($3g$ apart) and therefore should be ordered by cluster B, could also obtain timestamps that differ by 2 ticks. Cluster B cannot decide whether or not to order events with a timestamp difference of 2 ticks. To resolve this situation, cluster A must generate a $1g/4g$ precedent event set. Cluster B will not order two events if their timestamps differ by $\leqslant 2$ ticks, but will order two events if their timestamps differ by $\geqslant 3$ ticks, thus reestablishing the temporal order that has been intended by the sender.

To arrive at a *consistent view* of the order of *non-sparse events* within a distributed computer system (which does not necessarily reflect the temporal order of event occurrence), the nodes must execute an *agreement protocol*. The first phase of an agreement protocol requires an information interchange among the nodes of the distributed system with the goal that every node acquires the differing local views about the state of the world from every other node. At the end of this first phase, every node possesses exactly the same information as every other node. In the second phase of the agreement protocol, each node applies a deterministic algorithm to this consistent information to reach the same conclusion about the assignment of the event to an active interval of the sparse time base — the commonly agreed value. In the fault-free case, an agreement algorithm requires an additional round of information exchange as well as the resources for executing the agreement algorithm.

Agreement algorithms are costly, both in terms of communication requirements, processing requirements, and — worst of all — in terms of the additional delay they introduce into a control loop. It is therefore expedient to look for solutions to the consistent temporal ordering problem in distributed computer systems that do not require these additional overheads. The sparse time model, introduced above, provides for such a solution.

Many processes in the technical and biological world are cyclic [354]. A cyclic process is characterized by a regular behavior, where a similar set of action patterns is repeated in every cycle. In the *cyclic representation of time*, the linear time is partitioned into cycles of equal duration. Every cycle is represented by a circle, where an instant within a cycle is denoted by the phase, i.e., the angular deviation of the instant from the beginning of the cycle. Cycle and phase thus denote an instant in a cyclic representation. In the cyclic representation of sparse time, the circumference of the circle is not a dense line, but a dotted line, where the size and the distance between dots is determined by the precision of the clock synchronization.

An extension of the cyclic representation is the *spiral representation of time*, where a third axis is introduced to depict the linear progression of the cycles.

2.4.5   State of a System

The notion of *state* is widely used in the computer science literature, albeit with meanings that are different from the meaning of state that is useful in a real-time system context. In order to clarify the situation, we follow the precise definition of Mesarovic [220, p. 45] which is the basis for our elaborations:

The state enables the determination of a future output solely on the basis of the future input and the state the system is in. In other words, the state enables a "decoupling" of the past from the present and future. The state embodies all past history of a system. Knowing the state "supplants" knowledge of the past. . . . Apparently, for this role to be meaningful, the notion of past and future must be relevant for the system considered.

The sparse time model introduced above makes it possible to establish the consistent system-wide separation of the past from the future that is necessary to define a *consistent system state* in a distributed real-time computer system.

In order to enable the dynamic reintegration of a node into a running system, it is necessary to design periodic *reintegration instants* into the behavior, where the size of a node's state at the reintegration instant contained is a small set of *state variables*. We call the state at the reintegration instant the *ground state* of a node and the temporal distance between two reintegration points the *ground cycle*.

The ground state at the reintegration point is stored in a declared *ground-state data structure*. Designing a minimal ground state data structure is the result of an explicit design effort that involves a semantic analysis of the given application. The designer has to find a periodic instant where there is a maximum decoupling of future behavior from past behavior. This is relatively easy in *cyclic applications*, such as in control applications and multimedia applications. In these applications a natural reintegration instant is immediately after the termination of one cycle and before the beginning of the next cycle.

2.5 Autonomous Control of Communication Networks

A time-triggered communication network is designed for the periodic transmission of state information. It initiates all communication activities at predetermined global points in time. Hence, the temporal behavior of the communication network is controlled solely by the progression of time.

2.5.1 Types of Temporal Control Signals

A trigger is a control signal that initiates an action in the embedded computer system, like the execution of a task or the transmission of a message. Depending on the source from which a trigger is derived, one can distinguish event triggers and time triggers.

2.5.1.1 Event Triggers

An event trigger is a control signal that is derived from an event, i.e., a significant state change. The event can originate either from activities within the computer system (e.g., termination of a task) or from state changes in the controlled object (e.g., alarm condition indicated by a sensor). In the latter case, the event trigger serves as a mechanism by which the controlled object delivers a service request to the embedded computer system. In general, such a service request will start a sequence of computational and communication activities.

2.5.1.2 Time Triggers

A time trigger is a control signal that is generated at a particular point in time of a synchronized global time base. Time triggers are solely derived from the progression of the global time, which is established by a clock synchronization service (cf. Chapter 4, Section 4.2). The set of time triggers is a subset of the set of event triggers, since time triggers correspond to a particular class of events, namely changes in the state of the global time. Time triggers are discriminated from other event triggers, since systems restricting control signals to time triggers offer properties that are desirable for distributed real-time systems. Among these properties are temporal predictability and composability [168].

2.5.2 Information Semantics

A time-triggered communication network is designed for the periodic exchange of messages carrying state information. Information with state semantics contains the absolute value of a real-time entity (e.g., speed of the car is 41 km/h). Since applications are often only interested in the most recent value of a real-time entity, state information allows the communication network to overwrite old state values with newer state values.

Messages with state information are called *state messages*. The self-contained nature and idempotence of state messages ease the establishment of state synchronization, which does not depend on exactly-once processing guarantees. Since applications are often only interested in the most recent value of a real-time object, old state values can be overwritten with newer state values. Hence, a time-triggered communication network does not require message queues.

Information with event semantics relates to the occurrence of an event. Event information represents the change in value of a real-time entity associated with a particular event. Messages containing event information are called *event messages* and transport relative values (e.g., increase of the speed of the car by 2 km/h). In order to reconstruct the current state of a real-time entity from messages with event semantics, it is essential to process every message exactly once. The loss of a single message with event information can affect state synchronization between a sender and a receiver.

2.5.3 Temporal Firewall

As depicted in Figure 2.6, the communication interface of a time-triggered communication network acts as a temporal firewall [182]. The sender can deposit information into the communication interface according to the information push paradigm, while the receiver must pull information out of the communication interface. A time-triggered transport protocol autonomously carries the state information from the communication interface of the sender to the communication interface of the receivers. Since no control signals cross the communication interface, temporal fault propagation is prevented by design.

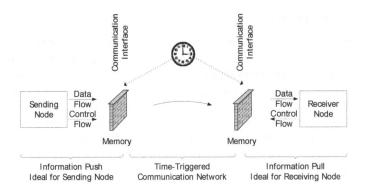

FIGURE 2.6
Data Flow (Full Line) and Control Flow (Dashed Line) of a Temporal Firewall Interface

The state messages in the communication interface memory form two groups. One group of messages is written by the host computer. The communication controller reads these messages and disseminates them during the slots reserved for the node via the underlying Time Division Multiple Access (TDMA) scheme. The messages of the second group are written by the communication controller and read by the host computer.

Consistency of information exchanged via the communication interface can be ensured by exploiting the a priori knowledge about the points in time when the communication network reads and writes data into the communication interface. The host computer performs *implicit synchronization* by establishing a phase alignment between its own communication interface accesses and the communication interface accesses of the communication controller. A different approach is the use of a protocol for *explicit synchronization*, such as the *Non-Blocking Write Protocol* [184].

2.5.4 Transport Protocols

The media access control strategy of a time-triggered communication network is TDMA. TDMA statically divides the channel capacity into a number of slots and assigns a unique slot to every node. The communication activities of every node are controlled by a time-triggered communication schedule. The schedule specifies the temporal pattern of message transmissions, i.e., at what points in time nodes send and receive messages. A sequence of sending slots, which allows every node in an ensemble of n nodes to send exactly once, is called a TDMA round. The sequence of the different TDMA rounds forms the cluster cycle and determines the periodicity of the time-triggered communication.

The a priori knowledge about the times of message exchanges enables the communication network to operate autonomously. The temporal control of communication activities is within the sphere of control of the communication network. Hence, the correct temporal behavior of the communication network is independent of temporal behavior of the application software in the host computer and can be established in isolation.

2.5.5   Flow Control

Time-triggered communication networks typically employ *implicit flow control* [169]. Sender and receiver agree a priori on the global points in time when messages are exchanged. Based on this knowledge, a component's ability for handling received messages can be ensured at design time, i.e., without acknowledgment messages. Implicit flow control is well-suited for multicast communication relationships, because a unidirectional data flow involves only a unidirectional control flow. Such an interface is called an *elementary interface* [170].

3

Properties of Time-Triggered Communication Systems

R. Obermaisser

University of Siegen

H. Kopetz

Vienna University of Technology

CONTENTS

3.1   Introduction

This chapter introduces fundamental properties of embedded systems based on time-triggered networks. The strengths of the time-triggered paradigm (e.g., composability, determinism, diagnosability, fault containment) are contrasted with weak points (e.g., average performance, flexibility). The discussion of these properties enables an application developer to decide when a time-triggered network is most suitable. In addition, the properties allow us to compare the different time-triggered protocols that will be introduced in the following chapters.

The chapter starts with the property of *composability*, which is a prerequisite of a component-based design framework. Composability enables the independent development of nodes and the smooth integration of these nodes using a time-triggered network. Composability rules out unintended interference between nodes and preserves the correctness of the nodes' services upon integration.

The following section addresses *determinism and predictability*. Determinism is important for the realization of fault-tolerance through active redundancy with exact voting. Determinism also facilitates the understanding and testing of complex embedded systems, because a causal chain between a cause and the consequent effect can be established.

The section on *diagnosability* analyzes the ability for monitoring the functionality and the performance of nodes. This information can be used for active diagnosis by taking actions at run-time such as the identification of faulty nodes and the execution of an online error recovery. In case of passive diagnosis, this information provides the basis for off-line maintenance decisions.

In safety-critical systems, *certifiability* is a key concern. In general, the design of a safety-critical real-time system must be approved by an independent certification agency in order to avoid danger to life, health, property or the environment. Time-triggered networks facilitate certification by their inherent determinism and predictability. In addition, time-triggered networks contribute to a constructive modular certification process where the certification of subsystems is done independently of each other.

The following section explains *fault containment and error containment* with an emphasis on time-triggered networks. Fault containment ensures that the immediate consequences of a fault are limited to a single node. Error containment prevents the propagation of an error through messages with incorrect values or timing.

Depending on the application, time-triggered networks need to satisfy given *performance attributes* (e.g., latency, variability of the latency, bandwidth) for different traffic classes such as periodic and sporadic messages.

The chapter closes with a view on *flexibility*. Many embedded systems must adapt to technological changes and new environmental contexts. The section about flexibility describes the ability to support different system configurations that evolve over time. A time-triggered network should accommodate changes without requiring a modification and retesting of the existing nodes that are not affected by the change.

3.2 Composability

In many engineering disciplines, large systems are built from prefabricated components with known and validated properties. Components are connected via stable, understandable and standardized interfaces. The system engineer has knowledge about the global properties of the components as they relate to the system functions and of the detailed specification of the component interfaces. Knowledge about the internal design and implementation of the components is neither needed nor available in many cases. *Composability* deals with all issues that relate to the component-based design of large systems.

3.2.1 Component-Based Design

Component-based design is a *meet-in-the-middle* design method. On the one side, the *functional and temporal requirements* on the components are derived top-down from the desired application functions. On the other side, the *functional and temporal capabilities* of the components are contained in the specifications of the available components (bottom-up). During the design process a proper match between component requirements and component capabilities must be established. If there is no component available that meets the requirements, a new component must be developed.

A prerequisite of any component-based design is a crystal-clear component concept that supports the precise specification of the services that are delivered and acquired across the component interfaces. In many non-real-time applications, a software unit is considered to form a component. In real-time systems, where the temporal properties of components are as important as the value properties, the notion of a software component is of questionable utility, since no temporal capabilities can be assigned to software without associating the software with a virtual or real machine. The specification of the temporal properties of the Application Programming Interface (API) between a software component and the machine is so involved that a simple specification of the temporal properties of the API is hardly possible. If the mental effort needed to understand the specification of the component interfaces is in the same order of magnitude as the effort needed to understand the internals of the component operation, then the utility of a component-based design methodology becomes questionable.

3.2.2   Component Interfaces

In the following we regard a node of a time-triggered cluster as a component. A node is a self-contained hardware/software unit with precisely specified interfaces. Each interface of a node should serve a *single well-defined purpose*. Based on the purpose, we distinguish between the following four message interfaces of a node:

- The *Linking Interface* (LIF) that provides the specified service of the node at the considered level of abstraction.

- The *Technology Independent Interface* (TII) that is used to configure and control the execution of the node.

- The *Technology Dependent Interface* (TDI) that is used to provide access to the internals of a node for the purpose of maintenance and debugging.

- The *Local Interface* that links a node to the outer world, that is the external environment of a set of related nodes, that we call a cluster.

The LIF and the local interface are *operational interfaces*, while the TII and TDI are *control interfaces*. The control interfaces are used to control, monitor or debug a node, while the operational interfaces are in use during the normal operation of a node.

3.2.2.1   Linking Interface

The *services* of a node are offered to the other nodes of the cluster at the Linking Interface (LIF) of the node. The LIF is an *operational message-based interface*. The LIF is thus the interface for the integration of nodes into the cluster. The LIF of a node abstracts from the internal structure and the local interfaces of the node. The specification of the LIF must be self-contained and cover not only the functionality and timing of the node itself, but also the semantics of its local interfaces. The LIF is *technology agnostic* in the sense that the LIF does not expose implementation details of the internals of the node or of its local interfaces. A technology agnostic LIF ensures that different implementations of nodes (e.g., host based on a general purpose CPU, FPGA, ASIC) and different local Input/Output subsystems can be connected to the node without any modification to the other nodes that interact with this node across its message-based LIF.

3.2.2.2 Technology Independent Interface (TII)

The technology independent interface is a *control interface* that is used to configure a node, e.g., assign the proper names to a node and its input output ports, to *reset*, *start* and *restart* a node and to monitor and control the resource requirements (e.g., power) of a node during run time, if so required. Furthermore, the TII is used to configure and reconfigure a node, i.e., to assign a specific *job* (i.e., software image) to a programmable node hardware.

The messages that arrive at the TII communicate either directly with the node hardware (e.g., *reset*) or with the node's operating system (e.g., *start a task*), but not with the application software. The TII is thus orthogonal to the LIF. This strict separation of the application-specific message interfaces (i.e., the LIF) from the system control interface of a node (i.e., the TII) simplifies the application software and reduces the overall complexity of a node.

3.2.2.3 Technology Dependent Interface (TDI)

The TDI is a *special control interface* that provides a means to look inside a node and to observe the internal variables of a node. It is related to the *boundary scan interface* that is widely used for testing and debugging large VSLI chips and has been standardized in the IEEE standard 1149.1 (also known as the JTAG Standard). The TDI is intended for the person who has a deep understanding of the internals of a node. The TDI is of no relevance for the user of the LIF services of the node or the system engineer who configures a node. The precise specification of the TDI depends on the technology of the node implementation and will be different if the same functionality of a node is realized by software running on a CPU, by an FPGA or by an ASIC.

3.2.2.4 Local Interface

The local interfaces establish a connection between a node and its outside environment, e.g., the sensors and actuators in the physical plant, the man-machine interface, or another computer system in another cluster. A node that contains a local interface is called a *gateway node* or an *open node*.

From the point of view of the LIF, only the *timing* and the *semantic content*, i.e., the meaning of the information exchanged across a local interface, is of relevance, while the detailed structure, naming and access mechanisms of the local interface are *intentionally left unspecified* at the cluster level. A modification of the local access mechanisms to the physical environment, e.g., the exchange of a CAN Bus by Ethernet, will not have any effect on the LIF specification, and consequently on the users of the LIF specification, as long as the semantic content and the timing of the relevant data items are the same.

A node that does not contain a local interface is a *closed node*. The distinction between open and closed nodes is important from the point of view of the specification of the semantics of a node. Only nodes that are closed nodes can be fully specified without knowing the *context of use*.

3.2.3   Linking Interface Specification

The *timed sequence of messages* that a node exchanges across an interface with its environment defines the *behavior* of the node at that interface. The *interface behavior* is thus determined by the properties of all messages that cross an interface. We distinguish between three parts of an interface specification: (i) the *transport specification* of the messages, (ii) the *operational specification* of the messages and (iii) the *meta-level specification* of the messages.

The *transport specification* describes all properties of a message that are needed to transport the message from the sender to the receiver(s). The transport specification covers the addressing and temporal properties of a message. If two nodes are linked by a communication system, the transport specification suffices to describe the requested services from the communication system. The communication system is *agnostic* about the contents of the data field of a message. For the communication system, it does not matter whether the data field contains multimedia data, such as voice or video, numerical data or any other data type.

In order to be able to interpret the data field of a message at the end points of the communication, we need the *operational* and the *meta-level specification*. The operational specification informs about the syntactic structure of the message that is exchanged across the LIF and establishes the *message variables*. Both the transport and the operational specification must be *precise* and *formal* to ensure the *syntactic interoperability* of nodes. The meta-level specification of a LIF assigns meaning to the *message variable names* introduced by the operational specification. It is based on an interface model of the user environment. Since it is impossible to formalize all aspects of a real-world user environment, the meta-level specification will often contain natural language elements, which lack the precision of a formal system. Central concepts of the application domains and applications can be specified using *domain specific ontologies*.

The meta-level LIF specification bridges the gap between the message variables, established by the operational specification, and the *user's mental model* of the service provided at the interface. Central to this meta-level specification is the LIF service model. The LIF service model defines the concepts that are associated with the message variable names contained in the operational specification. These concepts will be qualitatively different for *closed nodes* and *open nodes*.

The LIF service model for a closed node can be formalized, since a closed node does not interact with the external environment. The relationship between the LIF inputs and LIF outputs depends on the discrete algorithms implemented within the closed node. There is no input from the outer environment that can bring unpredictability into the node behavior. The sparse time-base within a cluster is discrete and supports a consistent temporal order of all events.

The LIF service model for an open node is fundamentally different since it must encompass the inputs from the outer environment, the local interfaces of the node, in its interface specification. *Only the operational specification of an open node can be provided without knowing the context of use of the open node.* Since a physical outer environment is not rigorously definable, the interpretation of the external inputs depends on human understanding of the natural environment. The concepts used in the description of the LIF service model must thus fit well with the accustomed concepts within a *user's internal mental model of the application domain*; otherwise the description will not be understood.

The LIF service model of an open node must meet the following requirements:

- *User orientation:* Concepts that are familiar to a prototypical user must be the basic elements of the LIF service model. For example, if a user is expected to have an engineering background, terms and notations that are *common knowledge* in the chosen engineering discipline should be utilized in presenting the model.

- *Goal orientation:* A user of a node employs the node with the intent to achieve a goal, i.e., to contribute to the solution of her/his problem. The relationship between user intent and the services provided at the LIF must be exposed in the LIF service model.

- *System view:* A LIF service user (the system architect) needs to consider the system-wide effects of an interaction of the node with the external physical environment, i.e., effects that go beyond the node. The LIF service model is different from the model describing the algorithms implemented within a node, since these algorithms end at the node's boundary.

3.2.4   Composition of Nodes

A node is a self-contained validated unit that can be used as a building block in the construction of larger systems. In order to enable a straightforward composition of a node into a cluster of nodes, the following four *principles of composability* should be observed.

3.2.4.1   Independent Development of Nodes

An architecture must enable the precise specification of the linking interface (LIF) of a node in the domains of value and time. This is a necessary prerequisite for the independent development of nodes on one side and the reuse of existing nodes that is based solely on their LIF specification on the other side. While the operational specification of the value domain of interacting messages is *state-of-the-art* in embedded system design, the temporal properties of these messages are often not considered with the appropriate care. The global time, which is available in time-triggered systems, is essential for the precise specification of the temporal properties of the LIF messages. Note that the transport specification and the operational LIF specification are independent of context of use of an open node, while the meta-level LIF specification of an open node depends on the context of use.

3.2.4.2 Stability of Prior Services

The stability of prior services principle states that the services of a node that have been validated in isolation (i.e., prior to the integration of the node into the larger system) remain intact after the integration.

3.2.4.3 Non-Interfering Interactions

If there exist two disjoint subgroups of cooperating nodes that share a common communication infrastructure, then the communication activities within one subgroup may not interfere with the communication activities within the other subgroup. If this principle is not satisfied, then the integration within one node-subgroup will depend on the proper behavior of the other (functionally unrelated) node-subgroups. These global interferences compromise the composability of the architecture. Time-triggered communication systems provide predictable message transport latency that are not influenced by the behavior of other, functionally unrelated messages.

3.2.4.4 Preservation of the Node Abstraction in the Case of Failures

In a composable architecture, the introduced abstraction of a node must remain intact, even if a node becomes faulty. It must be possible to diagnose and replace a faulty node without any knowledge about the node internals. This requires a certain amount of redundancy for error detection within the architecture. This principle constrains the implementation of a node, because it restricts the implicit sharing of resources among nodes. If a shared resource fails, more than one node can be affected by the failure. In time-triggered communication systems, the communications system contains redundant information about the permitted temporal behavior of nodes and disconnects nodes that violate their temporal specification in order to avoid error propagation from a faulty node (a babbling idiot) into the communication system.

3.3   Determinism and Predictability

3.3.1   The Concept of Determinism

The analytical-rational problem solving subsystem of humans excels in reasoning along causal chains. Causality refers to the unidirectional relationship that connects an *effect* to a *cause*. If this relationship is one of *entailment*, we speak of *determinism*, which we define as follows: *A system behaves deterministically if, given an initial state at instant* t *and a set of future timed inputs, the future states and the values and instants of all future outputs are entailed* (can be predicted without a doubt). In a distributed real-time computer system, the consistent definition of the *initial state* is only possible if all nodes agree on the instant *now* that separates *past events* from *future events*. The sparse time model, introduced in Chapter 2, Section 2.4.4 provides the basis for a consistent definition of the initial state in a distributed real-time system.

Deterministic behavior of a component is desired for the following reasons:

- The unconditional connection between initial state, input, output and the progression of time makes it easy to understand the real-time behavior of a component.

- Two replicated components that start from the same initial state and receive the same timed inputs will produce the same results at about the same time. This property is important if the results of a faulty component are to be masked (outvoted) by the results of correct components.

- The testability of the component is simplified, since every test case can be reproduced, eliminating the appearance of *spurious Heisenbugs* in the software.

In order to realize a deterministic behavior of a distributed real-time computer system, we must ensure that

- the *initial state* of all components is *defined consistently* in the distributed system. Furthermore, all inputs to replicated components must be presented to all replicas simultaneously. This is simplified if a global sparse time-base is available, since the sparse model of time solves the problems of the *uncertainty of simultaneity* in replicated channels that exists if a *dense time model* or a *discrete time model* is used.

- the computations are *certain*, i.e., there are no program constructs that produce arbitrary results (e.g., a random number generator) and that the final result of a computation will be delivered before an established instant in the future. This requires that the worst-case execution time (WCET) of the involved computations can be determined.

- the message transport system among the components is *predictable*, i.e., the instants of message delivery are known in advance and the temporal order of the received messages is the same as the temporal order of the sent messages at all independent channels.

According to Section 2.4.5, the *state* of a component can only be defined if there is a consistent separation of *past events* from *future events*. The sparse time model, introduced in Section 2.4.4, provides for such a consistent separation of past events from future events and makes it possible to define the instants where the initial state is defined. Without a sparse global time, the establishment of a consistent initial state of the components in a distributed system is difficult.

The implementation of a *certain computation* requires not only that the WCET of all involved programs are known but also that temporal consequences of resource sharing are bounded and known.

The implementation of a *predictable message transport system* requires that the temporal order of messages on different channels can be determined consistently and that the communication system will not reorder or arbitrarily delay messages. A time-triggered transport service provides these characteristics.

3.3.2   Replica Determinism

Replica determinism is a desirable relation among a set of replicated components that are introduced in order to mask the failure of one of the components of the set. A set of replicated RT components is *replica determinate if all the members of this set start from the same initial state and produce the same output messages at about the same instants*. If the behavior of one member of the set of replicated components deviates from the specified behavior, then this component has failed. In a properly configured fault-tolerant system, such a failure is masked by the other *replica determinate components*. The software design must ensure that a deviation of the specified behavior is caused by random physical faults and not by software instructions with unpredictable outcome. An example of such a software instruction is a *semaphore-wait-operation* that protects a resource from the concurrent access by two concurrent processes. The outcome of such a semaphore-wait-operation is uncertain. To one of the replicated components the access may be granted, while to another component the access may be delayed, resulting in a non-replica determinate behavior. From the point of view of fault tolerance, any non-replica determinate behavior is tantamount to a fault and leads to a loss of the further capability to tolerate a fault.

In a fault-tolerant system, the term *about* in the above definition of replica determinism relates to the time interval that it takes to replace a missing output message or an erroneous output message from a node by a correct message from the redundant replicas. This time interval must be derived from the dynamics of the application. If, in a time-triggered system, the components start from the same initial state (i.e., the ground state – see Section 2.4.5) and produce the same output messages at the same global ticks of their local clocks, then an upper bound for the time interval *about* is given by the precision of the global time.

The basic causes of replica non-determinism are: differing inputs, a difference between the progress of the computation and that of the local clocks in the replicas, differing oscillator drifts caused by the physical variations of the resonators and programming constructs that have *uncertain* results.

3.3.2.1 Differing Inputs

Whenever a value that is defined over a continuous value domain is mapped onto a discrete value domain, a digitalization error occurs. The physical RT entities in the controlled object, e.g., temperature and pressure, are defined over continuous value domains. The analog-to-digital transformation at the computer interfaces maps these values into discrete domains, causing a potential digitalization error of one bit. The same phenomenon occurs in the temporal domain: external time is dense, while internal time within a computer is discrete. If events that occur on a dense time-base are observed in a different order by two replicas, then, significantly different computational trajectories could develop.

3.3.2.2 Deviations of Computational Progress Relative to Real Time

In many computers, the same resonator drives the CPU and the real-time clock and no clear distinction is made between *real time* and *execution time*. One would there-fore assume that the progress of the local physical time is in synchrony with the progress of the computation. This assumption is not generally valid, since, to correct a randomly occurring transient error, many processors provide hardware-controlled instruction-retry mechanisms that take physical time without resulting in computa-tional progress. If, in two replicas, a different number of instruction retries are exe-cuted, the computational progress can diverge from the progress of the local physical time.

The differences in the progress between real time and the execution time lead to consequences whenever a program reads the local clock. Two replicas read different clock values at the same point of the computation, possibly resulting in different decisions.

3.3.2.3 Oscillator Drift

The control signals for the CPU originate from a physical oscillator, a quartz crystal. Because the mechanical dimensions of any two physical quartz crystals are slightly different, no two physical oscillators have the same drift. These slight differences in the drift of the oscillators of replicated nodes can lead to a non-determinate outcome for those decisions that involve, in one way or another, the local time. A prime exam-ple is the local use of time-outs. The same time-out value that is defined abstractly by a time-out value in a replicated program will lead to time-out intervals of slightly dif-fering physical lengths at the replicas. If a significant event, e.g., the expected arrival of an acknowledgment message that is monitored by a local time-out, occurs after the local time-out event in one replica, but before the same time-out event in another replica, then, remarkably different computational trajectories may develop in the two replicas.

3.3.2.4   Preemptive Scheduling

If dynamic preemptive scheduling is used, the points in the computations where an external event (interrupt) is recognized may differ at the different replicas. Consequently, the interrupting processes see different states at the two replicas at the point of interruption. They may reach different results at the next major decision point.

3.3.2.5   Nondeterministic Language Features

The use of a programming language with nondeterministic language constructs, such as the SELECT statement in an ADA program, can lead to the loss of replica determinism. Since the programming language does not define which alternative is to be taken at a decision point, it is left up to the implementation to decide the course of action to be taken. Two replicas may also take different decisions in the case of programming constructs with uncertain results. For example, a semaphore wait operation can also give rise to non-determinism, because of the uncertain outcome regarding the process which will win the race for the semaphore. Communication protocols that resolve a media-access conflict by reference to a random number generator, such as the Ethernet protocol, also suffer from replica non-determinism. The same argument applies to communication protocols that resolve the access conflict by relying on the outcome of non-determinate temporal decisions, such as ARINC 629 [7] or CAN [149].

3.3.3   Building a Replica Determinate System

The construction of replica determinate nodes requires careful design of the software system so that all the causes of replica non-determinism that have been discussed in the previous section are properly addressed.

3.3.3.1   Sparse Time-Base

A sparse global time-base makes it possible to assign a significant event to the same global clock tick at all the replicas without the execution of an agreement protocol. Any reference to the local real-time clock of a node (without the execution of an agreement protocol) can lead to replica non-determinism. This means that no local time-outs may be used in any part of the software, including the application software, the operating system and the communication software.

3.3.3.2   Agreement on Input

Whenever a redundant observation of an RT entity outside the sphere of control of a fault-tolerant computer system is performed, an agreement protocol must be executed among all replicas of the observing Fault-Tolerant Unit (FTU) to reach a common view of the exact digital value of an observation, and the exact instant on the sparse time-base when the observation was taken. The agreement on the time is the basis for establishing a consistent system-wide order of all observation events.

3.3.3.3 Static Control Structure

The implementation of a data-independent static control structure that can be validated independently of the data inputs is a safe choice for the implementation of replica determinate software. All inputs from the control object are periodically sampled by a trigger task, and no interrupt from the controlled object is allowed to occur. If the application timing requirements are so stringent (less than 1 msec response time) that a process interrupt causing a dynamic task preemption cannot be avoided, then all the possibilities of task preemption must be statically analyzed in the application context to ensure that replica determinism is maintained. Non-preemptive dynamic scheduling avoids the problems of unpredictable task interference.

3.3.3.4 Deterministic Algorithms

In the algorithmic section of an implementation, all constructs that could lead to non-determinate results must be avoided. Special attention must be paid to any dynamic synchronization construct that relies on the unpredictable resolution of a race condition, such as a semaphore wait operation. If software diversity is implemented, exact arithmetic must be performed to avoid the consistent comparison problem.

3.3.3.5 Deterministic Communication System

The communication system that transports replicated messages using independent redundant channels must be deterministic, such that no unplanned delay or reordering of messages occurs.

3.4 Diagnosability

Diagnosis aims at ascertaining the impact and location of perturbations by monitoring the functionality and performance of nodes. The reliable identification of failed nodes can be used for the autonomous recovery in case a failure is transient, and guide a maintenance engineer in the physical replacement of defective Field Replaceable Units (FRUs) in case the failure is permanent.

Depending on how diagnostic information is used, one can distinguish between *passive diagnosis* and *active diagnosis*. In passive diagnosis, the diagnostic information is stored and analyzed for maintenance and engineering feedback without taking on-line actions in the system. In case of active diagnosis, on the other hand, the diagnostic information is used to achieve fault-tolerance by directly intervening in the system behavior by means of reconfiguration (e.g., migration of services to spare nodes, graceful degradation). Since for safety-critical systems active diagnosis is directly related to safety properties, the diagnostic subsystem must be trusted and certified to the highest criticality level in the given application.

Both active and passive approaches involve distinct services for detection, analysis and use of diagnostic information, which will be explained in the following, along with the relationship to time-triggered communication protocols.

3.4.1 Detection of Errors and Anomalies

It is the purpose of the detection services to identify discrepancies between the current state and the intended state. In a time-triggered system, consistent information about the current state is available during the inactivity intervals of the sparse timebase (cf. Chapter 2, Section 2.4.4). The knowledge about the intended state of a system can be available in two ways:

- *A priori knowledge:* This part comprises knowledge about the properties of correct states, the timing of states and state changes, as well as the semantics of state variables. In the value domain, information about the code space can be used for error detection. For example, CRC codes or parity bits extend the hamming distance and permit the identification of erroneous code words. In the time domain, a priori knowledge can comprise restrictions on the time intervals in which certain states may occur (e.g., send instant of a message), the duration of states and the sequence of states. Furthermore, knowledge about the semantics of state variables can be used to formulate plausibility checks that express interrelationships between different state variables based on mathematical models of the system (e.g., using the laws of physics or based on past experience). This knowledge is also denoted as analytical redundancy [247]. Analytical redundancy defines a set of variables denoted as residuals with one or more residual generation filters. These residuals are ideally zero under no-fault conditions and sensitive to a selective set of faults [137].

- *Spatial and temporal redundancy:* Error detection can exploit spatial redundancy by comparing the information from redundant communication channels or redundant computational units. N-Modular Redundancy (NMR) as introduced in Section 3.6 is an example for the exploitation of replicated computational units to detect and mask the failure of replicas. In case of time redundancy, a single node performs redundant computations during different execution time intervals. The states of successive executions are compared to detect discrepancies. Time redundancy permits the detection of errors that are caused by transient faults with a duration that is shorter than the time between the execution time intervals.

In the following examples, error detection capabilities are given, which are located at different parts of a time-triggered cluster.

- *Self-Checking of Nodes.* Many embedded nodes execute Built-In Self-Tests (BISTs) such as memory tests [336], logic tests [198], sensor tests [154] and test computations with known results (also called challenge/response mechanisms). In addition, the communication controller can serve as a watchdog that monitors a regular lifesign from the host. The absence of the lifesign indicates a crash failure of the host. Using output assertions, a priori knowledge about the produced messages can be expressed and checked (e.g., message syntax). Such an assertion is a predicate on values of the message, and relevant state variables, that evaluates to "TRUE" if the message is correct [221, p. 112].

- *Mutual Checking of Nodes.* In analogy to the output assertions of sender nodes, receiver nodes can execute input assertions on messages. The benefit of the input assertions is that the error detection occurs in a different FCR than the production of the message. Self-checking of nodes is prone to leading to fault negatives, because the error detector can be affected by the same fault that caused the error of the rest of the node. Mutual checking of nodes can occur in a distributed fashion or using a dedicated node (e.g., called diagnostic electronic control unit in [211]).

- *Checks by the Communication System.* A time-triggered communication system possesses a priori knowledge about the temporal behavior of nodes. The communication infrastructure of a time-triggered communication system can be programmed with this knowledge (e.g., using central and local guardians [26, 106]) for the detection of node failures in the time domain (e.g., late message failures, early message failures, omission failures).

As outlined in [252], the knowledge about the intended state is often insufficient to perform immediate error detection. The state can indicate an error or be part of a rare but correct operational situation. If the distinction cannot be performed with certainty at a given time, we speak of an *anomaly*. In such a case, the state needs to be analyzed over time and at different nodes in order to discriminate between errors and intended (but unlikely) states.

3.4.2 Decision Making – Analysis of Diagnostic Information

In order to decide on a suitable reaction, the analysis of diagnostic information evaluates information about errors and anomalies from different nodes and performs correlation in the value and time domains. The types of decisions depend on whether active or passive diagnosis is realized. For passive diagnosis, a service technician needs to be supported in performing maintenance decisions. For example, no maintenance action is required in case of an external fault that has no permanent effect on the functionality of the FRUs. An example for such an external fault is Electromagnetic Interference (EMI). In case of an internal fault, which originates from within the FRU boundaries (e.g., crack in the PCB), examples of maintenance actions are the replacement of an FRU affected by a hardware fault or a software-update of an FRU with a software fault. So-called borderline faults are the class of faults that cannot be judged to be external or internal with respect to the FRU boundaries. An example of such a fault is a connector fault, where a connector consists of two parts, one attached to the FRU and the other attached to the cable loom. This class is responsible for a significant number of system failures [320]. A connector fault requires an inspection of the FRUs and the wiring harness.

For active diagnosis, the diagnostic subsystem decides on an action to recover from a fault. After a transient fault, a suitable action is a reset of the affected node with a subsequent restoration of the state (cf. Section 3.6.4). After a permanent fault, the services can be migrated to suitable spare nodes. Restrictions on the migration of services result from the types of nodes (e.g., deployed processing cores, performance, memory) and the management of physical inputs and outputs. In particular, inputs and outputs introduce a strong coupling between application service and specific nodes [92].

For decision making in active and passive diagnosis, different types of analysis technologies are available, such as solutions based on Bayesian networks [273], neural networks [357] or inference [36]. The choice of the analysis technology has a significant impact on the maximum time to reach a decision and the ability to provide guarantees (e.g., error detection coverage). Temporal predictability and guaranteed results are especially important for active diagnosis, where a failure of the diagnostic services has the potential to cause a failure of the application services and is thus safety-relevant. Therefore, safety-critical applications that need to be certified (e.g., in the avionic domain [277]) typically use static configurations or switch between a small set of pre-qualified static configurations [92].

3.4.3   Use of Diagnostic Information and Analysis Results

In case of active diagnosis, nodes contain local enactors [50] that accept commands via the TII from the decision making units and perform the computed actions. For example, after a transient fault an enactor can perform a reset of the node and acquire the restart state from the replicas.

Diagnostic information and the results of the analysis can also be fed back to the nodes in order to enable application-level fault tolerance. A typical example is a membership service [68]. The membership service provides binary information about each node, denoting whether a node is operational or non-operational. In addition, the membership service provides consistency by guaranteeing that all correct nodes have the same view on the operational state. For this purpose, algorithms for the processor-group membership problem with suitable agreement algorithms are used.

In case of passive diagnosis, service technicians retrieve the diagnostic information and the proposed maintenance decisions. This information can be stored locally in the nodes where the detection occurred or in a central database. Independently of the storage location, filtering and preprocessing of diagnostic information can help to reduce the amount of diagnostic data. The retrieval typically occurs via a diagnostic access port that is part of the node's TDI or TII. For example, in the automotive industry the On-Board Diagnosis (OBD) standard defines the diagnostic connectors, the electrical signalling protocols, the messaging formats and significant vehicle parameters [113]. OBD supports a unified data communication, which supports different underlying communication protocols such as ISO14230 (KWP-2000) or SAEJ1850. Solutions for the retrieval of diagnostic information can be classified depending on whether the retrieval of diagnostic information occurs off-line (i.e., in a special maintenance mode) or during normal operation. The retrieval during normal operation is desirable in systems with a high cost of down time (e.g., factory automation).

3.5 Certifiability

Legislatures and the public have decided that certification agencies – such as the Federal Aviation Administration (FAA) for commercial transport aircraft – must monitor the design of safety-critical applications that are crucial to the preservation of human life. Likewise, certification is required in military systems, the railway industry, the nuclear industry and in industrial control systems.

In accordance with the safety standards of the respective domains, certification agencies demand the construction of a safety case. *A safety-case is a documented body of evidence that provides convincing and valid argument that a system is adequately safe for a given application in a given environment* [34].

A safety case ensures that all justifiable precautions have been taken in order to minimize the risk to the public. Adversely, the safety case establishes a shared responsibility between the certification agency and the developer in case of an accident.

Certification is a significant cost factor in the development of embedded systems, e.g., in the avionic domain [128]. Consequently, there is a need for systems that are designed for validation in order to simplify the certification process. Design for validation [155] occurs by devising a complete and accurate reliability model, by avoiding design faults, and by minimizing parameters that have to be measured. The construction of the reliability model has to be based on a detailed understanding of failure modes and fault-tolerance mechanisms. The reliability model must be based on parameters that can be accurately measured, e.g., failure rates of nodes.

3.5.1 Safety Case

The safety case provides documented arguments and evidence in order to justify that a system is sufficiently safe for deployment. A safety case includes diverse evidence and arguments that accumulate as the project proceeds. The following two types of evidence can be distinguished:

- *Process evidence:* The focus of process evidence lies on management and is gained by checking the quality of the life cycle process and the quality assurance organization. The quality and experience of the personnel is also an important factor. In addition, the process evidence includes results of reviews, tests and compliance with the plans.

- *Product evidence* investigates the product as such. This evidence consists of experience with the operational system (e.g., testing by fault injection), results from simulations and formal verification. Experience with similar designs (e.g., field failure rates) and architecture properties also contributes to the product evidence.

Today, emphasis is placed on process-related evidence in the field of computer systems. The reliability requirements of safety-critical applications are orders of magnitude higher than what can be validated experimentally by using measurements and testing. Therefore, certification must be based on the life-cycle processes of the development, reviews and analysis of the system and experience gained with similar systems.

The main elements of a safety case are claims, evidence, inference and arguments [34]. A *claim* is a statement about a property of a subsystem. Examples of claims are statements about reliability, availability, performance and security. The *evidence* provides the basis for the justification of safety and includes facts, assumptions and further subclaims. The *arguments* link the evidence with claims, and the transformational rules for the arguments are called *inference*. Depending on the claim and the available evidence, arguments can be deterministic, probabilistic or qualitative. Deterministic arguments (e.g., exhaustive test, worst-case timing analysis, formal proof of a property) are preferred over quantitative arguments (e.g., probabilistic statistical reasoning). Also, quantitative arguments are preferred over qualitative. Qualitative arguments reflect an indirect effect of evidence on the claimed attributes such as the satisfaction of design rules and the qualifications of the developers.

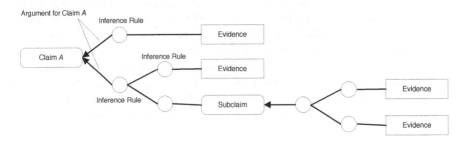

FIGURE 3.1
Hierarchical Structure of a Safety Case with Subclaims

Figure 3.1 depicts the relationship between the elements of a safety case. The safety case can be structured hierarchically if another claim serves as evidence.

The safety case is typically a living document that evolves throughout a project. Along with the refinement of requirements and the design, the safety case is progressively refined into a *layered safety case*. The traceability of subsystems, refined requirements and design features to the top-level requirements is essential in order to maintain the link to the claimed safety attributes.

3.5.2 Modular Certification

Traditionally, certification of a system (e.g., an aircraft) is considered as an indivisible whole. The goal of modular certification is the separation of the certification of platform services from applications and aims at independent safety arguments for different components.

Ideally, the supplier of a component describes and analyzes significant properties in isolation. In order to certify the whole, the manufacturer integrates these properties with the properties of the rest of the system. For example, [228] proposes a model for software certification where suppliers provide test certificates in a standard portable form.

Such an approach reduces the efforts for the manufacturer of the overall system, because certification arguments from suppliers are available. In addition, the reuse of components in different systems is facilitated. The supplier can develop a certification argument once and use it in many systems.

The main challenge in modular certification is the need for a component and interface model, which is also valid in the presence of faults. Certification is mainly concerned with faults and the analysis of the hazard of a component in the context of the global system. Reference [284] explains the problem of a disintegrating tire, which can penetrate the wing tanks in a Concorde. However, this hazard might not exist on other aircraft.

Although a time-triggered communication protocol cannot solve the analysis of such hazards at the application level, the temporal and spatial partitioning of a time-triggered communication protocol (see Section 3.6) provides an important baseline for modular certification.

At higher levels, modular certification can exploit assume-guarantee reasoning where assumptions and guarantees take into account both normal and abnormal conditions. Temporal and spatial partitioning ensures that components respect interfaces after failures (e.g., no modification of private data in another component) and interact only based on the respective assumptions and guarantees. Assume-guarantee reasoning in this context has been formally examined in [284].

3.5.3 Certification in Application Domains

Certification typically occurs according to safety standards, which define a safety life cycle ranging from an initial specification with safety requirements to the design, implementation and deployment of safety-critical systems.

For example, IEC 61508 is a universal safety-standard for different application domains including automotive, industrial control, powerplans, medical devices and railway systems. The focus of IEC 61508 is the identification of potential hazards and the evaluation of the impact and occurrence probability. Four Safety Integrity Levels (SILs) are distinguished in order to reflect the acceptable risk. Each SIL is associated with quantitative and qualitative requirements for the development process and the safety life cycle management. The most critical systems belong to SIL4 and a probability of a critical failure of 10^{-9} is demanded (i.e., ultra dependability).

Domain-specific safety standards have been developed based on IEC 61508. For example, in the automotive domain the functional safety standard ISO 26262 is currently under development. This standard is entitled "Road vehicles – Functional safety."

In the aerospace domain, Quality Assurance (QA) is critical due to the relatively small production numbers and potentially large impact of failures on safety of operation. QA stretches through development, production and operation and maintenance phases of an aircraft. This section addresses especially the use of formal methods for the development of algorithms that may have an impact on the safety of an aircraft.

The regulations driving the safety of an aircraft are Federal Aviation Regulations (FAR) 25 Paragraph 1309 (or internationally are reflected in Joint Aviation Regulations [JAR]). For the methods of compliance with the FAR and JAR 25 requirements for a new system design, five methodologies are generally adopted, some of which are described in more detail in ARP4754 [321] and ARP4761 [322]:

1. Analysis including engineering analysis, stress analysis, system modeling and similarity modeling.

2. Failure analysis including FMEA (Failure Mode and Effects Analysis), FTA (Fault Tree Analysis) and safety analysis (including Functional Hazard Assessment (FHA), (Preliminary) System Safety Assessment ([P]SSA), and Common Cause Analysis [CCA]).

3. Laboratory tests including component tests, qualification tests, system tests through an integrated systems test rig.

4. Ground Tests–On aircraft ground tests.

5. Flight Tests–On aircraft flight tests.

3.5.4 Time-Triggered Communication Protocols and Certification

In an distributed system, the services of the communication protocol (e.g., message exchange, time services, fault isolation) are among the most critical parts of the system, since any design fault in these services is likely to result in correlated failures in multiple functions. *The core algorithms and architectural mechanisms in fault-tolerant systems are single points of failure: they just have to work correctly* [284, p. 27]. Consequently, high emphasis should be placed on validating the services of the communication protocol, e.g., by employing formal verification techniques.

Therefore, Chapter 5 provides results on the validation by means of fault injection and formal verification along with the explanation of the time-triggered communication protocols.

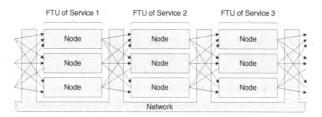

FIGURE 3.2
Incoming Voting

3.6 Fault Containment and Error Containment

An FCR has been introduced in Chapter 2, Section 2.3.2 as the boundary of the immediate impact of a fault. In conformance with the fault-error-failure chain introduced by Laprie [21], one can distinguish between faults that cause the failure of an FCR (e.g., design of the hardware or software of the FCR, operational fault of the FCR) and faults at the system level. The latter type of fault is a failure of an FCR, which could propagate to other FCRs through a sent message that deviates from the specification. If the transmission instant of the message violates the specifications, we speak of a *message timing failure*. A *message value failure* means that the data structure contained in a message is incorrect.

Such a failure of an FCR can be tolerated by distributed fault-tolerance mechanisms. The masking of failures of FCRs is denoted as error containment [187] because it avoids error propagation by the flow of erroneous messages. The error detection mechanisms must be part of different FCRs than the message sender [172]. Otherwise, the error detection mechanism may be impacted by the same fault that caused the message failure.

For example, a common approach for masking node failures is *N-Modular Redundancy (NMR)* [194]. *N* replicas receive the same requests and provide the same service. The output of all replicas is provided to a voting mechanism, which selects one of the results (e.g., based on majority) or transforms the results to a single one (average voter). The most frequently used *N*-modular configuration is triple-modular redundancy (TMR).

We denote three replicas in a TMR configuration a *Fault-Tolerant Unit (FTU)*. In addition, we consider the voter at the input of a node and the node itself as a self-contained unit, which receives the replicated inputs and performs voting by itself without relying on an external voter. We call this behavior *incoming voting* (see Figure 3.2).

In the following, the prerequisites for error containment through FTUs are discussed.

3.6.1   Independent Fault Containment Regions

Common-mode failures are failures of multiple FCRs, which are correlated and occur due to a common cause. Common-mode failures occur when the assumption of the independence of FCRs is compromised. They can result from replicated design faults or from common operational faults such as a massive transient disturbance. Common-mode failures of the replicas in an FTU must be avoided, because any correlation in the instants of the failures of FCRs significantly decreases the reliability improvements that can be achieved by NMR [130].

The definition of FCRs needs to take into account shared resources that could be impacted by a fault. Typically, shared resources include the computing hardware, the power supply, the timing source, the clock source and the physical space. For example, if two subsystems depend on a single power supply, then these two subsystems are not considered to be independent and therefore belong to the same FCR. Since this definition of independence allows that two FCRs can share the same design, design faults are not part of this fault-model. Diversity is a method to avoid common design faults in FCRs by using different designs that perform the same function.

3.6.2   Strict Control on Node Interactions

A distributed system uses shared networking resources in order to support the interaction between nodes. In order to preserve the independence of FCRs, strict control on the node interactions and the use of these shared resources is required. Based on the specification of the permitted behavior of a node at its linking interface, a time-triggered communication protocol can isolate nodes that violate the linking interface specification. Thereby, the time-triggered communication protocol realizes *temporal partitioning* and *spatial partitioning* [281] at the network-level:

- *Temporal partitioning*. Media access control is concerned with the assignment of time intervals to each node for the transmission of the node's messages. A node that sends untimely messages has the potential of disrupting the communication abilities of the other nodes by delaying their messages. The media access control mechanisms of a time-triggered communication system can be designed to prevent this interference between nodes in the temporal domain. A time-triggered communication system is autonomous and has a priori knowledge of all intended message sent and receive instants. This knowledge can be used by local or central guardians [26, 106] in order to block untimely messages.

- *Spatial partitioning*. Spatial partitioning ensures that one node cannot alter the code, private data or messages of another node. Spatial partitioning must ensure that no messages with corrupted data or wrong addresses are delivered. A message with a wrong address or identification – also called a masquerading failure – could cause correct messages of other nodes to be overwritten at the recipient. Faults occurring at the communication channel can be detected using CRC checks. For faults that occur within the sender, an unforgeable authentication mechanism is necessary. Depending on the failure assumptions, an authentication mechanism can be implemented with simple signature schemes or cryptographic mechanisms [114].

3.6.3 Replica Determinism

Replica determinism has to be supported by the architecture to ensure that the replicas of an FTU produce the same outputs in defined time intervals. As discussed in Section 3.3.2, a time-triggered communication system addresses key issues of replica determinism. In particular, a time-triggered communication system supports replica determinism by exploiting the sparse global time base in conjunction with preplanned communication and computational schedules. Computational activities are triggered after the last message of a set of input messages has been received by all replicas of an FTU. This instant is a priori known due to the predefined time-triggered schedules. Thus, each replica wakes up at the same global tick and operates on the same set of input messages. The alignment of communication and computational activities on the sparse global time-base ensures temporal predictability and avoids race conditions.

3.6.4 Recovery and Repair

A time-triggered communication protocol should support adequate recovery and repair mechanisms in order to reestablish the original reliability of an FTU after the failure of a replica. Although a failure of an FCR is masked by an FTU and is thus not visible to the users, a failure of an FCR nevertheless reduces or eliminates any further fault-masking capability.

After a permanent fault, the masked failures are reported to a diagnostic system in order to perform a repair action at the next maintenance point. In case of a transient fault where the hardware is still operational, the recovery of an FCR can occur by restarting the FCR with a valid state. Part of the state can be restored from the environment by performing a complete scan of all the sensors and resynchronizing the node with the external world. In addition, output data that is in the control of the computer can be enforced on the environment using a *restart vector* [169, p. 136]. State that can be neither read from the environment nor enforced onto the environment must be recovered from a node-external source such as another node that has stored this information redundantly.

The instant for the reintegration of an FCR can be selected to minimize the state that needs to be recovered (cf. ground state in Chapter 2, Section 2.4.5). In cyclic systems (e.g., control loops, many multimedia systems), an ideal instant for the reintegration of a node is at the beginning of a new cycle.

The recovery from transient faults is of particular importance, because transient faults are, at least by a factor of 1000, more likely than permanent faults. Hence, the time needed to recover after a transient failure is an important input for a reliability model. After the recovery, the FCR is able to tolerate another transient or permanent fault.

3.7 Performance

This section discusses the types of communication loads that need to be handled by time-triggered communication networks. Based on these communication loads, performance attributes are presented.

3.7.1 Periodic, Sporadic and Aperiodic Messages

In a cluster with nodes interconnected by a time-triggered network, each node is assigned one or more messages that are repeatedly sent by the node. A node requests the dissemination of a message m at points in time $\rho_{k,m} \in \mathbb{R}^+$ ($k \in \mathbb{N}$), where $\rho_{k,m}$ are stochastic variables.

The parameter m identifies the message as well as the corresponding node. The parameter k counts the instances of the message at that node. Every message m is characterized by two parameters, a minimum interarrival time $d_m \in \mathbb{R}^+$ and a random interval offset $\delta_{k,m} \in \mathbb{R}^+$.

$$\forall k \in \mathbb{N}: \quad \rho_{k+1,m} - \rho_{k,m} = d_m + \delta_{k,m}$$

d_m specifies an a priori known minimum interval of time between two transmission requests of m. The stochastic variables $\delta_{k,m}$ cover the random part in the time interval between two transmission requests of m. For a particular message m the stochastic variables $\delta_{k,m}$ possess a corresponding probability distribution in the interval $[0, u_m]$. u_m is called the random interval length.

In conjunction with knowledge about the parameter b_m denoting the length of message m in bits, the average bandwidth usage of a message m is as follows:

$$\text{bandwidth}(m) = \frac{b_m}{d_m + \mathbb{E}\left[\delta_{k,m}\right]} \tag{3.1}$$

where $\mathbb{E}\left[\delta_{k,m}\right]$ is the expected value of the random interval length.

The two message parameters d_m and u_m employed in this message model allow us to distinguish between different message types. For a *sporadic message*, the transmission request instants are not known, but it is known that a minimum time interval exists between successive transmission requests.

$$m\,\text{sporadic} \;\leftrightarrow\; \left(\forall k\; 0 \leqslant \delta_{k,m} \leqslant u_m\right) \wedge \left(d_m \in \mathbb{R}^+\right)$$

For a *periodic message*, the transmission request instants are known and the random interval is 0.

$$m\,\text{periodic} \;\leftrightarrow\; \left(\forall k\; \delta_{k,m} = 0\right) \wedge \left(d_m \in \mathbb{R}^+\right)$$

For an *aperiodic message*, no minimum time interval between successive transmission requests is known.

$$m\,\text{aperiodic} \;\leftrightarrow\; \left(\forall k\; 0 \leqslant \delta_{k,m} \leqslant u_m\right) \wedge \left(d_m = 0\right)$$

3.7.2 Performance Attributes

Important performance attributes in real-time communication networks are the bandwidth, the network delay and the variability of the network delay (i.e., communication jitter).

The bandwidth is a measure of the available communication resources expressed in bits/second. The bandwidth is an important parameter as it determines the types of functions that can be handled and the number of messages and nodes that can be handled by the communication network.

The network delay denotes the time difference between the production of a message at a sending node and the reception of the last bit of the message at a receiving node. At some instant $t_{request}$ the sending node requests the transmission of a message by invoking a send operation at the node's communication controller. Depending on the communication protocol and the current traffic on the communication channel, the transmission of the message will start after the *access delay* d_{access} at the send instant t_{send}. After the transmission delay $d_{transmission}$ the message arrives at the receiver node at instant $t_{receive}$.

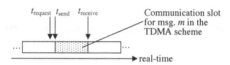

FIGURE 3.3
Phase-Alignment between Request Instant and Send Instant of a Periodic Message

Depending on whether the communication protocol is time-triggered or event triggered, the access delays exhibit different characteristics. In a time-triggered system the send instants t_{send} of all nodes are periodically recurring instants, which are globally planned in the system and defined with respect to the global time-base. The access delay d_{access} of a message in a time-triggered system is thus locally determined at the sending node. The access delay is independent of the traffic from other nodes and depends solely on the relationship between the request instants and the preplanned send instants. Furthermore, since the next send instant t_{send} of every node is known a priori, a node can synchronize locally the production of periodic messages $t_{request}$ with the send instant t_{send} and thus minimize the access delay of a message. This phase-alignment between the request instant of a periodic message with the communication schedule of the time-triggered communication protocol is depicted in Figure 3.3.

In addition, the TDMA scheme of a time-triggered communication protocol can contain periodic communication slots for the transport of sporadic and periodic messages. In case of a transmission request, the message is transmitted during the next communication slot in the TDMA scheme. The period of the communication slots can be determined by the maximum permitted network delay, the bandwidth requirements and knowledge about the message interarrival times (e.g., minimum and average values). As depicted in Figure 3.4, sporadic and aperiodic messages incur a delay depending on the (stochastic) time of the request instant relative to the preplanned send instants of the time-triggered communication protocol. In the worst case, the transmission of a sporadic or aperiodic message is requested immediately after a send instant, thereby incurring a delay equal to the period of the communication slot. At the cost of additional communication resources, this network delay can be reduced by increasing the frequency of the communication slots reserved for sporadic and aperiodic messages. In order to reduce the consumption of communication resources, several time-triggered communication protocols have been extended with means for the event-triggered communication of sporadic and aperiodic messages such as time intervals with dynamic arbitration as described in Chapter 4, Section 4.4.

Period of slot for message m in the TDMA schedule

FIGURE 3.4
Worst-Case Relationship between Request Instant and Send Instant of a Sporadic Message

In an event-triggered system the access delay d_{access} of a message depends on the state of the communication system at the instant $t_{request}$. If the communication network is idle at the instant $t_{request}$ the message transmission can start immediately leading to an access delay close to zero. If the channel is busy at the instant $t_{request}$ then the access delay d_{access} depends on the media access strategy implemented in the communication protocol. For example, in the CSMA/CD protocol of Ethernet [140], nodes wait for a random delay before attempting transmission again. In the CSMA/CA protocol of CAN [149], the access delay of a message depends on its priority relative to the priorities of other pending messages. Hence, in an event-triggered network the access delay of a message is a global property that depends on the traffic patterns of all nodes.

A bounded network delay with a minimum variability is important in many embedded applications. For example, achievement of control stability in real-time applications depends on the completion of activities (like communicating sensor and control values) in bounded time. Hard real-time systems ensure guaranteed response even in the case of peak load and fault scenarios. Guaranteed response involves assurance of temporal correctness of the design without reference to probabilistic arguments. Guaranteed response requires extensive analysis during the design phase such as an off-line timing and resource analysis [17]. An off-line timing and resource analysis assesses the worst-case behavior of the system in terms of communication delays, computational delays, jitter, end-to-end delays, and temporal interference between different activities.

In hard real-time systems, missed deadlines represent system failures with the potential of consequences as serious as in the case of providing incorrect results. For example, in drive-by-wire applications, the dynamics for steered wheels in closed control loops enforce computer delays of less than 2 ms [131]. Taking the vehicle dynamics into account, a transient outage-time of the steering system must not exceed 50 ms [131]. In the avionic domain, variable-cycle jet engines can blow up if correct control inputs are not applied every 20 to 50 ms [187].

While control algorithms can be designed to compensate a known delay, delay jitter (i.e., the difference between the maximum and minimum value of delay) brings an additional uncertainty into a control loop that has an adverse effect on the quality of control [169]. Delay jitter represents an uncertainty about the instant a real-time entity was observed and can be expressed as an additional error in the value domain. In case of low jitter or a global time-base with a good precision, state estimation techniques allow us to compensate for a known delay between the time of observation and the time of use of a real-time image. State estimation uses a model of a real-time entity to compute the probable state of the real-time entity at a future point in time.

4

Core Algorithms

M. Paulitsch
EADS Innovation Works

W. Steiner
TTTech Computertechnik AG

R. Obermaisser
University of Siegen

C. El Salloum
Vienna University of Technology

CONTENTS

4.1 Introduction

The focus of this chapter are the algorithms that provide the core of a time-triggered communication protocol.

Clock synchronization is concerned with aligning the local clocks of all nodes in the cluster with a given precision. The resulting system-wide global time-base is a prerequisite for the realization of a time-triggered network, which performs the temporal coordination of all communication activities using the global time.

Clock synchronization is based on the assumption of initially synchronized nodes. The core algorithms for *startup* serve for establishing this initial synchronization by negotiating and agreeing on the initial synchronization point. If a node joins a cluster with already synchronized nodes, this process is called *integration*.

Many communication protocols support both *event-triggered* and *time-triggered communication*. Event-triggered control excels with respect to flexibility (e.g., strong migration transparency, no need to change a communication schedule when adding messages) and resource efficiency through the sharing of bandwidth between nodes. In event-triggered communication, messages exchanges are initiated when a significant event occurs. In contrast to time-triggered systems, these events are not necessarily derived from a clock tick but can include arbitrary state changes in the computer system or the environment (e.g., interrupt from a sensor).

The section on *diagnosis* introduces diagnostic services of time-triggered networks. After explaining basic error detection mechanisms, which provide a local view of the operational state of other nodes, the establishment of a globally consistent membership vector is described.

The last section describes algorithms for *interactive consistency*, which ensure that all non-faulty nodes of a cluster receive a consistent value for any communicated message. Interactive consistency reduces the complexity of developing fault-tolerant embedded systems, since developers would otherwise have to realize agreement at the application level.

4.2   Clock Synchronization

Clock synchronization in its most general meaning comprises startup, integration and keeping clocks synchronized during continuous operation. In this book, we differentiate between synchronization of clocks during continuous and during initial synchronization. The reason for this approach lies more in the literature and the approaches to synchronization of clocks than in the problem of synchronization itself. This section describes the problems and approaches of continuous synchronization of clocks and Section 4.3 describes the initial synchronization.

Since any clock drifts, the clock times of an ensemble of clocks will drift apart if they are not periodically re-synchronized with respect to each other.[1] We call the clocks of a subsystem *local clocks*. Clock synchronization is concerned with bringing the time of local clocks in close relation with respect to each other. A measure for the quality of synchronization is the *precision* described in detail in Chapter 2, Section 2.4 and defines the maximum difference of local clocks.

Internal Clock Synchronization

The process of mutual resynchronization of an ensemble of clocks to maintain a bounded precision is called *internal clock synchronization*. Internal clock synchronization does not necessarily mean synchronization to real time, as all clocks of an ensemble can drift with respect to real time. Internal clock synchronization is defined as *optimal* by Srikanth and Toueg [304], if the drift rate of the synchronized clock time of an ensemble of clocks (with respect to real time) is smaller than or equal to the drift rate of the largest drift rate of the ensemble of clocks.

Mathematically, the problem of internal clock synchronization can be formulated as follows (using the notion of Chapter 2, Section 2.4):

Property 1 *Internal clock synchronization: For any two local clocks j and k and all microticks mt^i*

$$|z(mt_k^j) - z(mt_k^j)| < \Pi \tag{4.1}$$

---

[1] "Re-synchronize clocks with respect to each other" means "to bring the different times of the clocks closer together."

where Π is the precision of the clock synchronization algorithm.

External Clock Synchronization

If an ensemble of clocks synchronizes their clock times to a distinguished set of clocks that are not part of this ensemble, this is called *external clock synchronization*. A quality measure for this kind of synchronization is the accuracy described and defined in Chapter 2, Section 2.4.

Mathematically, the problem of external clock synchronization can be formulated as follows:

Property 2 *External clock synchronization: For any local clock j, reference clock r representing the reference time to which clocks synchronize to, and all microticks mt^i*

$$|z(mt_k^j) - z(mt_k^r)| < \alpha^r \qquad (4.2)$$

where α^r is the accuracy of the clock synchronization algorithm to the reference clock.

Clock Synchronization Versus Synchronization of Clock Rates

Clock synchronization (i.e., synchronization of the times of clocks) is not the same as clock rate synchronization [201]. Clock synchronization does *not* mean only synchronizing the rates of clocks, because synchronizing rates of clocks leaves the initial state of a clock open. Synchronizing clock times, however, automatically implies synchronization of clock rates. Synchronizing rates is computationally less intensive compared to synchronizing times and may suffice for certain applications [201].

State Correction Versus Rate Correction

In order to synchronize local clocks, the local clocks' time values must be corrected. There are two different approaches to correcting local clocks: state correction and rate correction. For *state correction*, a node applies a computed correction value at once and immediately after calculation. For *rate correction*, the rate of the clock is accelerated or decelerated by changing the number of hardware clock ticks over an interval. The sum of the number of local clock time changes in the interval equals the computed correction value.

4.2.1 Principle of Operation of Clock Synchronization

As mentioned above, the goal of clock synchronization is to bring or maintain the local clock times of the synchronizing clocks in close relation to each other or to a reference clock. In order to achieve this goal, each node cyclically performs the following three phases:

Phase 1: Collection of clock time values. In phase 1, a node collects information about the clock time values of other clocks actively participating in the synchronization. This can either be the difference to its own clock time or the actual clock time value.

Phase 2: Calculation of correction value. After performing phase 1, each node computes a correction value using (some or all) of the collected clock time values in phase 2. The computation is performed using the convergence function, which should bring clock times closer together.

Phase 3: Clock correction. In phase 3, a node corrects its clock time using the calculated correction value of phase 2.

Schneider argues in [297] that basically all internal clock synchronization algorithms are solutions to the following three subproblems:

Resynchronization. What is the event that causes the resynchronization of clocks?

Remote clock time readings. How does a clock obtain the clock time values of other clocks?

Convergence function. How are the correction values of different clocks computed in order to keep clocks synchronized within a bounded interval (the precision) while preserving the monotonicity of the clock times and keeping the drift of the clocks better than or equal to the hardware clock drifts?

Schneider's argument is also valid for the external clock synchronization algorithm. For all clock synchronization problems, the correction approach (state correction or rate correction) is another sub-problem to be addressed.

4.2.1.1 Resynchronization Initiation

The trigger for resynchronization can be a key differentiator in clock synchronization algorithms. Often periodic resynchronization, so-called *round-based* algorithms, are used. The problem is when a round starts initially. The most common assumption is initial synchronization leaving the problem of initial synchronization to another algorithm (like startup). Another approach is the use of message exchanges to initiate resynchronization.

4.2.1.2 Remote Clock Time Readings

There are two major approaches identified by [5]: *Time Transmission* (TT) and *Remote Clock Reading* (RCR). In algorithms using the TT technique, local clocks send their clock value based on local time. Alternatively, a processor (or node) estimates the remote time based on a message received from the remote clock. The RCR technique has been introduced by [67] to circumvent the absence of knowledge of an upper bound of communication delays. In the RCR technique, a processor p_j willing to estimate the clock of a remote processor p_i sends a request message to p_i and waits for a certain amount of time for p_i's response. The estimate of the remote clock is obtained using the round trip times and the distribution of communication delays and the drifts of p_i and p_j during the estimation process. The result can be an interval or a value out of this interval. The key difference between RCR and TT is who initiates the process of remote clock reading.

4.2.1.3   Convergence Functions

In the literature, clock synchronization approaches often differentiate themselves via the convergence function. The convergence function uses the different remote clock estimates and calculates a single new clock value which the local clock uses to correct itself to. Schneider [296] and Anceaume and Puaut [4, 5] list different convergence functions with the following being the most often referenced ones.

Convergence Averaging Techniques

In the following, some kind of averaging on clock estimates characterizes the convergence functions. Some functions tolerate potentially faulty clock estimates, while others don't. For the overview, $f(p_i, x_1, ..., x_n)$ identifies a convergence function, where p_i is the processor (or node) requesting the convergence function and $x_1, ..., x_n$ are the estimated clock values. For the convergence function f_{mm} and f_{im}, x_i is an interval otherwise an integer value.

Interactive convergence function. [188] uses this function, which is also called *egocentric average function* and denoted by f_e. $f_e(p_i, x_1, ..., x_n)$ returns the average of the arguments modified in the following: x_j, $1 \leq j \leq n$, stays x_j if $|x_j - x_i| < \bar{\omega}$ (i.e., if x_j is not further than $\bar{\omega}$ away from x_i) otherwise x_j is replaced with x_i. $\bar{\omega}$ has to be chosen appropriately, but be at least $\bar{\omega} > \pi$, π being the precision. For f_e no sorting algorithm needs to be applied to the remote clock estimates, which is an advantage.

Fast convergence function. f_{fc} denotes this function and is used in [212, 115]. f_{fc} returns the average of all arguments x_1 to x_n that are within $\bar{\omega}$ of at least $n - f$ other arguments. f_{fc} yields a high-quality precision, but is computationally quite complex.

Fault-tolerant midpoint function. Denoted f_{ftm} and used in [352] returns the midpoint of the range of values spanned by arguments x_1 to x_n after the f highest and f lowest values have been discarded.

Differential fault-tolerant midpoint. Denoted f_{dftm} and used in [101, 102] is optimal with respect to the best precision achievable and best drift rate achievable for logical clocks. f_{dftm} is defined as $\frac{min(T-\Theta,x_l)+max(T+\Theta,x_u)}{2}$, where $x_l = x_{h_{f+1}}, x_u = x_{h_{n-f}}$ with $x_{h_1} \leq x_{h_2} \leq x_{h_n}$, $h_p \neq h_q$; $1 \leq h_p$, $h_q \leq n$ where T is p's logical time and Θ is the maximum reading error of a remote clock.

Sliding window function. This function selects a fixed size window w that contains the larger number of clock estimates. This function is proposed in [257] and proposed to convergence functions differing by the way a window is chosen when multiple windows contain the same number of clock estimates and differing by the way the correction term is computed once the window has been identified. The first function, f_{mean}^{det}, chooses the first window and returns the mean of the clock values contained in the window instance. The second function, f_{median}, chooses the window containing clock estimates having the smallest variance and returns the median of all clock estimates within the selected window. The main interest of sliding window convergence functions is that logical clocks closeness degrades gracefully when more failures are assumed to occur.

Minimization of the maximum error interval-based function. Denoted f_{mm} and takes for each x_i an interval $[L_{p_i}(t) - e_{p_i}(t), L_{p_i}(t) + e_{p_i}(t)]$, where $e_{p_i}(t)$ is the maximum error on p_i's clock estimate and returns an interval for the corrected clock value. f_{mm} is used in [217].

Intersection of the maximum error interval-based function. Denoted f_{im} and also used in [217] is similar to f_{mm} in the sense that it also takes intervals representing clock estimates as arguments. f_{im}, however, returns an intersection of the intervals of the clock estimates.

Convergence Non-Averaging Techniques

Convergence non-averaging techniques compute a new clock value based on the fact that a fixed number of estimates of remote clocks have been received to compute a new clock value. The number of expected clock estimates depends on the type and number of tolerated failures. When all required clock estimates are received, the local clock is corrected to the value that is computed using the respective algorithm.

Reference [70] uses only one estimate as only performance failures are tolerated and the logical clock is corrected with kR, where k is the round number and R is the round duration. Similarly, [124] also only corrects kR. In [304], the logical clock is corrected with the value $kR + \pi$, where π is constant selected large enough for clocks not to be set backwards. In [347], the clock is corrected with the value of one of the clock estimates contained in the clock estimation message.

4.2.2 Classifications of Clock Synchronization Algorithms

The following paragraphs further classify clock synchronization algorithms.

Internal and External Clock Synchronization Algorithms

Clock synchronization algorithms can be classified by the clocks used as reference clocks. If the clocks used as reference are not part of the synchronizing ensemble, this is commonly called *external clock synchronization*. Examples of external clock synchronization algorithms are [71, 217]. If the clocks used as reference are also synchronizing clocks, this is called *internal clock synchronization*. There are a number of different internal clock synchronization algorithms protocols, which all basically differ by the different approaches to the three subproblems described by Schneider [297]. Examples for internal clock synchronization algorithms are [22, 101, 124, 183, 189, 188, 209, 212, 304, 272, 350, 70]. Fetzer and Cristian combined the internal and external clock synchronization [102, 71].

Sometimes internal clock synchronization is also additionally characterized as symmetric and asymmetric. In *symmetric* algorithms all clocks participating play the same role, whereas in *asymmetric* algorithms one or more predefined local clock(s) play a specific role often referred to as *master(s)*. Symmetry influences the ability of the algorithm to support failures and requirements in terms of message exchanges needed to achieve synchronization within a sought precision [4].

Deterministic, Probabilistic and Statistical Clock Synchronization Algorithms

Internal clock synchronization algorithms can further be classified by assumptions about the transmission delay of the communication system in *deterministic*, *probabilistic* and *statistical* clock synchronization algorithms [4].

- *Deterministic* clock synchronization algorithms assume that an upper bound on transmission delay exists. If this assumption holds, a certain quality level of clock synchronization (in terms of a guaranteed precision) can be guaranteed.

- *Probabilistic* and *statistical* clock synchronization algorithms do not assume a strict upper bound on the transmission delay. Instead, *statistical* clock synchronization algorithms assume that the first and second moment of the distribution of the transmission delay and – sometimes – that the distribution is known. As a consequence, a node does not know the precision of the clock synchronization at a given time. *Probabilistic* clock synchronization algorithms make no assumption about the distribution of the transmission delay. Instead it is assumed that with probability p, $p < 1$, the transmission delay will be smaller than a guaranteed constant maximum transmission delay. As a consequence, all clocks know whether they are synchronized or not at any time [67].

Hardware Versus Software Implementation

Clock synchronization algorithms that are implemented in hardware and use specialized hardware components, such as [299, 183], achieve tight synchronization. On the other side, implementations in software, such as [67, 101, 124, 304, 209, 188, 212, 257, 347], do not achieve synchronization as tight as hardware algorithms, but use commercial-off-the-shelf components. Lately, the trend towards support of some critical clock synchronization elements being supported by hardware enables tighter precision values for clock synchronization. An example of this trend is driven by the standardization activities around IEEE 1588 [142] effectively requiring some hardware support to be efficiently possible.

4.2.3 Limits in and Performance of Clock Synchronization Algorithms

As the remote clock readings are estimates of local clocks by other nodes and contain some unknown variances, clock synchronization can never be perfect. In other words, local clocks cannot be re-synchronized perfectly, so that at a given time the local clock readings are equal at all clocks. For a symmetric clock synchronization algorithm and a given uncertainty ε and n local clocks, Lundelius and Lynch have shown in [210] that the bound that a symmetric clock synchronization algorithm can achieve immediately after resynchronization cannot be smaller than $\varepsilon(1 - 1/n)$. For asymmetrical clock synchronization and one master, the bound is ε, the uncertainty of the remote clock reading of the master reference clock.

Krause et al. show in [185] that there is a relationship between precision and drift of the global time. The more frequently an algorithm synchronizes, the tighter the precision, but frequent resynchronization may lead to a larger drift of the global time. This is easily explained by the fact that any error terms in remote clock readings are integrated and increase the drift.

Anceaume and Puaut provide an overview of mostly internal clock synchronization algorithms in [4, 5] and describe the key parameters of these algorithms such as precision, failure modes tolerated, assumptions and drift rate of the global time (called accuracy by Anceaume). Table 4.1 provides an overview and classification of some clock synchronization algorithms. Table 4.2 provides an overview of the key parameters of the deterministic clock synchronization algorithms presented in Table 4.1. Table 4.3 provides a similar overview for statistical and probabilistic clock synchronization algorithms.

4.2.4 Related Work on Clock Synchronization Algorithms

Halpern et al. [124] present a fault-tolerant clock synchronization algorithm that works in arbitrary networks. Approaches to clock synchronization are refined for the employment in large distributed systems [218, 295, 294, 347]. Rushby and von Henke have formally verified clock synchronization algorithms [286]. Schedl simulated several clock synchronization algorithms [292].

Most of the presented clock synchronization algorithms use replication as fault tolerance strategy. Dolev, Welch and Papatriantafilou present approaches that achieve fault tolerance using self-stabilization [77, 76, 246]. Analyses of self-stabilizing clock synchronization algorithms can be found in [208, 59].

Measurement and control applications are increasingly using distributed system technologies such as network communication, local computing and distributed objects. As these measurement and control applications are based on distributed embedded systems, clock synchronization has become an important area for standardization. In 2001, the IEEE standard organization started a standardization process for clock synchronization in measurement and control applications. In 2002, the standardization committee published a draft standard [141] for clock synchronization, called "Precision Time Protocol" (IEEE P1588). This standard was revised and extended in 2008 and is called "IEEE Standard for a Precision Clock Synchroniza-

TABLE 4.1

Classification of clock synchronization algorithms [5].

Reference	Type	Failures C	Failures L	Failures P	Structure	Synchronization Event Detection	Remote Clock Estimation	Clock Correction
[70]	D	R	P	O/P	SYM-FLOOD	MSG	TT	NAV
[124]	D	B	P	B	SYM-FLOOD	MSG	TT	NAV
[304]	D	B	R	B	SYM-FLOOD	MSG	TT	NAV
[352]	D	B	R	B	SYM-FLOOD	SYNC	TT	$AV - f_{ftm}$
[188] (CSM)	D	B	R	B	SYM-FLOOD	SYNC	TT	$AV - f_e$
[347]	D	B	O	C/P	SYM-FLOOD	MSG	RCR	NAV
[257]	D	B	R	B	SYM-FLOOD	SYNC	TT	$AV - f_{sw}$
[116] (Tempo)	D	T	P	C	ASYM-MAST	not required	RCR	$AV - f_{fc}$
[212]	D	B	R	B	not imposed	not imposed	not imposed	$AV - f_{fc}$
[100]	D	B	R	B	not imposed	not imposed	not imposed	$AV - f_{dftm}$
[102]	D	B	R	B	not imposed	not imposed	not imposed	$AV - f_{dftm}$
[188] (CON)	D	B	R	B	not imposed	not imposed	not imposed	$AV - f_e$
[188] (COM)	D	B	R	B	not imposed	not imposed	not imposed	$AV - f_e$
[217] (MM)	D	R	R	B	not imposed	not imposed	RCR	$AV - f_{mm}$
[217] (IM)	D	R	R	B	not imposed	not imposed	RCR	$AV - f_{im}$
[240] (TT)	S	R	-	C	SYM-RING	not imposed	TT	not considered
[240] (RCR)	P	R	-	C	SYM-RING	not imposed	RCR	not considered
[12]	S	R	-	R	ASYM-SLAV	not imposed	TT	$AV - f_{id}$
[67]	P	T	-	P	ASYM-SLAV	not imposed	RCR	$AV - f_{id}$

Note: D stands for deterministic, P for probabilistic, S for statistical.

Components are labeled C (Clock), L (Link), and P (Processor). R means Reliable, C Crash, B Byzantine, O Omission, P Performance, and T Timing. Probabilistic and statistical algorithms do not assume an upper bound on message delays, which makes omission and performance failures irrelevant as indicated by "-".

SYM stands for symmetric, ASYM for asymmetric. FLOOD, RING, MAST, and SLAV represent flooding-based, ring-based, master-based, and slave-controlled schemes, respectively.

NAV is used for non-averaging and AV for averaging techniques. AV also contain the name of the convergence function.

SYNC or MSG are used when rounds are detected thanks to initially synchronized clocks or message exchanges, respectively.

RCR stands for "remote clock reading," TT for "time transmission."

TABLE 4.2
Properties of deterministic clock synchronization algorithms [5].

Reference	Type	Redundancy	Precision	Drift Rate	Msg.
[70]	I	none	$(\delta+\varepsilon)(1+\rho)+2\rho[R(1+\rho)+(\delta+\varepsilon)]$	ρ	n^2
[124]	I	none	$(1+\rho)(\delta+\varepsilon)+2\rho(1+\rho)R$	$(f+1)2\rho$	n^2
[304]	I	$2f+1$	$(\delta+\varepsilon)[(1+\rho)^3+2\rho](1+\rho)R$	ρ	n^2
[352]	I	$3f+1$	$5\varepsilon+4\rho\varepsilon+4\rho R$	$\rho+\frac{\varepsilon}{R_{min}}$	n^2
[188] (CSM)	I	$2f+1$	$(f+6)\varepsilon+6\rho S+2\rho R$	$(2f+12)\varepsilon+10\rho S+2\rho R$	n^{f+1}
[347]	I+E	$(f_0+1)(f_p+1)$	$\Gamma_t+2\rho\Gamma_a+2\rho R$	$\frac{\rho R-(1-\rho)\Gamma_t}{R+(1-\rho)\Gamma_t}$	$3n$ bcasts
[257]	I	$4f+1$	$\frac{n^2+n-f^2)(\delta+\varepsilon)+2R\rho(n-f)(n-f+1)}{n^2-5fn+n+4f^2-2f}$	not given	n^2
[116] (Tempo)	I	$2f+1$	$4D(1+2\rho)-4(\delta-\varepsilon)+2\rho R$	not given	$3n$
[212]	I	$3f+1$	$\frac{(n+f)\Theta+2f(\pi+\Theta)}{n}+2\rho R$	not given	-
[100]	I	$3f+1$	$4\Theta+4\rho R+2\rho\beta$	ρ	-
[102]	I+E	$2f+1$	$\varphi=\Delta+\Theta+\rho R$	ρ	-
[188] (CON)	I	$3f+1$	$max(\frac{n}{n-3f}\Theta+2\rho(R+2S^{\frac{n-f}{n}}),\beta+2\rho R)$	not given	-
[188] (COM)	I	$3f+1$	$(6f+4)\Theta+2\rho S(4f+3)+2\rho R$	$(12f+8)\Theta+(8f+5)2\rho S+2\rho R$	-
[217] (MM)	I	$3f+1$	$2E_m(t)+2(\delta+\varepsilon)+2\rho(R+2(\delta+\varepsilon))$	not given	-
[217] (IM)	I	$3f+1$	$\delta+\varepsilon+2\rho R$	not given	-

Note: Although not required, [5] assumes a fully connected network for [70, 124, 304] in order to be able to compare the algorithm precision with the one of other algorithms. The worst case precision for [352] given here is obtained when taking $\beta=4\varepsilon+4\rho R$ as suggested by the authors. R_{min} is the minimum duration of a round. For [188] (CSM), S stands for the time interval during which clock estimates are obtained (last S seconds of R). In [347] at most f_0 omissions are supported during each synchronization round and there can be at most f_p fault clock-node pairs per round. Γ_t is maximum difference between message received at different nodes. Γ_a is the maximum duration of an agreement protocol. For [116], D equals $\delta+\varepsilon$. For comparison reasons, external time masters do not drift and reference time servers approximate real time with an a priori given error Δ for [102]. For [217](MM), $E_m(t)$ stands for the smallest clock error in the system.

TABLE 4.3

Properties of probabilistic and statistical clock synchronization algorithms [5].

Reference	Type	Precision	Messages
[240] (TT)	I	$\int_{-\infty}^{\infty} dy \int_{-\infty}^{w+y+n(\delta-\epsilon)} f_{max}(y,m) f_{min}(x,m) dx$	not mentioned
[240] (RCR)	I	$P_{\gamma<\gamma_{max}} = erf\left(\frac{\gamma_{max}\sqrt{2m}}{\sqrt{n}\mu}\right)$	$m = 2\frac{n\mu^2}{2\gamma_{max}^2} erf^{-1}\left(P_{\gamma<\gamma_{max}}\right)^2$
[12]	I	$\gamma_{max} = 2\left(\gamma_{synch} + (R+2\epsilon)\rho\right)$	$\frac{2\sigma_d^2(erfc^{-1}(p))^2}{\epsilon_{max}^2}$
[67]	I	$U - \delta + \epsilon + \rho k(1+\rho)W$	$\frac{2}{1-p}$

Note: For [240] (TT), f_{min} and f_{max} stand for the density function of the lower and upper bound on the skew interval, respectively, generated by a message under the assumption that endpoints of the interval can be modeled as independent random variables. The constant $w/2$ is the sought remote clock reading error. The estimate in column *Precision* is the probability a remote clock is estimated with an error $w/2$ using m messages.

For [240] (RCR), the value given in the *Precision* column is the probability a remote clock is estimated with an error lower than a given constant γ_{max} using m messages. Function erf - defined as $erf(u) = frac2\sqrt{\pi} \int_0^u e^{-u^2} dy$. Variable γ is the difference between any two clocks assuming that the average of the transmission delay δ is known, γ_{max} the maximum value the estimate is allowed to vary from the true value. μ^2 is the variance of a single hop on the ring. The value in column *Messages* is the average number of messages exchanged per resynchronization.

For [12], the *Precision* value is the best precision achievable when a precision of γ_{synch} in obtained just after resynchronization. Variable p stands for the probability a precision of γ_{synch} is obtained, σ^2 is the standard deviation of the message delay. $erfc^{-1}(p)$ is the inverse of the complementary error function defined as $erfc(u) = 1 - erf(u)$.

For [67], the *Precision* column value is the best precision achievable assuming no error occurs. Variable k is the number of successive reading attempts performed by the master, and W is the delay elapsed between each reading attempt. U is the maximal round trip delay accepted by the master to consider a slave response, beyond which, the master discards a reading attempt. The figure given in the *Messages* column is the average number of messages required in a fail-free environment. p is the probability that a process observes a round trip delay greater than $2U$.

tion Protocol for Networked Measurement and Control Systems" [142]. This standard addresses the needs of measurement and control systems: microsecond to submicrosecond accuracy; administration free; and most importantly, accessible for both high-end devices and low-cost, low-end devices. This standard decouples the application and communication task by introducing an application and a communication layer and an isolation layer that isolates application activities from communication activities and provides time provision services. The communication layer of the protocol requires all communicating nodes to follow a time-division multiple access (TDMA) strategy for the communication medium in order to achieve low latency jitter. The TDMA strategy is achieved by implementing a master/slave protocol [142].

4.2.5 Time Standards and Sources

Time standards are an agreed origin and representation of time. This section describes the two time standards *Internal Atomic Time* and *Universal Time Coordinated*. Clocks that are synchronized to time standards are potential sources for references for external clock synchronization. That is, they can be used as relational clocks. We call systems that provide clocks that are synchronized to time standards *time sources*.

4.2.5.1 Time Standards

International Atomic Time

In 1967, the Bureau International de l'Heure (BIH) specified the *Temps Atomique International* (TAI) in order to provide a time standard that can be produced in a laboratory, but is in agreement with the second derived from astronomical observations. The TAI defines the second as the duration of 9192631770 periods of the radiation of a specified transition of the cesium atom 133. The TAI is a strictly monotonic timescale (also called a *chronoscopic* time scale). That is a timescale without any discontinuities.

Universal Time Coordinated

The *Universal Time Coordinated* (UTC) is a time standard that is in close relation to the time derived from astronomical observations of the rotation of the earth relative to the sun. UTC is a time standard used for business relations and is a basis for widely used synchronized clocks, such as wall clocks. There is an offset between TAI and UTC due to the deceleration of the rotation speed of the earth, which is adjusted in second intervals by the BIH whenever necessary. UTC time is always kept within ±0.9 seconds of the time derived from astronomical observations (including polar wander effect corrections) by the insertion of an extra second (positive leap second) as needed. While it is theoretically possible to have to remove an extra second (negative leap second), it has not happened so far. From the last statements it can be concluded that UTC is not free of discontinuities.

4.2.5.2   Time Sources

Global Positioning System

The U. S. Department of Defense funds, controls and operates the *Global Positioning System* (GPS). GPS is a dual-use, satellite-based system that provides accurate location and timing data world-wide. Provision is done via specially coded satellite signals that can be processed in a GPS receiver, enabling the receiver to compute position, velocity and time. The time at GPS satellite is synchronized to UTC time with an accuracy that is better than 15 *ns* (there is an offset between GPS time and UTC to provide monotonicity). The accuracy of the timing information of commercially available GPS receivers using special measurement methods is about 50 *ns*. GPS is an accepted and widely used time source for military and civil application [197, 133, 72].

Global Navigation Satellite System

The *GLObal NAvigation Satellite System* (GLONASS) is – similar to GPS – a dual-use, satellite-based navigation system that enables global-wide positioning, velocity measuring and timing information. The GLONASS system is managed by the Russian Space Forces for the Russian Federation Government. The GLONASS system provides two types of navigation signals with different precision levels. The time of GLONASS satellites is synchronized to UTC (with a constant offset to provide chronoscopic behavior) with approximately 15 *ns* [192].

Galileo

Galileo is the planned European satellite navigation and time transferring system. The European Union launched the GALILEO project due to the following concerns regarding GPS and to a degree GLONASS [164]: Both systems are under the unilateral control of a foreign national defense authority, the absence of guarantees of service, priority on military needs, intentionally degraded use to civil needs, poor availability in urban areas and unpredictable gaps in coverage. Galileo should overcome these concerns and become fully operational in 2008.

Radio-Controlled Clocks

In several regions of the world, time information is broadcast via radio services with a UTC timecode modulation (examples are DCF77, WWVH, WWV, and HBG). Timecode receivers enable the reception of a time signal with an accuracy with respect to UTC in the order of 10 *ms*. Receivers, however, are subject to occasional gross errors due to propagation and equipment failures [197, 298, 224]. Terrestrial radio stations allow for accuracies of timing information in the range of 50 *ms* to 10 μs [199].

4.2.6   Time Aspects from an Application-Specific View

Time can serve different aspects for an application. For a real-time control application, time must have properties such as chronoscopic behavior. For applications that use time as reference for users for synchronizing their wall clocks and watches, time must be synchronized to UTC and chronoscopic behavior is thus impossible. This section describes different aspects of time.

Optimal Representation of Time for Real-Time Systems

For time in real-time systems, a representation should be chosen that has the following properties:

- No overflow occurs in system life time

- The precision of the time is smaller than the granularity of the representation

- Chronoscopic behavior

The time should be accessible via reliable sources world-wide and the jitter at receivers with respect to the reference time should be low. The GPS time is a continuous representation of time with high availability and low jitter. Yet, GPS time overflows every 19.6 years because the GPS week is represented by a 10 bit value. This shortcoming of GPS is overcome by the time format [241] described in Figure 4.1. This representation guarantees no overflow during the lifetime of the product (horizon of more than 10,000 years) and the full second can be easily accessed due to binary representation of a second overflow. The granularity of this time representation is 59.6 *ns*, which is in the order of the accuracy of common GPS receivers. Using this time representation and a satellite-based time provision system enables the above-mentioned requirements.

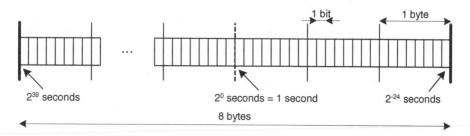

FIGURE 4.1
Optimal Representation of Time in Real-Time Systems [241]

4.3 Startup and Restart

4.3.1 Introduction and Overview

The initial synchronization of distributed local clocks in a network of computing nodes is an essential prerequisite for successful time-triggered communication. Indeed, clock synchronization algorithms for most applied time-triggered protocols are based on the assumption of initially synchronized local clocks as established by a startup algorithm. While both algorithms can be interwoven, in principle, the main motivation of a dedicated startup algorithm is in achieving a simple clock synchronization algorithm as it is continually executed during mission time. Typical periods of clock synchronization are in the order of some tens of milliseconds, resulting in several millions of repetitions in a transatlantic flight for example. In contrast, the startup algorithm is ideally executed only once. Hence, we isolate complexity in the startup algorithm as a simple clock synchronization algorithm contributes to low runtime and resource requirements as well as robust and precise synchronization.

The success of a startup algorithm ideally does not depend on the individual power-on times of the participating nodes. Therefore, most startup algorithms are subdivided in a so called "coldstart" and "integration" phase. Coldstart is the phase when nodes negotiate and agree on the initial synchronization point. Nodes that are powered on late will then integrate to this synchronized timebase. Coldstart and integration are discussed in more detail in Section 4.3.2.

Once synchronization is established, clock synchronization maintains the synchronization quality. Clock synchronization algorithms can be designed in a fault-tolerant way such that they even sustain faulty nodes and faulty channels. In rare scenarios, when the underlying fault hypothesis is exceeded, it may happen that the synchronization is lost. Therefore, startup algorithms are often used for the resynchronization after a network-wide synchronization failure. In this case we speak of a restart of the system. Before restart, a node has to detect that it has lost synchronization. For this purpose, time-triggered protocols use clique avoidance or clique detection algorithms. We discuss restart in Section 4.3.3.

The startup problem is often associated with the leader election problem. However, it is the task of the startup to reach an agreement on a point in time rather than to reach an agreement on a particular node which would then in turn propose a starting point in time. One reason to have a dedicated startup algorithm rather than to use a leader election algorithm is that the leader just elected may become faulty and a new election phase is required.

Figure 4.2 depicts boxes which represent common phases of operation in a time-triggered protocol: Integration, coldstart and the synchronized phase. Arrows between boxes denote transitions between two phases. Arrows marked with (A)..(C) represent the initial phase entered after power-on. Note that initially entering the synchronized phase (C) demands an external synchronization event, such as the simultaneous power-on of all nodes or an external time reference like GPS. The algo-

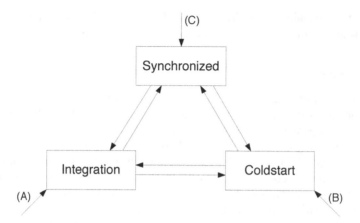

FIGURE 4.2
The Startup/Restart Triangle: A Time-Triggered Protocol Usually Distinguishes between the Integration, the Coldstart, and the Synchronized Phase

rithms realizing the individual phases as well as the transitions and the initial phase are key characteristics of a particular instance of a time-triggered protocol.

In one example protocol, a node may start in the integration phase after initial power-on (entry point A). In the integration phase, the node listens to the network to discover whether there exists already a synchronized timebase, or not. The presence of a synchronized timebase may be indicated by the periodic synchronization messages used in the clock synchronization algorithm. When the node receives such synchronization messages, it may use these messages for integration and enter the synchronized phase. In a case when a node does not receive such synchronization messages for a sufficient long duration, it is safe to conclude that there does not exist a synchronized timebase. In this case, the node transits to the coldstart phase and aims to establish a new timebase. As more and more nodes become powered-on the coldstart phase will succeed in establishing a timebase and the nodes may move on to the synchronized phase.

While the clock synchronization problem has been studied thoroughly since the early 1980s, the startup problem got relatively little attention: Claesson et al. present a strategy for the solution of the startup problem that is based on unique message lengths and full-duplex communication links [60]. Lönn discusses startup algorithms in [204] and formally verifies with Pettersson a particular startup algorithm in [206]. Startup algorithms based on unique timeouts are similar to the startup algorithm of the Token Bus protocol [139]. Krüger introduces such an algorithm for the startup of time-triggered protocols [186]. The TTP/C startup algorithm as specified in [171] is based on this principle. More recent studies of the startup problem have been conducted by Widder [353] and Steiner and Kopetz [312]. Starting time-triggered communication in ring topologies has been researched by Paulitsch and Hall [249].

In the remainder of this chapter, we continue discussing the phases of the startup triangle and its transitions in more detail.

4.3.2   Startup

As introduced in Chapter 2, components in a system can be characterized by their functionality: Nodes are components that access sensor values, perform computations and/or operate actuators. Channels are components that connect nodes to each other. Depending on the network topology, channels may be pure wiring only in case of a multi-drop bus topology or may consist of several nodes such as routers, switches, hubs or repeaters in case of a star or tree-based topology. Sometimes, a component may be both, node and channel, e.g., a FireWire node acts as computing node and also relays messages from other nodes.

A node that has access to an oscillator, e.g., a quartz crystal, can use the regularity of this oscillator to implement a clock. A clock is a hierarchical set of cyclical counters and has state and rate. The state of a clock at some point in real-time is the current assignment of all its counters. The rate of a clock is the update rate of these counters. The state of a clock, therefore, changes with the progress of real-time in relation to the frequency of the oscillator. According to [282]: Let $\mathcal{C}$ be the *clocktime*, that is time represented within a node by its counters, and let $\mathcal{R}$ be the *real-time*; then the clock of node p is represented by the function:

$$C_p : \mathcal{R} \to \mathcal{C} \tag{4.3}$$

meaning that at each point in real-time t there exists a corresponding assignment of a node p's counters that represent the node's local view of time $C_p(t)$.

Formally stated, a startup algorithm ensures that there exists a point in time t_0, such that the difference of the local clocks in any two nodes p and q that are powered-on and non-faulty is less than the precision Π.

$$\exists t_0 : |C_p(t_0) - C_q(t_0)| < \Pi \tag{4.4}$$

Property 1 *Timely Startup: Whenever at least a minimum configuration of nodes and channels is powered-on, a startup algorithm establishes synchronous communication of the non-faulty nodes within an upper bound in time.*

The minimum configuration is specific to a time-triggered protocol. The "whenever" factor in the property specification is highly important, since it does not specify an upper bound in time until a minimum configuration is powered-up. The timeliness property is stronger than a "liveness" property: In contrast to liveness properties, timeliness properties require a known upper bound on the duration after which a property has to be established.

Property 2 *Safe Startup: When the startup algorithm terminates, all correct nodes and channels that communicate synchronously are synchronized to each other.*

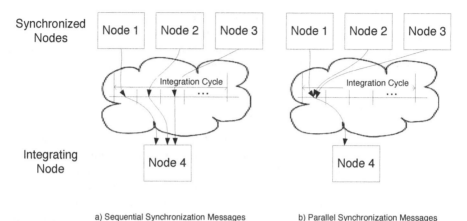

a) Sequential Synchronization Messages   b) Parallel Synchronization Messages

FIGURE 4.3

Integration Messages Must Be Periodically Communicated to Allow Integration; The Integration Messages May Be Sent in Sequence or in Parallel

The safety property ensures that the startup algorithm will not produce multiple "cliques," that are disjoint subsets of nodes that communicate synchronously within the subset but not with nodes in different subsets. However, cliques can be formed temporarily during the startup process.

4.3.2.1 Integration

Nodes in the synchronized phase periodically exchange integration messages. An integrating node determines from these integration messages the number of synchronized nodes. When this number is sufficiently high, the integrating node p uses the synchronization messages to initialize its local clock C_p. In this section, we discuss this integration process and alternative realizations.

Figure 4.3 depicts a network of four nodes. Nodes 1..3 are already synchronized, which means that they execute the off-line configured access schedule to the network. The network is represented by a cloud and may allow only sequential communication, like in a multi-drop bus, or allow also parallel communication, like a star/tree-based network. As a consequence, the nodes either have to send the integration messages in sequence or may send them in parallel. Many time-triggered protocols use the clock synchronization messages also as integration messages. However, this is an implementation choice rather than a necessary paradigm. In principle it is possible to use different sets of messages with potentially different message periods. However, to increase bandwidth efficiency, one type of message is typically used for both, integration and clock synchronization.

From a purely functional perspective, integration of a node requires the reception of a single integration message only. The receiver knows the state of the local clock in the sender C_S at the scheduled dispatch point in time $t_{dispatch}$. Furthermore, *latency* and *jitter* of the received integration message are bounded. Upon reception of the integration message, at $t_{receive}$, the receiver can set its local clock C_R to max(*latency*): $C_R(t_{receive}) = $ max(*latency*). As a consequence, at $t_{receive}$, C_R and C_S are synchronized (where ρ is the maximum drift rate of the local clocks in the system):

$$C_R(t_{receive}) - C_S(t_{receive}) \leq jitter + (\text{max}(latency) \times \rho) \qquad (4.5)$$

The network jitter will be typically the dominant factor in this equation, especially in networks that share the physical communication links with unsynchronized traffic. For example, in the TTEthernet chapter we discuss the "transparent clock" mechanism as a means to measure the individual jitter of a message in the network and the "permanence function" to transform the measured jitter into network latency.

The minimum content of an integration message is a type field which distinguishes the integration message from all other messages in the system. This field may even be encoded in other fields present in the message, as for example an Ethernet frame with a particular Ethernet MAC destination address. Other typical content of an integration message is as follows:

- Sender identifier: An identifier of the sender of the integration message.

- Membership vector: A bitvector with a fixed relation of bits to nodes in the system, for example to the set of nodes that are allowed to send integration messages. In case of parallel synchronization messages, the channels in the network may form a single integration message out of a group of integration messages. In order to preserve the information about the original senders, the channels can set the respective bits in the membership vector of the newly generated frame.

- Integration cycle number: The length of the overall communication cycle is determined by the least common multiple of all frame periods in the system. Long communication cycles may require multiple integration points, such that a node may not only integrate at the beginning of the communication cycle but also at periodic points in between. The integration cycle number identifies these points.

- Global time value: A time-triggered protocol may provide a global time service with a time-horizon that exceeds the communication cycle. The global time value is then used to integrate to this global time.

Integration algorithms become slightly more complex, when faulty integration senders have to be assumed. Such faulty senders may send integration frames at faulty points in time or with fault message contents. Hence, the reception of an integration message from a single sender is in general insufficient for a fault-tolerant integration process. Instead, integration messages from a group of nodes have to be used. The size of this group follows the classic fault-tolerance results: $k + 1$ senders to tolerate k fail-silent faulty senders, $2k + 1$ for fail-consistent failures, and $3k + 1$ for

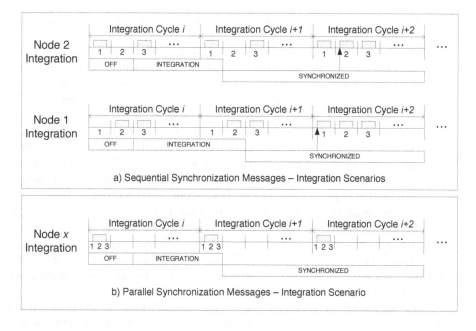

FIGURE 4.4
The Determination of the Number of Synchronized Nodes Can Be Done by "Counting" the Number of Received Integration Messages or by Checking the Contents of an Integration Message

arbitrarily-faulty senders. When error-containment measures are implemented, like central guardians, the failure modes refer to the failure manifestation on the interface to the error-containment unit, e.g., the output as forwarded or blocked by the central guardian.

As different networks allow different synchronization message transmissions, the process of deducing the number of senders from the received integration messages differs. Figure 4.4 shows some example scenarios on how the information from multiple senders may be combined in the integration process.

The top two examples in Figure 4.4 depict the fault-tolerant integration process when the integration messages are only sequentially transmitted. Here we assume a system of three nodes where two nodes are synchronized and the third node is integrating. In the first scenario, Nodes 1 and 3 are synchronized and Node 2 is powered-on just after the slot of Node 2. It receives the integration message of Node 3 and continues listening for integration messages with matching contents and timing. In the next integration cycle, Node 2 receives the integration message from Node 1, which matches the previously received integration message. Node 2 received two integration messages from different senders. In case of a fail-consistent failure model, this is sufficient for Node 2 to integrate. The second scenario is similar, with Node

1 being the integrating node. Again, Node 1 needs integration messages from two independent sources for integration.

When parallel integration messages are possible, all synchronized nodes may send their integration message at the same scheduled dispatch point in time. The integration messages are then sent to a dedicated set of "compressing nodes" rather than broadcast to all nodes. Each compressing node generates a new compressed integration message in which the membership vector identifies the original integration message senders. This compressed integration message is then broadcasted in the network. An integrating node can deduce the quality of the compressed integration message by the number of bits set in the membership vector. When this number is sufficiently high, the integrating node can use the integration message for integration. The third example in Figure 4.4 depicts a scenario in which parallel integration messages are used. Here, a Node x is able to integrate after the reception of a single compressed integration message with a sufficiently high number of bits set in the membership vector.

Dedicated compression nodes have been introduced to reduce the overall number of broadcast messages and to implement two convergence phases in the clock synchronization process.

The integration phase can terminate "successfully" or "unsuccessfully": When the node has received a sufficiently long sequence of messages, it is able to synchronize to a running system and the integration terminates successfully. If the node is not able to integrate for a given duration, the node terminates the integration phase unsuccessfully and transits to the coldstart phase.

4.3.2.2 Coldstart

When a node perceives no synchronized communication or synchronized communication of a too little number of nodes, it may enter the coldstart phase. A time-triggered protocol may be configured such that only a dedicated set of nodes, the "core system," is allowed to enter the coldstart phase while the remaining nodes enter the synchronized phase only by integration. In the coldstart phase, a node aims to establish a new synchronized time-base rather than integrating to an existing one. As the node is not synchronized yet, its input/output behavior to the network is uncoordinated with other nodes. In particular, a node in the coldstart phase will asynchronously send periodic coldstart messages and will asynchronously receive coldstart messages.

Coldstart messages are signals for the nodes to start synchronized communication. There are different options to construct coldstart messages [305]:

- Noise: any kind of activity other than the idle state of the network channels. This form of coldstart message depends on the physical layer.

- Semantic-Free Coldstart Message: A well-formed message with minimum content only identifying the message as a coldstart message, as for example a type field.

- Semantic-Full Coldstart Message: A well-formed message with content exceeding the minimum; for example, information on where to start in the communication schedule similar to that discussed for the integration messages.

Similar to the integration process, the reception of a coldstart message is used to set the local clock in the receiver C_R such that it holds approximately the same value as the local clock in the sender C_S (see Equation 4.5).

In a fault-tolerant system, it is necessary to configure more than one node to send coldstart messages to ensure the presence of a starting node despite failures. The asynchronous transmission of coldstart messages leads to a contention problem at the network when two or more nodes send their coldstart message within a short time interval. Depending on the network properties, the contention may result in a physical collision, a logical collision or network buffering.

Physical collisions appear when half-duplex communication links are used or in multi-drop network topologies, like the traditional bus network topology. Here, two or more senders access the same communication link at the same point in time. Thereby, the electro-magnetic waves in the physical communication links overlay. Physical collisions may have different effects in different time-triggered protocols. When noise is used as coldstart message the collision itself may be still sufficient as a starting signal for time-triggered communication. In case of semantic-free coldstart messages, the physical collision may destroy the message consistently for all nodes or inconsistently only for some nodes. However, as there is no more content transported in the coldstart message, it is guaranteed that all nodes that receive the semantic-free coldstart message will set their local clocks to about the same position in the communication cycle. The physical collision of semantic-full coldstart messages is more critical. Again, the physical collision of semantic-full coldstart messages may be consistent or inconsistent. In the inconsistent case, different nodes may receive different semantic-full coldstart messages. Hence, different nodes potentially set their local clocks to different points in the communication cycle.

In summary, the physical collision of coldstart messages may result in an insufficient quality of synchronization (when using noise as a coldstart message), no initial synchronization at all (when using semantic-free coldstart messages) or inconsistent initial synchronization (when using semantic-full messages). To resolve these problems, a contention-resolving algorithm can be used. In particular, a deterministic contention resolving algorithm guarantees that there exists an upper bound in time, when the access of at least one node will not result in a contention:

Property 3 *If several nodes have produced a contention at their n-th access to the shared medium, there exists an x such that the $(n+x)$-th access of at least one node in this set will not result in a contention.*

This property is essential for the coldstart phase in networks with bus-based topology since it guarantees that even if there are more nodes sending their coldstart messages at approximately the same point in time, there exists an upper bound in time when one node will send its coldstart signal without a contention.

The contention problem naturally arises in communication networks based on a

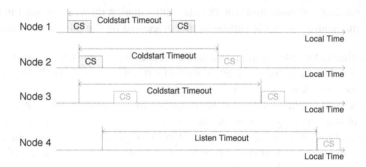

FIGURE 4.5

TTP Coldstart Scenario: Nodes 1 and 2 Produce a Collision, Which is Resolved by Unique Timeouts

shared communication medium and, hence, communication protocols have to provide solutions for this problem. A summary of contention resolving algorithms of well-established communication protocols (such as Avionics Full-Duplex Ethernet, Token Bus, etc.) can be found in [305, p.35 ff.]. Such protocols usually use priority-based algorithms where the priorities are realized as different unique timeouts; e.g., a re-try timeout – if a contention occurs, the one node with the shortest timeout will be the first to re-transmit its message. An example contention-resolving algorithm based on unique timeouts is the TTP coldstart procedure [313] depicted in Figure 4.5.

The coldstart procedure assigns each node a unique coldstart timeout which defines the period in which the node dispatches coldstart messages. In the scenario depicted in Figure 4.5, nodes 1 and 2 send their coldstart messages at approximately the same point in time, which results in a collision. Node 1 has the shorter coldstart timeout assigned and is, hence, the first one to send a succeeding coldstart message. Receivers of this coldstart message may start synchronized operation. Node 3 would send a coldstart message before the second coldstart message of node 1, but resets its coldstart timeout when it perceives the collision. Also, node 4, which is powered-on late and does not perceive the collision, will wait an initial listen timeout which is sufficiently long that its first coldstart message would not collide with coldstart messages from nodes that already produced a collision.

Logical collisions are a result of the replication of the shared medium where the replicas are controlled by mutually independent instances, e.g., central guardians. Each of these instances guarantees a transmission free of physical contentions on one replica. However, since these instances are independent of each other, nodes that start to broadcast at approximately the same time may occupy only a subset of the replicas each. A receiver, therefore, will receive messages from different senders on the replicas of the communication medium. Logical collisions may occur in networks with star topology and there is similar potential for inconsistent initial synchronization when semantic-full coldstart messages are used as discussed in the context of physical collisions. Contention-resolving algorithms can also be used to resolve logical collisions.

A network capable of buffering coldstart messages in case of a contention does not have to realize a contention-resolving algorithm. In case of a contention, the coldstart messages can be buffered in the network nodes, like switches, and be relayed sequentially. Again, measures like the transparent clock and the permanence function (discussed in the context of TTEthernet) can be applied to mitigate the additional transmission jitter on the coldstart messages.

The coldstart will terminate "successfully," if the node finds a sufficient number of nodes/channels synchronized. The node then transits to the synchronized phase. The coldstart terminates "unsuccessfully" if the number of nodes/channels is not sufficient. When the coldstart phase terminates unsuccessfully, another coldstart phase may be started immediately or the integration phase may be re-started, which allows the node to check whether a different set of nodes reached the synchronized phase already. Determining whether a set of nodes/channels is sufficient or not is done at design time and can be bound to the following numbers:

1. The numbers of nodes/channels necessary to guarantee the correct operation of the application, or

2. The numbers of nodes/channels necessary to allow an unsynchronized node to integrate into a synchronized system (this is the approach used within this chapter).

4.3.3 Restart

Network startup addresses the problem of initial synchronization after power-on of the nodes and channels in the network. Once initial synchronization is established, the clock synchronization routine takes over and maintains the synchronized timebase. Clock synchronization algorithms for time-triggered protocols are typically also fault-tolerant and ensure synchronization even in the presence of failures. The restart procedure allows faulty nodes to re-synchronize to the network once their error state has been corrected, for example after a node reset. On the other hand, in rare situations, like an airplane hit by a lightning stroke, the number and mode of failures may exceed the fault hypothesis such that global synchronization is lost. In a good case, this synchronization loss is consistent throughout the network; in a bad case, disjoint subsets of nodes may remain synchronized within their subset but the synchronization between the subsets is lost. In the latter case, the subsets are called "cliques." The restart procedure also addresses these rare situations.

Figure 4.6(a) depicts the general concept behind a restart procedure. The concentric circles depict the sets of possible states of a system. The center circle forms the set of system states covered by the fault hypothesis. We say this is the set of safe states. The other circles depict system states that violate the fault hypothesis. These circles form a classification: inner circles represent unsafe states from which recovery can be done by simple procedures; the more outward the more complex the

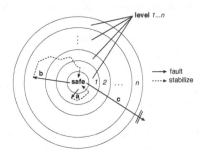

(a) General levels of safe-state recovery

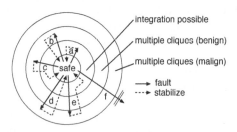

(b) Particular levels of safe-state recovery in the TTA [314]

FIGURE 4.6
Different Levels of Safe State Violations and Their Recovery Actions (with kind permission from Springer Science + Business Media [314])

recovery procedure may be. The unsafe system states in time-triggered communication protocols are typically different types of clique formations. A fault affecting multiple nodes may cause a transition from a safe system state to an unsafe one. Fault b in Figure 4.6(a), for example, is a more severe failure than fault a. The restart procedure of a time-triggered protocol implements algorithms to stabilize the system to the safe system states. Figure 4.6(b) depicts the restart scheme for the time-triggered architecture (TTA). For the different classes of unsafe states, different algorithms, in particular clique detection algorithms, can be implemented. Once a clique is detected, the startup algorithm is used to bring the system back to the safe state.

4.3.3.1 Clique Detection Algorithms

Clique detection algorithms are diagnosis algorithms that keep track of the health state of the synchronized global time. They may only use the regularity and irregularity of the synchronization message reception as symptoms for their conclusion on the health state of the synchronized global time. We can distinguish three types of clique detection functions: Synchronous clique detection, asynchronous clique detection and relative clique detection.

The synchronous clique detection algorithm counts the number of nodes that are currently synchronized to each other. When a node detects this number to be smaller than a threshold T^{sync}, then the synchronous clique detection algorithm has detected a clique scenario. The quantification of T^{sync} is influenced by several design decisions such as the number of faulty nodes allowed in the system or the number of nodes required for the integration process: when a node discovers too few nodes synchronized with itself than would be necessary to integrate, the synchronous clique detection algorithm can decide to reset the node. Furthermore, T^{sync} may be statically configured and dynamically adapted. In the latter case, T^{sync} can be a function of the overall number of powered-on nodes in the system. Hence, the threshold may change during mission time. When more and more nodes become powered-on the thresholds increase and they are decreased when nodes are powered-down.

Asynchronous clique detection algorithms count the number of nodes that are powered-on, but are not synchronized with each other. A node that counts equal or more than T^{async} unsynchronized nodes has detected a clique by means of asynchronous clique detection.

The relative clique detection algorithm simply compares the T^{sync} with T^{async}. When $T^{sync} \leq T^{async}$ then the relative clique detection algorithm has detected a clique.

The counting processes in the clique detection algorithms depend on the underlying physical network properties. In bus-based networks, the receiver of a synchronization message can associate a particular node with the received synchronization message. When the receiver receives the synchronization message of a node x within a defined acceptance window around an expected receive point in time, then the receiver qualifies the synchronization message as temporally valid, otherwise temporally invalid.

We know that the receive point in time of the synchronization message is a symptom from which the receiver can deduce the state of the local clock in the sender $\mathcal{C}_S$. Hence, when $|\mathcal{C}_S - \mathcal{C}_R| \leq \Pi$ then the receiver classifies the sender as synchronized with itself. When there exists a one-to-one relation between nodes and received synchronization messages, an acceptance window of size $2 \times \Pi$ is necessary and sufficient to qualify sender and receiver to be either synchronized or unsynchronized.

In star-based networks, the receiver may associate multiple nodes with a single synchronization message. In TTEthernet, for example, switches will compress the synchronization messages from the senders into a single synchronization message that is then forwarded. The receiver knows from a bitvector within the received synchronization message how many nodes are related to this message. In this case, the counting process of the clique detection functions use this number to qualify against the configured thresholds T^{sync} and T^{async}. When multiple failures have to be tolerated, the compression of synchronization messages from multiple senders into a single synchronization message imposes an interesting corner case for the clique detection functions. As the received synchronization message represents the local clock of multiple receivers, a failure in one of the senders influences the perception of the

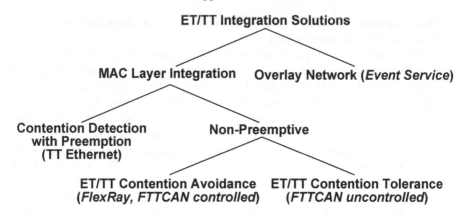

FIGURE 4.7
Solutions for Integration of Event-Triggered and Time-Triggered Communication

receiver of all local clocks in the senders. Hence, when the sender to received synchronization messages is a multiple-to-one relation, the influence of the faulty node and/or faulty channel has to be reflected when selecting the size of the acceptance window.

4.4   Integration of Event-Triggered and Time-Triggered Communication

Due to the respective advantages of the paradigms of event-triggered and time-triggered control, solutions for the integration of both communication approaches have been developed. The rationale behind these integrated communication protocols is the effective covering of mixed criticality systems, in which a safety-critical subsystem exploits time-triggered communication services and a non-safety-critical subsystem can exploit event-triggered communication services. Thereby, the communication protocol can support different, possibly contradicting requirements from different application subsystems.

For this reason, several communication protocols integrating event-triggered and time-triggered control have been developed [105, 251, 234, 179]. These protocols differ at the level at which the integration takes place (i.e., either at the Media Access Control (MAC) layer or above), and the basic operational principles for separating event-triggered and time-triggered traffic.

Communication protocols for the integration of event-triggered and time-triggered control can be grouped into two classes, depending on whether the integration occurs on the Media Access Control (MAC) layer or through an overlay network (see Figure 4.7).We further subdivide the MAC layer solutions depending on whether contention between event-triggered and time-triggered messages is resolved by message preemption or non-preemptively. In the latter case, contention along with the resulting communication jitter can either be tolerated or contention can be avoided by enforcing constraints on the transmission start instants of messages.

4.4.1 Integration of Event-Triggered and Time-Triggered Communication at MAC Layer

This class of protocols employs a MAC layer that supports both event-triggered and time-triggered message transmissions. The start and end instants of the periodic time-triggered message transmissions, as well as the sending nodes are specified at design time. For this class of messages, contention is resolved statically. All time-triggered message transmissions follow an a priori defined schedule, which repeats itself with a fixed round length, which is the least common multiple of all time-triggered message periods. Within each round, the time intervals that are not consumed by time-triggered message exchanges are available for event-triggered communication. Consequently, time is divided into two types of slots: event-triggered and time-triggered slots. The time-triggered slots follow from successive time-triggered message transmissions, while event-triggered slots are located in between the time-triggered slots. In event-triggered slots, message exchanges depend on external control and the start instants of message transmissions can vary. This difference with respect to the start instants of event-triggered and time-triggered slots is depicted in Figure 4.8. Furthermore, event-triggered slots can be assigned to multiple (or all) nodes of the system. For this reason, the MAC layer needs to support the dynamic resolving of contention when more than one node intends to transmit a message. During event-triggered slots a sub-protocol (e.g., CSMA/CA, CSMA/CD) takes over that is not required during time-triggered slots in which contention is prevented by design.

While time-triggered messages can always be fit into the respective slots at design time, on-demand event-triggered messages require support by the MAC protocol for protecting time-triggered slots. The mechanism for the delimitation of event-triggered and time-triggered slots allows us to further classify protocols with an integration of event-triggered and time-triggered communication at the MAC layer: Contention avoidance protocols, preemptive protocols and contention tolerant protocols.

4.4.1.1 Event-Triggered and Time-Triggered Communication — Contention Avoidance

Contention avoidance protocols reserve at the end of each event-triggered slot a time interval, in which no message transmissions may be started. The length of this time interval is equal to the maximum message transmission duration of an event-triggered message. Consequently, it is ensured that an event-triggered message transmission can always be completed before the next time-triggered slot starts.

An example of a communication protocol realizing this solution is FlexRay [105]. FlexRay denotes event-triggered slots as dynamic segments, while time-triggered

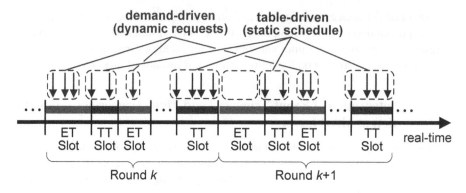

FIGURE 4.8
Event-Triggered and Time-Triggered Communication Slots

slots are named static segments. The static segments realize a strict TDMA scheme, while the dynamic segments employ an event-driven mini-slotting sub-protocol. The fixed duration of the dynamic segment is subdivided into mini-slots that identify potential start times of message transmissions. Each node is assigned a unique mini-slot, in which the node can start a message transmission in case the medium is idle. Consequently, an earlier mini-slot gives a node a higher priority compared to nodes with later mini-slots. Due to this demand driven access pattern, the reserved bandwidth of a dynamic segment can be shared between nodes. In FlexRay, the time interval reserved for contention avoidance at the end of an event-triggered slot is part of the dynamic slot idle phase.

Flexible Time-Triggered (FTT–CAN) in controlled mode is another example of a protocol integrating event-triggered and time-triggered communication at the MAC Layer with contention avoidance. Flexible Time-Triggered CAN (FTT-CAN) [251] is a CAN-based master/slave protocol for building a predictable time-triggered communication service on top of CAN, while also permitting event-triggered CAN communication. Time-triggered slots in FTT-CAN are used for the so-called *synchronous traffic*, which consists of periodic time-triggered messages. Time intervals not used by these periodic messages are available for the *asynchronous traffic* (i.e., event-triggered slots). Contention within the asynchronous traffic is resolved by the CSMA/CA arbitration mechanism of CAN. Furthermore, in *controlled mode* a node may only send an asynchronous message if the remaining time interval before the next synchronous message has a sufficient length for preventing any interference of synchronous and asynchronous messages.

4.4.1.2 Event-Triggered and Time-Triggered Communication — Contention Detection with Preemption

An example of a protocol that performs the integration of event-triggered and time-triggered communication based on contention detection with preemption is the Time-Triggered Ethernet (TTE) protocol [179]. TTE defines the coexistence of standard event-triggered Ethernet traffic and time-triggered traffic that is temporally guaranteed. For the time-triggered traffic, a static or dynamic scheduler has to define the conflict-free periods of the messages. The scheduler ensures that no conflict between TTE messages occurs. If an event-triggered Ethernet message comes into conflict with a time-triggered Ethernet message, then the TTE switch [316] preempts the transmission of the event-triggered message. After the completion of the transmission of the time-triggered Ethernet message, the switch autonomously retransmits the preempted event-triggered Ethernet message.

4.4.1.3 Event-Triggered and Time-Triggered Communication — Contention Tolerance

Contention tolerant protocols neither restrict transmission start instants within an event-triggered slot nor preempt ongoing event-triggered message transmissions. All event-triggered message transmissions are permitted to finish, thus leading to potential perturbations of the boundaries of time-triggered slots.

An example of this protocol type is FTT-CAN in *uncontrolled mode* [251]. In analogy to the controlled mode of FTT-CAN, both event-triggered communication (i.e., asynchronous traffic) and time-triggered communication (i.e., synchronous messages) are supported. Every node participating in time-triggered communication is equipped with a local table that contains information about the time-triggered messages transmitted and received during each communication round. Nodes can start with the transmission of event-triggered messages at arbitrary instants. Through assigning higher priorities to time-triggered messages, it is ensured that time-triggered messages always win in the arbitration process. Nevertheless, the non-preemptive nature of CAN results in transmission jitter of time-triggered messages, in case an event-triggered message is being transmitted when a time-triggered message transmission is scheduled. In the worst-case, a time-triggered slot with one or more time-triggered messages is delayed by the maximum transmission duration of an event-triggered message.

4.4.2 Event-Triggered Overlay Networks

Event-triggered overlay networks based on a time-triggered communication protocol are a solution for the integration of the two control paradigms by layering event-triggered communication on top of time-triggered communication. The MAC protocol is TDMA, i.e., time is divided into slots and each slot is statically assigned to a node that exclusively sends messages during this slot. A subset of the slots in each communication round is used for the construction of event-triggered overlay networks. This solution is similar to layer 2 Virtual Private Networks (VPNs), which emulate a point-to-point layer 2 connection over layer 3 (e.g., IP/MPLS networks) [334].

Event-triggered overlay networks have been established for different event-triggered protocols [234, 30]. A generic *event service* exploits time-triggered slots in order to support event-triggered on-demand message transmissions. This generic event service can then be used for the realization of layer 2 protocols (e.g., CAN [234]) or higher protocol layers (e.g., TCP/IP [30]).

4.4.3   Generic Event Service

The generic event service maps an event-triggered protocol to the sparse time-base [166] of the time-triggered communication protocol. Although message transmission requests can occur at arbitrary instants, the dissemination of the messages on the underlying time-triggered network is always performed at the predefined global instants of the time-triggered slots.

The generic event service is based on a temporal subdivision of the communication resources. Time Division Multiple Access (TDMA) of a time-triggered physical network statically divides the channel capacity into a number of slots and assigns to each node a unique slot that periodically reoccurs at a priori specified global points in time. Each node's slot is subdivided into two subslots, namely a slot for time-triggered communication and a slot for the event-triggered dissemination of messages (see Figure 4.9).

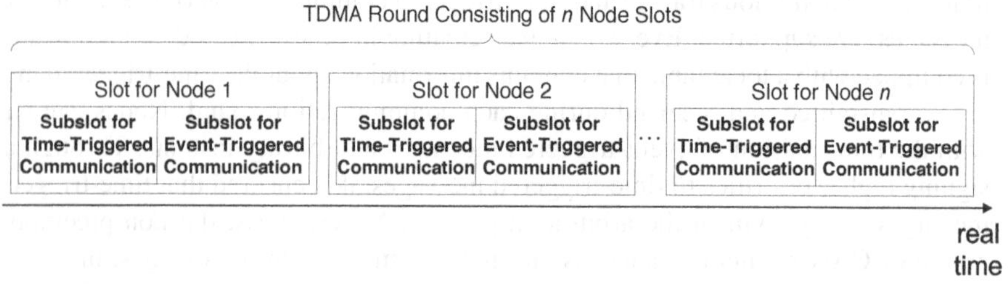

TDMA Round Consisting of *n* Node Slots

Slot for Node 1		Slot for Node 2			Slot for Node *n*	
Subslot for Time-Triggered Communication	Subslot for Event-Triggered Communication	Subslot for Time-Triggered Communication	Subslot for Event-Triggered Communication	···	Subslot for Time-Triggered Communication	Subslot for Event-Triggered Communication

real time

FIGURE 4.9
Subdivision of Communication Slots for an Event-Triggered Overlay Network

Since in the time-triggered protocol each slot is exclusively written by a single node, a particular slot enables only a single node to broadcast messages to all other nodes. In order to support a general communication topology, in which each node can transmit event-triggered messages, a system with *n* nodes employs *n* event-triggered slots as an input to the event service.

In each node, outgoing messages are buffered in message queues until the respective node's subslot for event-triggered communication occurs in the TDMA scheme (cf. Figure 4.10). Also, the queuing of messages handles bursts during which the bandwidth consumption of outgoing messages exceeds the bandwidth that is available via event-triggered subslots. In every node, the event service performs a fragmentation of outgoing messages into packets that can be placed in the node's event-triggered subslot. In addition, the event service reassembles messages out of received packets.

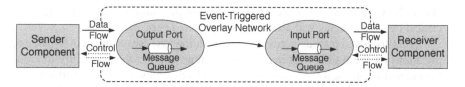

FIGURE 4.10
Event-Triggered Overlay Network

4.4.3.1 Higher Protocols: CORBA Internet Inter-ORB Protocol

The Common Object Request Broker Architecture (CORBA) specification of the Object Management Group (OMG) [238] provides a solution for distributed object computing based on a flexible middleware for integrating applications in heterogeneous environments. CORBA offers interoperability between components on different platforms and written in different programming languages.

The Object Request Broker (ORB) is the central building block in the CORBA architecture and hides the internal details of the execution environment (e.g., operating system, communication network). An ORB communicates with the application and with other ORBs using network connections and inter-process communication mechanisms. CORBA offers location transparency, since there are no differences for a client in accessing local and remote objects. In case of remote objects, the Internet General Inter-ORB Protocol (GIOP) establishes the interoperability between different nodes. The Internet Inter-ORB Protocol (IIOP) is a specific mapping of the GIOP onto TCP/IP. The protocol layers of IIOP are depicted in Figure 4.11.

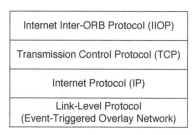

FIGURE 4.11
Commonly Used Protocol Layers for CORBA

Reference [207] describes a realization of GIOP using an event-triggered overlay network. At the link level, an implementation of the generic event service serves as the basis for the IP, TCP/IP and IIOP.

4.4.3.2 Higher Protocols: Controller Area Network (CAN)

The automotive industry is on the verge of deploying computer systems not only for safety-related and comfort functionality, but for safety-critical by-wire systems.

While the CAN protocol is prevalent in present day automotive networks, safety-critical by-wire systems will also use time-triggered networks.

Event-triggered overlay networks are a solution to reuse existing CAN-based applications on these time-triggered networks, thus offering the possibility to reduce the number of physical CAN networks, which leads to cost reductions and reliability improvements. However, for the reuse of CAN-based legacy applications, it is important to note that time-triggered networks and event-triggered overlay networks can exhibit a different temporal order of received messages than a physical CAN network for the same sequence of message transmission requests. While this difference may not be a concern in many newly developed applications, it poses a problem for the reuse of CAN-based legacy software. Substantial re-testing or adaptations of existing code would be required to ensure a correct behavior of legacy software despite the different temporal message order compared to the platform the application software has been developed for.

Therefore, a CAN protocol emulation middleware was introduced in [235] that establishes in an overlay network the same temporal order of the receive instants as in a physical CAN network. The CAN protocol emulation aims at the reuse of legacy applications with a minimum of redevelopment and retesting efforts.

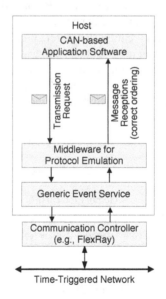

FIGURE 4.12
Node with Middleware for CAN Protocol Emulation

As depicted in Figure 4.12, in every node a protocol emulation middleware is executed, which is located in-between the generic event service and the application software. The protocol emulation middleware simulates at run-time a physical CAN network in every node and takes into account the message request instants, the message priorities and the message lengths [151]. The input of this simulation process are both messages for which a transmission has been requested at the same node and messages received via the generic event service from other nodes. Based on these inputs, the CAN emulation computes for each message the send instant when the message would have been sent on a physical CAN network. Messages are passed to the application in the order of ascending message send instants. Due to the non-preemptive nature of CAN, this strategy ensures ascending message receive instants and thus the correct temporal message order.

Before a message received from an event-triggered overlay network is forwarded to the application, the message goes through the following three steps:

- **Pending Non-Permanent.** A message m with a transmission request instant $t_{request}$ is non-permanent, if future messages, i.e., messages that have not yet arrived at the protocol emulation, can exhibit an earlier request instant than m. The notion of message permanence is based on the definition in [169]. Since messages with an earlier request instant can precede m in the temporal message order, message m must become permanent before it can be used in the simulation.

- **Pending Permanent.** After passing the *permanence test* (described below) a pending message becomes permanent. For a *permanent message*, it is ensured that all future messages will possess later request instants. The simulation determines the temporal message order based on the pending permanent message as its input.

- **In-Order.** After a pending message has been sent in the simulation of the physical CAN bus, the message is denoted as *in-order*. The message is revealed to the application software.

A message m_1 is *permanent* at instant t_p, if it is known that no message m_2 with an earlier or equal request instant $(m_2.t_{request} \leq m_1.t_{request})$ can be received at a later instant t $(t > t_p)$ via the event-triggered overlay network.

For determining permanence, we exploit the fact that transmission request timestamps of messages received from any particular node through the event-triggered overlay network are monotonically increasing.

In order to determine the permanence of messages, the protocol emulation maintains a vector $\overrightarrow{t_{latest}}$, which contains a timestamp for every sender. The ith element of this vector contains the message request instant of the most recent message received from sender i. Since the request instants of messages from a particular sender are monotonically increasing, the element associated with the sender in $\overrightarrow{t_{latest}}$ represents a temporal bound for subsequent messages, i.e., all future CAN messages must contain a later request instant.

A sufficient condition for the permanence of a message is that its request instant is earlier than all temporal bounds for message request instants of correct nodes in the vector $\overrightarrow{t_{latest}}$:

$$\text{permanence test:} \quad \bigvee_{i=1}^{n} \left(m.t_{request} < t_{latest,i} \right) \rightarrow m \text{ permanent} \qquad (4.6)$$

where n is the number of nodes in the system.

The temporal ordering of messages occurs through a simulation of a physical CAN network, where simulated message transmissions represent the simulation steps. The current simulation time is specified by the instant $t_{\text{idlestart}}$. $t_{\text{idlestart}}$ is a special instant that separates the messages which have been sent on the simulated CAN bus from those that have not. $t_{\text{idlestart}}$ marks the beginning of idleness on the simulated CAN bus. The message transmissions before $t_{\text{idlestart}}$ are already fixed, i.e., no later transmission requests can result in a modification of the sequence of message transmissions. Consequently, $t_{\text{idlestart}}$ also separates the ordered messages from the non-ordered ones.

In case the simulation time lies before the minimum request instant of a future timestamped CAN message ($t_{\text{idlestart}} < \min_i(t_{\text{latest},i})$) and one or more pending permanent messages are available, a simulation step can be taken. Out of the set of pending permanent messages, the protocol emulation chooses a message for the next simulation step based on the request instants and the message priorities. After the simulation step, the selected message becomes in-order and is transferred from the protocol emulation to the application software. Simulation steps are executed until no more pending permanent messages are available or a future timestamped CAN message can exhibit an earlier request instant than the current simulation time.

The protocol emulation has been validated in [235] using a communication matrix from a real-world automotive application, as well as synthetic message patterns.

4.5   Diagnostic Services

This section gives an overview of diagnostic services for time-triggered communication protocols. First, we describe basic error detection services, that enable a node to establish a local view of the operational state of other nodes in the system. In the next step, we introduce *membership agreement*, which is a service that provides a consistent view of the system's health state among multiple nodes.

4.5.1   Error Detection

In order to achieve the required level of safety, it is required for many applications, that errors in the systems are reliably detected and isolated in bounded time. In this section we will see, that many error detection mechanisms are facilitated in time-triggered systems due to their deterministic nature and the availability of a priori knowledge.

In general, error detection mechanisms can be implemented at the architectural level, or within the application itself. The advantage of implementations at the architectural level is that they are built in a generic way and thus, can be verified and certified once and for all. Furthermore, they can be implemented directly in hardware (e.g., within the communication controller) which can reduce the error detection latency and relieves the host CPU from computational overhead for error detection. Nevertheless, according to Saltzer's *end-to-end argument* [290] every safety-critical system must contain additional end-to-end error detection mechanisms at the application level, in order to cover the entire controlled process.

In the following, we describe error detection mechanisms based on *syntactic checks* and *semantic checks* that are implemented at the protocol level in many time-triggered protocols, and error detection by *active redundancy* which is usually implemented at a higher level.

4.5.1.1 Error Detection by Syntactic Checks

The syntactic checks are targeted on the syntax of the received frames. Protocols usually check for the satisfaction of specific constraints defined in the frame format. Examples are start and end sequences of the entire frame, or specific sequences within the frame like the *byte start sequence* that is mandatory before the transmission of every byte in the FlexRay protocol.

Another important category of syntactic checks are checksums, which protect the integrity of data with respect to accidental faults. *Cyclic Redundancy Checks* (CRCs) are employed in many protocols for this purpose. Often, there is a dedicated CRC that protects the data in the header of the frame and another CRC that protects the payload. The generation and the checking of the CRCs is usually implemented in hardware for performance reasons.

4.5.1.2 Error Detection by Semantic Checks

Semantic checks can be implemented very efficiently in time-triggered protocols, due to the a priori knowledge of communication patterns. Examples of semantic errors that can be detected by using this knowledge are:

Omission or Crash Failures: These failures can be detected, since the periodic points in time when a node should send are known.

Invalid Frame ID: Invalid frame IDs can be detected, since the allocation of frame IDs to periodic time instances is known.

Invalid Sender ID: Invalid sender IDs can be detected in topologies with active star couplers, since a star coupler knows which sender is attached to which port.

Invalid Temporal Information in the Frame Header: These failures can be detected by comparing the timing information included in the frame with the global time.

Violation of Slot Boundaries: Can be detected via the global time and the predefined schedules.

4.5.1.3   Error Detection by Active Redundancy

Value errors in a frame's payload cannot be detected by the above-mentioned mechanisms, if the payload CRC is valid. Such errors can happen, when the sending node computed an erroneous value at the application level before the CRC is generated. Value errors can be systematically detected by *active redundancy*, which means that the sending node is replicated, and the values generated by the redundant senders are compared at the receivers.

The most convenient way to systematically apply active redundancy is *bit-exact voting*, where the outputs of the redundant senders are bit-wise compared without the need of interpreting the values. Bit-exact voting is only possible if the redundant components are *replica deterministic* [169]. The time-triggered architecture provides, by the *sparse time-base*, an optimal platform for building replica-deterministic components, and thus for performing error detection in the value domain by active redundancy.

In order to make the voting process transparent to the user, a voting layer is included in major software communication stacks like OSEK-FTCOM or the communication stack of AUTOSAR.

4.5.2   Membership Agreement

The error detection mechanisms presented in the former section are performed locally on each node. Therefore, they establish a node's local view of the health state of the other components in the system. Due to *asymmetric faults*, the local view of different nodes is not necessarily the same. The goal of membership agreement protocols is to establish a consistent view of the system's health status among all correct nodes or processes of the system.

Two major properties of each membership algorithm are the following [68]:

Agreement: All correct nodes compute the same membership view

Timeliness: A faulty process will be removed from the member process group in a bounded time

The basic functionality of many membership agreement protocols is similar, and consists of the following steps:

1. In each communication round, each node observes the statically assigned slots of all other nodes.

2. Each node records for every slot whether a correct frame was received. The discrimination between correct and incorrect frames can be done with the error detection mechanisms introduced above.

3. Based on the observations, each node can build its local view on the health state of the other nodes. Due to asymmetric failures, the local opinion of different nodes can vary.

4. Each node disseminates its local option to all the other nodes. The dissemination can occur in statically defined slots in each round. Another option is to transmit the local opinion only when it has changed in a dynamic slot (this is only possible if the communication protocol provides dynamic arbitration like in FlexRay or TTEthernet). The local opinion can be sent explicitly as a vector with one bit for each node in the network, or implicitly as it is done in TTP/C where the local view is included in the calculation of the frame CRC, but the vector itself is not part of the frame.

5. Having received the local options of all other nodes, each node can construct a matrix containing the system view of all nodes in the system.

6. Based on the matrix, the nodes can execute a decision algorithm like majority voting.

The realization of membership agreement differs from protocol to protocol. In TTP/C, the membership service is an architectural service realized in hardware. Other protocols like FlexRay, provide no membership at the architectural level. In these cases, the membership service can be realized in software as proposed in [31] and [161].

5

Time-Triggered Protocol (TTP/C)

R. Obermaisser

University of Siegen

CONTENTS

5.1 Protocol Overview

The Time-Triggered Protocol (TTP) is a communication protocol for distributed fault-tolerant real-time systems. It is designed for applications with stringent requirements concerning safety and availability, such as avionic, automotive, industrial control and railway systems. TTP was initially named TTP/C and later renamed TTP. The initial name of the communication protocol originated from the classification of communication protocols of the Society of Automotive Engineers (SAE), which distinguishes four classes of in-vehicle networks based on the performance (see Figure 5.1). TTP/C satisfies the highest performance requirements in this classification of in-vehicle networks and is suitable for network classes C and above.

Network Class	Examples of Protocols	Bandwidth	Typical Latencies	Examples of Automotive Applications in this Class
Class A	Local Interconnect Network (LIN)	< 10 kbps	10-100ms	sensor/actuator access
Class B	Controller Area Network (CAN)	10kbps-125kbps	10-100ms	comfort domain
Class C	Controller Area Network (CAN)	125kbps-1Mbps	5ms	powertrain domain
Class D	Time-Triggered Protocol (TTP), FlexRay	> 1 Mbps	5ms	multimedia, X-by-wire

FIGURE 5.1
SAE Network Classes

The design of TTP has been driven by the following six design principles [177]:

- *Global time.* The global time is based on the metric of the physical second and provides the control signals for all communication and computational activities. In addition, the global time is also used to monitor the temporal accuracy of real-time data.

- *Temporal firewalls.* The communication interface in TTP is a unidirectional data-sharing interface with state-data semantics, where the communication network accesses this interface according to an a priori known schedule. Since no control signals cross the communication interface, it is a temporal firewall (cf. Chapter 2, Section 2.5.3) and supports fault and error containment.

- *Unification of interfaces.* The temporal firewall of TTP unifies the interface properties between host and communication system, as well as the gateway interfaces between multiple TTP clusters. In addition, the temporal firewall hides the local interfaces of a node (e.g., sensors, actuators, man-machine interface). Therefore, uniform interaction mechanisms are available between nodes regardless of the implementation technology within a node for interfacing the controlled object or other communication networks.

- *Two-phase design methodology.* TTP supports a two-phase design methodology for the component-based development of large distributed real-time systems. In the first phase, the message interaction patterns among nodes are designed as temporal firewall interfaces. During the second phase, the nodes are implemented taking the specification of the temporal firewall interfaces from the first design phase as constraints. TTP ensures that no unintended side effects occur during system integration at the communication network.

- *Real-time database.* TTP is designed to support the realization of a real-time database in each node, which consist of temporally accurate images of the relevant real entities [169].

- *Scalability.* TTP supports scalability with multi-cluster systems. Nodes can be expanded into gateway nodes, which implement two communication interfaces to two TTP networks. Each cluster can be understood independently by analyzing the temporal firewall interfaces within the cluster and the gateway node. The autonomous temporal behavior of a cluster does not depend on the operation of other clusters.

TTP provides a *consistent distributed computing base* [173] in order to ease the construction of reliable distributed applications. Given the assumptions of the fault hypothesis, TTP guarantees that all correct nodes perceive messages consistently in the value and time domains. In addition, TTP provides consistent information about the operational state of all nodes in the cluster. For example, in the automotive domain these properties would reduce the efforts for the realization of a safety-critical brake-by-wire application with four braking nodes. Given the consistent information about inputs and node failures, each of the nodes can adjust the braking force to compensate for the failure of other braking nodes. In contrast, the design of distributed algorithms becomes more complex [190], if nodes cannot be certain that every other node works on the same data. In such a case, the agreement problem has to be solved at the application level.

This chapter is organized as follows. Section 5.2 introduces the protocol services of TTP. At the core of TTP is a time-triggered communication service that builds on the fault-tolerant global time base. The protocol services of TTP are parameterized by a static data structure called the message descriptor list, which includes the time-triggered communication schedule with the points in time of all message transmissions and receptions. The elements of this data structure and the message schedule are explained in Section 5.3. Section 5.4 describes the interface, which is used by a host application in a node to access the services of TTP. This interface contains the messages as well as control and status variables. Finally, the chapter addresses the results of validation and verification efforts for TTP in Section 5.5 and example configurations and implementations in Section 5.6.

5.2 Protocol Services

The TTP protocol provides the services of a time-triggered communication protocol as introduced in Chapter 4. The *communication services* support the predictable message transport with a small variability of the latency. The *fault-tolerant clock synchronization* maintains a specified precision and accuracy of the global time-base, which is initially established by the *restart and startup services* when transiting from asynchronous to synchronous operation. The *diagnostic services* provide the application with feedback about the operational state of the nodes and the network using a consistent membership vector. The diagnostic services in conjunction with the a priori knowledge about the permitted behavior of nodes is the basis for the *fault isolation services* of TTP. Finally, *configuration services* offer flexibility by switching between predefined modes or programming new communication schedules into the system.

5.2.1   Communication Services

5.2.1.1   Temporal Structuring of Communication

The smallest unit of transmission and media access control on the TTP network is a *TDMA slot*. A TDMA slot is a time interval with a fixed duration (as defined by the Message Descriptor List (MEDL) in Section 5.3.1) that can be used by a node to broadcast a frame to all other nodes. A frame is a transmission of defined length on a TTP channel containing both application and protocol data. The application data within a frame represents a message, which possesses corresponding semantics for the host (e.g., a speed value in a control loop).

A sequence of TDMA slots is called a *TDMA round*. The *cluster cycle* defines a pattern of periodically recurring TDMA rounds. Although the sequence and the length of the TDMA slots in every TDMA round are equal, the frame communicated in a TDMA slot can differ between TDMA rounds. A TDMA slot in two TDMA rounds can be used for the exchange of frames with different sender nodes, content and size. The fixed duration of the TDMA slot limits the maximum size of a frame.

While a TDMA slot is defined within the TDMA round, a so-called *round slot* uniquely identifies a slot within the entire cluster cycle. A round slot possesses a unique sender node. In addition, the content and size of the frame communicated in the round slot is known.

If two TDMA rounds have two different sender nodes for a TDMA slot, the nodes are called *multiplexed nodes*. Multiplexed nodes can improve the use of communication bandwidth. If no multiplexed nodes are used, TTP possesses a unique mapping between TDMA slots and sender nodes. Otherwise, the sender is only known at the level of round slots.

Figure 5.2 depicts the structuring of the communication activities on a TTP network. This TDMA scheme is used after the completion of the startup and the establishment of synchronous operation.

A node that does not send during any slot is called a *passive node*. A node can be passive by design, if it is only required to receive frames (e.g., a bus sniffer for diagnosis). A failure can also lead to a passive node, such as the absence of the timely update of the life-sign or isolation performed by the bus guardian.

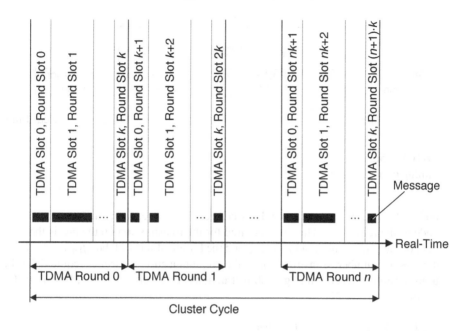

FIGURE 5.2
TDMA Scheme

5.2.1.2 Timing of a TDMA Slot

The constituting elements of a TDMA slot are depicted in Figure 5.3. The TDMA slot begins with the *transmission phase*, which is used for sending a frame on the bus. The *post-receive phase* is reserved for the protocol execution, e.g., updating the membership vector and writing the status information in the CNI. The *idle interval* allows us to stretch a TDMA slot in case a specific duration of TDMA slots and TDMA rounds is required in an application. The *pre-send phase* serves for the preparation of the transmission phase of the next TDMA slot. In this time interval, the communication controller loads the information about the next slot from the MEDL.

The durations of the post-receive phase and the pre-send phase depend on the controller implementation, namely the number of instructions for the protocol execution, the clock rate of the communication controller and the memory access times of the CNI memory.

The transmission phase duration in macroticks is computed using the following formula:

$$\text{transmission phase} = \left\lceil \frac{4 \cdot \Pi + d_{\text{correction}} + n_{\text{frame}}/f_{\text{bitrate}} + d_{\text{processing}}}{g} \right\rceil + 1$$

As introduced in Chapter 2, Π is the precision of the clocks in the cluster and g is the granularity of the global time-base. The computation of the transmission phase must take into account a maximum deviation of Π between the clocks of different TTP

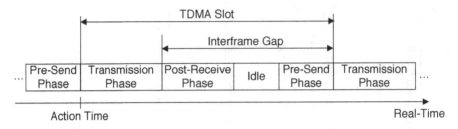

FIGURE 5.3
Timing of a TDMA Slot

nodes. $d_{correction}$ is the maximum delay correction of all receiving nodes to the sender node of the round slot. The time required for the transmission on the bus is the quotient of the frame length in bits n_{frame} and the bitrate of the TTP bus $f_{bitrate}$. $d_{processing}$ denotes the number of instructions in the communication controller required for the transmission or reception (e.g., 280 instructions at a clock rate of 40 MHz in the C2 controller [337, p. 21]).

5.2.1.3 Frame Types and States

The TTP protocol distinguishes two main types of frames (cf. Figure 5.4): *coldstart frames* are sent during the unsynchronized operation at startup, while *normal frames* are used in the synchronous operation. In addition, *download frames* can be used to parameterize the TTP protocol or to write an application image into a node.

Normal Frame

Frame Type	Mode Change Request	Controller State (Optional)	Application Data	CRC

Coldstart Frame

Frame Type	Global Time	Round Slot of Sender	CRC

FIGURE 5.4
Frame Formats

A coldstart frame contains a frame identifier which marks the frame as a coldstart frame, the global time of the sender, the current round slot of the sender and a Cyclic Redundancy Code (CRC). With the knowledge about the current global time and round slot, receiving nodes are enabled to integrate and transit from asynchronous to synchronous operation.

A normal frame consists of a frame type field, information about mode change requests, an optional controller state, application data and a CRC. If the controller state is included, one speaks of a frame with explicit controller state. Such a frame serves for the integration of nodes during synchronous operation (i.e., no coldstart frames are exchanged).

The controller state encompasses internal variables of the TTP communication controller, which are required to be globally consistent among all correct nodes in the TTP cluster. The controller state consists of the global time, the round slot position, the current mode of the cluster, information about pending mode changes and the membership vector.

Regardless of the inclusion of the controller state in a frame, the controller state is always used in the CRC calculation of the frame. The TTP protocol introduces the notion of *frames with an implicit controller state*, if the state is not contained in a frame. An implicit controller state reduces the protocol overhead and allows a more efficient use of the communication bandwidth at the cost of higher computational complexity.

A divergence of the controller state between receiver and sender is always detected by a CRC error. In particular, this information is used to establish agreement on a consistent view concerning the membership.

After each round slot configured for reception, a node classifies the received frames depending on detected errors:

- *Null frames.* If a receiver observed no transmission activity on a communication channel, the expected frame is called a null frame. A null frame occurs in case of a crash failure of a node.

- *Invalid frames.* A frame is received within the expected time interval of the round slot and the coding rules are satisfied. For example, the Modified Frequency Modulation (MFM) encoding, which is supported by existing TTP communication controllers, imposes limits on the minimum and maximum number of 0 bits that may appear between consecutive 1 bits. A frame that violates these constraints is called an invalid frame.

- *Incorrect frames.* An incorrect frame is a valid frame with an incorrect CRC. The CRC error indicates a disagreement of the controller states or a transmission error (e.g., bit flip during the transmission).

- *Tentative frames.* As long as the controller state agreement and acknowledgment (cf. Section 5.2.4.2) is in progress, a frame is tentative. In order to perform this agreement, the membership views of successor nodes in the TDMA round are evaluated.

- *Correct frames.* A correct frame is a valid frame, which has passed the CRC check and the acknowledgment.

5.2.2 Clock Synchronization

The Fault-Tolerant Average (FTA) algorithm [183] is used for clock synchronization in TTP. The FTA algorithm computes the convergence function for the clock synchronization within a single TDMA round. It is designed to tolerate k Byzantine faults in a system with N nodes. Therefore, the FTA algorithm bounds the error that can be introduced by arbitrary faulty nodes. These nodes can provide inconsistent information to the other nodes.

Each node collects the $N - 1$ measured time differences between the node's clock and the clocks of the other nodes. The time differences are determined by the difference of the actual arrival time of a frame and the expected arrival time of frames (as defined by the time-triggered communication schedule). These time differences, which indicate the deviations of the local times of sender and receiver, are sorted by size. The k largest and the k smallest time differences are discarded. The average of the remaining time differences is the correction term for the node's clock.

Using the FTA algorithm, a faulty time value is discarded if it is larger or smaller than the other time values. Otherwise, a faulty time value must be within the precision window. As discussed in [183] the worst-case scenario occurs if all correct clocks are at opposite ends of the precision window and the Byzantine clock is seen at different corners by two nodes. In this case, each Byzantine clock will cause a difference of $\Pi/(N-2k)$ in the calculated averages at two different nodes in an ensemble of N clocks. In the worst case, a total of k Byzantine errors will thus cause an error term of $k\Pi/(N-2k)$.

Considering the jitter ε of the synchronization frames and the drift offset Γ, the convergence function of the FTA algorithm is as follows:

$$(\varepsilon + \Gamma) \cdot \frac{N - 2k}{N - 3k} \tag{5.1}$$

In the TTP protocol, the MEDL controls which TDMA slots are used for clock synchronization. The slots at which new correction terms are calculated are marked consistently at all receivers in the MEDL in order to ensure that all nodes correct their clocks at the same time.

In addition to internal clock synchronization, the TTP protocol supports external clock synchronization using a time reference such as GPS. For this purpose, a host acts as a time gateway and possesses a connection to GPS or to another cluster with a different time base. The time gateway periodically computes a correction term and provides this term to its communication controller using the external rate correction field in the CNI. The external rate correction term denotes the number of microticks that need to be corrected in the next synchronization interval.

If the absolute value of the external correction or the absolute value of the total correction term is larger than $\Pi/2$, a node freezes due to a synchronization error.

5.2.3   Restart, Re-Integration, Integration

Cluster start-up is the process of establishing a synchronized cluster after power-on. After power-on, a node listens on the communication channels for a *coldstart frame*. A coldstart frame consists of a frame type field (identifying the frame as a coldstart frame), the global time at the send instant of the frame, a sender round slot position in the MEDL and a CRC. If the frame is received, then the node sets its controller state accordingly. The node adopts the global time and the round slot position in the MEDL.

If no coldstart frame is received within the *listen timeout* and the coldstart allowed flag is set in the MEDL of the node, then the node sends a coldstart frame itself. The sending of a coldstart frame can be repeated until a maximum number of allowed coldstarts is reached (as defined in the MEDL). This limit is introduced in order to prevent a node with an incoming link fault to interfere with the synchronization of the other nodes in the cluster.

During startup the nodes perform asynchronous access to the communication medium controlled by the *startup timeout*. This parameter is unique for each node in a cluster. For a given node, it denotes the number of TDMA slots prior to the sending slot of the node. The listen timeout introduced above is the sum of the node's controller startup timeout plus two TDMA rounds. Hence, the duration between two coldstarts is always shorter than the listen timeout.

In case of a collision of the coldstart frames of two nodes, TTP performs the so-called *big bang mechanism*. In case of large propagation delays, a collision can be perceived inconsistently. Only a subset of the nodes could receive a correct coldstart frame, thereby leading to the formation of cliques. All nodes can detect this situation based on the transmission phases of two coldstart frames in relation to the sum of the maximum propagation delay and the frame duration. In order to prevent cliques, nodes will not integrate on coldstart frames once a big bang scenario has been detected. Thereby, coldstarting nodes will not detect traffic and they will restart their startup timeouts again. Because of the unique startup timeouts, no second collision will occur between the nodes.

If a node joins a cluster, which is already synchronized, this process is called the *integration* of the node [313]. In order to support integration, the MEDL must contain at least one frame with the controller state within the minimum listen timeout (i.e., two TDMA rounds). This constraint ensures that an integrating node does not initiate a coldstart.

In order to avoid the integration on faulty frames, an integrating node maintains an *integration counter*. The MEDL contains a parameter called *minimum integration value*, which specifies the number of correct frames that need to be received before a node considers itself integrated and may start to send.

5.2.4 Diagnostic Services

5.2.4.1 Life-Sign

The TTP protocol detects the crash failure of a host based on a periodic *host life-sign*. In every TDMA round, the host of a node must provide a life-sign to the communication controller. More precisely, a host needs to set the life-sign after the start of the node's transmission slot and before the beginning of the pre-send phase of the node's transmission slot in the next TDMA round. The communication controller verifies whether the host has set the life-sign during the node's transmission slot.

If the life-sign is not set by the host, the communication controller does not send frames and transits into passive mode. Frame transmission is only continued when the host updates the life-sign again. The life-sign is used both during normal operation and at startup. The set life-sign is also a prerequisite for sending coldstart frames to startup the cluster.

In addition to the host life-sign, the updating of the global time and schedule position by the communication controller serves as a *controller life-sign*. Thereby, the host can react to a crash failure of a communication controller in an application-specific way (e.g., enter a safe state).

5.2.4.2 Membership Service

The TTP protocol informs nodes about the operational state of every other node in the TTP cluster using a membership vector. The membership vector is a vector with a bit for every node, denoting whether the respective node is operational.

A node A considers another node B as operational, if node A has correctly received the frame that was sent by node B prior to the membership point. In case redundant communication channels are used, the reception on one of the channels is sufficient in order to consider a sender to be operational.

The delay between the failure of a node and the indication in the membership vector is bounded by the duration of two TDMA rounds. The points in time for establishing the membership information are called *membership points*. In TTP, the post-receive phase (PRP) of a sending node serves as a membership point.

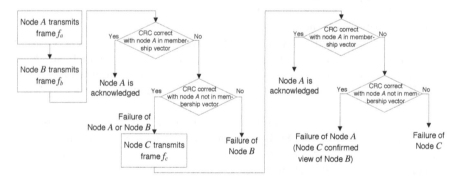

FIGURE 5.5
Acknowledgment Scheme (based on information in [338])

The agreement of the membership is part of the controller state agreement based on CRC calculation (cf. Section 5.2.1.3). If the membership vectors of sender and receiver of a frame diverge, the receiver will detect a CRC error on the received frame. The reason for the CRC error is that the membership vector is part of the controller state and the sender's controller state is appended to the frame before the calculation of the CRC. The receiver appends its own controller state before the calculation of the frame's CRC.

As depicted in Figure 5.5, the successor nodes of a sender in the time-triggered communication schedule determine whether the sender has sent a correct frame and is therefore classified as operational:

- The node 'A' sets its membership flag to '1' before starting to send frame f_A. The membership vector with node A classified as operational is used in the calculation of the CRC of frame f_A.

- The successor of the sender node 'A' is the node 'B,' which will send the next frame f_B according to the time-triggered communication schedule. Node 'B' sets the membership flag of 'A' to 1, if frame f_A is correctly received on one of the redundant channels. Otherwise, the membership flag of node 'A' is set to 0. The new membership vector with the operational state of node 'A' is used to send frame f_B.

- Node 'A' receives the frame f_B and learns about the view of node 'B' on the operational status of node 'A.' If the CRC is correct, node 'A' knows that node 'B' classified node 'A' as correct. If the CRC is incorrect, node 'A' determines whether the CRC would be correct in case of node 'A' classified as non-operational. If the CRC is now correct, nodes 'A' and 'B' disagree on the operational state of node 'A.' This means that either node 'A' is faulty (e.g., outgoing link failure) or node 'B' is faulty (e.g., incoming link failure). To distinguish between the two cases, node 'A' evaluates the frame of the next node 'C' in the time-triggered communication schedule. If the CRC remains incorrect despite the assumption of node 'A' being non-operational, node 'B' is considered to be non-operational by node 'A' and node 'C' becomes the successor node.

- Node 'A' receives the frame f_C and learns about the view of node 'C' on the operational status of node 'A.' If the CRC of f_C is correct, node 'A' resolves the disagreement between nodes 'A' and 'B' by classifying itself as operational. The reason for this decision is that a majority of two nodes classify node 'A' as operational, namely nodes 'A' and 'C.' If the CRC of f_C is incorrect, node 'A' determines whether the CRC would be correct in case of node 'A' classified as non operational. If so, node 'A' resolves the disagreement with node 'B' by classifying itself as non-operational. There is now a majority of two nodes classifying node 'A' as non-operational, namely nodes 'B' and 'C.' If the CRC remains incorrect despite the assumption of node 'A' being non-operational, node 'C' is considered to be non-operational by node 'A' and the next node (i.e., node 'D') becomes the second successor node.

The design principle that a node 'A' assumes to be correct until at least two other nodes indicate the failure of node 'A' is denoted as the *self-confidence principle*. As long as this disagreement between two nodes persists (i.e., until the disagreement is resolved by a third node), receivers classify a frame as tentative in the frame status field (cf. Section 5.2.1.3). Thereby, the application can delay the use of the frame until it is known whether the sender is operational.

If a node learns that it was classified as non-operational by two receivers, the node increases a counter of acknowledgment failures. In case this acknowledgment failure counter exceeds a configurable threshold value, an acknowledge error is raised and the controller enters the freeze state (cf. Section 5.5).

5.2.4.3   Clique Detection

The clique detection and avoidance in TTP [27] has the goal of avoiding the partitioning of a cluster into cliques that are not able to communicate with each other. The clique avoidance algorithm selects the largest partition (clique) as a winner, while the nodes of other partitions are shut down by entering the freeze state.

In every TDMA round, the communication controller determines whether it is in agreement with the majority of the other nodes concerning the controller state. For this purpose, the communication controller counts the number of round slots where the frame status is correct, as well as the number of round slots where the frame status is incorrect or invalid. In the pre-send phase of its own transmission slot, a node checks whether the value of the failed slots counter is larger than the agreed slots counter. In this case, a clique error is detected and the communication controller transits into the freeze state.

5.2.4.4   Communication System Blackout Detection

A communication system back out is detected, if only null frames are received during a TDMA round. This means that no correct transmission from other nodes occurred. The communication controller raises a communication blackout error and enters the freeze state.

5.2.5   Fault Isolation

The TTP protocol was designed to isolate and tolerate an arbitrary failure of a single node during synchronized operation [178]. After the error detection and the isolation of the node, a consecutive failure can be handled. Given fast error detection and isolation mechanisms, such a single fault hypothesis is considered to be suitable in many safety-critical systems [237]. The fault hypothesis assumes an arbitrary failure mode of a single node. TTP does not guarantee to tolerate two independent node failures, i.e., a second failure before the detection and isolation of the first failure. Such a scenario is considered very unlikely and addressed by the so-called *Never-Give-Up (NGU) strategy* [174]. If failures outside the fault hypothesis are detected, the communication system informs the application. Depending on the application, a safe-state can be entered (e.g., setting all signals to red in a railway application). Assuming transient faults, a restart of the TTP cluster can be performed in a fail-operational system.

In order to tolerate timing failures, a TTP cluster uses local or central *bus guardians*. In addition, the bus guardian protects the cluster against slightly-off-specification faults [1], which can lead to ambiguous results at the receiver nodes.

A *local bus guardian* is associated with a single TTP node and can be physically implemented as a separate device or within the TTP node (e.g., on the silicon die of

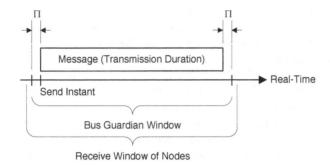

FIGURE 5.6
Bus Guardian Window and Nodes' Receive Window

the TTP communication controller or as a separate chip). The local bus guardian uses the a priori knowledge about the time-triggered communication schedule in order to ensure fail-silence of the respective node. If the node intends to send outside the preassigned transmission slot in the TDMA scheme, the local bus guardian cuts off the node from the network. In order to avoid common mode failures of the guardian and the node, the TTP protocol suggests the provision of an independent external clock source for the local bus guardian.

The *central bus guardian* is always implemented as a separate device, which protects the TDMA slots of all attached TTP nodes. An advantage compared to the local bus guardians is the higher resilience against spatial proximity faults and the ability to handle slightly-off-specification faults.

In safety-critical systems, a TTP cluster is deployed with two independent bus guardians for the two redundant TTP channels. The failure mode of a central bus guardian is assumed not to be arbitrary. According to the fault hypothesis of TTP, the failure of a central guardian only leads to the transmission of frames that are detectably faulty at the receivers. In the time domain, a failure of a guardian can lead to untimely frames that are perceived at the receiving nodes outside the slots defined by the TDMA scheme. In the value domain, a faulty guardian can produce frames with an invalid CRC. According to the fault hypothesis, the guardian may not generate incorrect frames with valid CRCs and correct timing. The reason for this assumption is that nodes would receive two different frames from the redundant communication channels with correct CRCs and timing. Hence, the receiving nodes would be unable to determine which frame is the correct one and should be provided to the application. In order to justify this fault assumption, implementations of the central bus guardian contain no logic for the generation of CRC codes. TTP addresses the replaying of old frames by including the global time in controller state, which is used together with the application data in the CRC calculation.

Both central and local guardians use a *bus guardian window* in order to ensure timely frames (cf. Figure 5.6). The bus guardian window enables access to the communication system for the node at the specified time and for the complete slot duration, but prevents any transmission from the node for the remaining duration of the TDMA round. The start instant and the end instant of the bus guardian window take into account the different views on the global time, which are bounded by the precision of the clock synchronization.

In order to avoid slightly-off-specification failures, the bus guardian uses a bus guardian window that is shorter than the receive windows used by the receiver nodes. This means that the bus guardian is more restrictive concerning the time of a frame transmission than any receiver node. The bus guardian limits the frame transmission in such a way that frames transmitted too early or too late are blocked or truncated, thus resulting in an invalid transmission for all receivers. Thereby, the bus guardian protects the communication system from a node transmitting a correct frame with a temporal deviation close to the precision. Such a transmission can result in an inconsistently perceived failure (i.e., so-called Byzantine fault) when the frame is received correctly by some nodes and incorrectly by other nodes. A detailed discussion of the dimensioning of bus guardian windows can be found in [283].

5.2.6 Configuration Services

TTP supports the switching between predefined static modes in order to adapt to changing environment conditions. Furthermore, TTP includes basic configuration capabilities in order to update the MEDL and download the application software.

5.2.6.1 Mode Changes

TTP supports the switching between predefined static configurations called *cluster modes*. The rationale for this protocol service is that many applications exhibit mutually exclusive modes of operation. For example, the flight control system of an airplane can support different modes such as on-ground, take-off, low-altitude and landing [40]. Likewise, cars can exhibit different modes of operation such as a normal mode and a limp-home mode [24, 333].

At any time, all nodes of the cluster must be in the same cluster mode. The cluster mode is part of the controller state, thus a divergence of cluster modes is detected by the CRC calculation. Every cluster mode must also possess the same sequence of TDMA slots.

A host can request a mode change using the control area of the Communication Network Interface (CNI). In the next pre-send phase, the communication controller checks this request against the mode change permissions in the MEDL. If the mode change is permitted by the MEDL, information about the new mode is included in the next sent frame.

Receiving nodes act on this mode change information in the post-receive phase. The communication controller of a receiver node also checks whether the sender is allowed to request the new mode according to the mode change permissions in the MEDL. If the request is permitted, the new mode will become active at the beginning of the next cluster cycle. The ongoing cluster cycle is not preempted.

If another mode change request arrives before the end of the cluster cycle, then the new one overwrites the previous one. Nodes can also cancel a pending mode change request by sending a special value for the mode change request. In this case, the mode of the current cluster cycle remains in place.

5.2.6.2 Boot Loader

TTP nodes use a boot loader [326], which supports the startup of the node and the download of a new MEDL or an application image. When the node is powered up, the boot loader determines whether download frames are exchanged on the network. If so, the boot loader switches the TTP controller to the await and download states.

The implementation of the boot loader depends on the type of the TTP controller. In case of a TTP controller with flash memory (e.g., C2), the MEDL is directly downloaded to the memory of the TTP controller. If a TTP controller without flash memory is used (e.g., C2NF), then the MEDL is attached to the application image during download. During each startup, the boot loader passes the MEDL from the host's flash to the RAM of the TTP communication controller.

During the download, the boot loader acquires Application Descriptor Blocks (APDBs), each containing a header with version information and an identification of the target node, as well as an application image and/or a MEDL. Since TTP nodes can support multiple applications, APDBs can also be linked (see Figure 5.7).

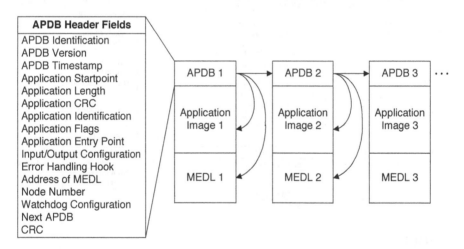

FIGURE 5.7
Application Descriptor Blocks

A TTP cluster uses a special TTP node called the *download master node* to send the download frames. Download master nodes with an Ethernet interface are available in order to acquire the application images and MEDLs [325].

5.3   Protocol Parameterization

5.3.1   Message Descriptor List

The MEDL is the central configuration data structure in the TTP protocol. Each
node possesses its own MEDL, which reflects the node's communication actions
(e.g., sending of frames, clock synchronization) and parameters (e.g., delays to other
nodes). At design time, TTP development tools [339] are used to temporally align
the MEDLs of the different nodes with respect to the global time-base. For example,
the period and phase of a frame transmission is aligned with the respective frame
receptions taking into account propagation delays and jitter.

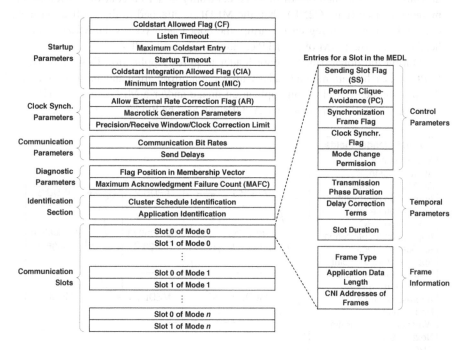

FIGURE 5.8

Layout of the Message Descriptor List

The contents of the MEDL are depicted in Figure 5.8. The actual memory layout
depends on the implementation of the communication controller and is not fixed by
the TTP protocol. For example, the memory layout of the TTP controller C2 can be
found in [337].

For each node, the MEDL includes several *start-up parameters*. The MEDL defines whether a node is allowed to send a frame during a coldstart (so-called coldstart
frames). If so, the listen timeout defines how long the node waits for frames on the
communication channels before starting with the transmission of a coldstart frame.

The maximum number of attempts of the node to send a coldstart frame is bounded by a threshold value: the maximum coldstart entry. The startup timeout provides a node-specific unique timeout value in order to avoid repeated collisions. In addition, the MEDL specifies whether a node is allowed to integrate on a coldstart frame. Otherwise, a normal frame with an explicit controller state is required for integration. A threshold value in the MEDL (called minimum integration count) denotes the minimum number of correct frames that need to be received before a node considers itself to be integrated and starts transmitting frames. This threshold prevents a node from integrating based on a single faulty frame.

Another group of entries in the MEDL are *clock synchronization parameters*. They define whether the external rate correction is enabled. Also, the microtick/macrotick ratio and the precision of the global time-base are specified. According to the reasonableness condition [169, p. 52] of a global time-base, the duration of a macrotick must be smaller than the precision in order to guarantee that the timestamps assigned to the same event at two nodes differ by at most one macrotick. The clock synchronization parameters are used to configure most of the TTP services, e.g., the duration of the receive windows or the maximum clock correction term.

Another group of entries in the MEDL are *communication parameters* such as the communication bitrate and the send delay. The send delay depends on the propagation delay and ensures that no correctly synchronized node receives a frame before the action time.

The *diagnostic parameters* define the node's position in the membership vector and the maximum number of acknowledgment failures before entering the freeze state.

An *identification section* consists of an identifier of the MEDL (called cluster schedule identification) and an application identification, which names the host application. The identification section can be used by the application to determine if a correct MEDL is deployed in the TTP node.

Thereafter, the MEDL provides a description of the slots of the TDMA scheme. The TDMA scheme consists of TDMA slots, each of which can be used by a TTP node for the transmission of a frame. TTP supports different modes, each of which can have a different layout of TDMA slots.

For each TDMA slot, the MEDL defines the required actions of a node, the temporal parameters of the slot and the frame that is exchanged during the slot.

- *Control parameter.* Possible actions in a slot are the sending of a frame, the execution of the clique avoidance algorithm and the synchronization of the clock. For the clock synchronization, the control parameters define whether the receive instant of a frame shall be used for synchronization and whether a correction term shall be computed in the slot. Finally, the mode change permission denotes if a mode change can be accepted in this slot.

- *Temporal parameters.* The maximum duration of the frame transmission in the slot is provided, while considering the delay on the communication channel and communication jitter. The sum of this transmission phase duration, the pre-send phase and the post-receive phase gives the minimum for the slot duration. The slot duration is the length of the TDMA slot in macroticks. If a node receives during the TDMA slot, a correction term is given in order to compensate for the propagation delay from the sending node.

- *Frame information.* This section of the MEDL describes the frame. Firstly, the frame type is given, such as a frame with implicit or explicit controller state. Also, the number of bytes of application data in the frame is specified. In order to enable the processing of frames, an address in the CNI is associated with the TDMA slot. In case of a frame reception, the application data is written to this CNI address. If a frame is transmitted, the application data is read from the CNI address.

5.4   Communication Interface

The communication interface between the host and the communication controller is called the Communication Network Interface (CNI) in TTP. The CNI is a memory area that is structured into three parts: a status area, a control area and a message area.

The host's access to the status and control area is constrained in order to ensure consistency (e.g., no transmission of a partially updated message). The host is allowed to access the status and control information in the CNI during the transmission phase and the idle phase. The message area of the CNI can be read or modified during the transmission phase except for the frame under transmission or reception.

5.4.1   Status Area

The *status area* provides access to the global time, diagnostic information and the status of the communication controller. The specific memory layout is not fixed by the TTP specification and depends on the implementation. For example, Figure 5.9 depicts the layout of the status area for the TTP communication controller C2 [338].

The status area contains the current state of the communication controller (i.e., the controller state) with the following information:

- *Global time:* The *global time* in the status area denotes the point in time of the next frame transmission in macroticks. The global time is updated during the pre-send phase before the next sending slot of a node.

- *Cluster mode:* The status area provides information about the *cluster mode* at which the TTP system is operating. In case a mode change is requested in the next cluster cycle, this condition is indicated by the field Deferred Pending Mode Changes (DMC).

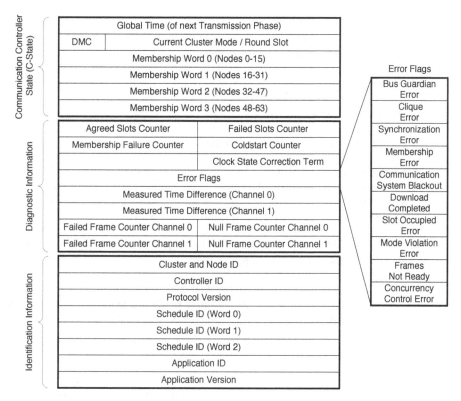

FIGURE 5.9
Layout of Status Area in the CNI of the TTP Communication Controller C2 [338]

- *Round slot position:* The status area contains the current position in the time-triggered communication schedule as defined in the MEDL.

- *Membership vector:* The membership vector provides consistent information about the operational state of the nodes (cf. Section 5.2.4).

In addition to the controller state information, the status area includes diagnostic information:

- *Agreed and failed slot counters:* The agreed slots counter counts in each TDMA round the number of nodes that have sent at least one correct frame. Adversely, the failed slots counter counts in each TDMA round the number of nodes sending at least one failed frame but no correct frame. The agreed slots counter and the failed slots counter are used for the clique avoidance algorithm.

- *Membership failure counter*: The membership failure counter (also called *acknowledgment failure counter*) is used by the acknowledgment algorithm and counts the number of successive acknowledgment failures. If a threshold value is exceeded, a node is considered to exhibit a permanent failure and terminates its operation. The threshold value is contained in the MEDL and called the *maximum acknowledgment failure count*.

- *Coldstart counter value:* The coldstart counter value is a value used by the startup algorithm (cf. Section 5.2.3). It counts the number of coldstarts executed by the communication controller. A node only sends coldstart frames, if the coldstart counter value does not exceed a threshold for the maximum permitted coldstarts as specified in the MEDL.

- *Clock state correction term:* The clock state correction term is computed by the clock synchronization algorithm. This field can be used for external clock synchronization (e.g., to GPS). The granularity of the clock state correction term are microticks of the communication controller.

- *Error flags:* The error flags denote internal protocol errors and host failures detected by the communication controller.

  - The *bus guardian error* shows that a local bus guardian detected a bus access violation such as an attempt to send outside the time interval specified in the time-triggered communication schedule.

  - The *clique error* occurs if the controller state of the communication controller is different to the majority of the cluster.

  - An error in the clock synchronization subsystem is indicated by a *synchronization error*. For example, a computed clock correction term larger than $\Pi/2$ (where Π is the precision of the global time) leads to such an error.

  - If the number of successive membership failures has exceeded the threshold in the MEDL (i.e., maximum membership failure count parameter), this condition is recorded in the *membership error field*.

  - A *communication system blackout* occurs in case no bus activity is perceived except for the own transmission of the node during the duration of a TDMA round.

  - The *download completed* flag informs the host that the download of the MEDL into the communication controller has been completed. This flag is usually used by the controller for a restart.

  - A *slot occupied error* is raised if the slot of a node is already used by another node (as indicated by the membership vector).

  - A mode change request by a host leads to *mode violation error* in case it violates the mode change permissions.

  - Prior to the transmission of a frame by the communication controller, the host must set a flag that is associated with the frame. Thereby, the host confirms that the frame is ready for transmission. The failure to set this flag leads to a *frames not ready* error.

- The *concurrency control error* records a violation of the non-blocking write protocol [184].

• *Time difference values:* For every communication channel, the status area also contains values denoting the difference between the expected arrival time and the actual arrival time of the most recent received frame. Due to the reasonableness condition of the global time-base [169, p. 52], this value must be between -2Π and $+2\Pi$ (where Π is the precision of the global time-base).

• *Null frames and failed frames:* Further diagnostic information consists of a counter for the number of null frames and failed frames (i.e., invalid or incorrect) for each communication channel during the last TDMA round.

The third part of the status area is comprised of identification information. This part provides an identification of the node, controller and communication schedule, as well as version information about the protocol and the application.

5.4.2 Control Area

The *control area* of the CNI is used by the host to control the operation of the communication controller. Firstly, the host can switch on and switch off the communication controller using the *controller on flag*. At power-on and after a critical failure of the communication controller the controller on flag is set to zero. Using the flag, the host can start or restart the operation of the communication controller.

The *control area* is also used by the host to provide the periodic life-sign. By writing the life-sign entry, the host demonstrates that it does not exhibit a crash failure (cf. Section 5.2.4). If the life-sign is not set by the host within a TDMA round, the communication controller does not send frames and transits into passive mode. Frame transmission is only continued when the host updates the life-sign again.

The *external rate correction field* of the control area serves for the implementation of external clock synchronization. External clock synchronization links the global time of the TTP cluster to an external time base. This synchronization is unidirectional and periodically adjusts the rate of the global time in the TTP cluster to bring it into agreement with the external time-base. The difference between the occurrence of a significant event in the external time-base (e.g., the start of the full second) and the occurrence of the related significant event in the global time of the TTP cluster, is measured by using the microticks of a time gateway. The necessary rate correction term is then computed and written into the external rate correction field of the control area. This field specifies the number of microticks that shall be added to the correction term during the next resynchronization interval.

The *mode change request field* can be used by a host to request a mode change to a new time-triggered communication schedule in all nodes of a cluster. Mode changes enable different operating modes and adaptability to different environmental contexts. A mode change request is broadcast to all other nodes of the cluster at the next scheduled send instant of the node.

The *timer field* enables the host to select a periodically recurring point in time at which a control signal is generated by the communication controller. When the global time reaches the value of the timer field, a timer interrupt is generated. Using such a programmable timer, which is synchronized to the global time of TTP, it is possible to periodically trigger the dispatcher of a time-triggered operating system [324]. On its behalf, the dispatcher can control the execution of processing tasks according to a time-triggered task schedule to establish synchronization with the communication activities of the TTP network. Thereby, the operating system implements implicit synchronization as introduced in Section 2.5.3.

The host can determine which events should trigger an interrupt by writing the *interrupt enable field*. Examples of supported events are the timer expiration as requested by the timer field and error conditions such as BIST errors (e.g., CRC error of protocol code), protocol errors (e.g., clock synchronization error) and host errors (e.g., no host life-sign). In addition, interrupts can indicate changes of the cluster's status such as an update of the membership vector or a mode change.

The final entry of the control area is the *time startup field*, which contains the controller state time that is broadcast during cluster start-up. This field is useful to establish initial synchronization to an external time-base. A time gateway can use this field to initialize the cluster with the current external time.

5.4.2.1 Message Area

The message area contains the frames that are sent and received at the TTP network. The structure of the message area depends on the time-triggered communication schedule. As depicted in Figure 5.10, each frame consists of application data, a status field and an optional diagnosis field. In case of a frame that is sent by the TTP node, the status field is a send status flag that denotes whether the application data of the frame is valid for transmission. This flag is called the Ready Status (RS) flag. For a frame that is received by the TTP node, the status field shows different types of detected communication failures. Possible failures are no traffic on the communication channel, coding errors, CRC errors, disagreement of the controller state and a request for a mode change that is not permitted according to the mode change permissions in the MEDL. The diagnosis field is optional and stores the CRC and the header of a received frame. The MEDL defines whether the diagnosis field is appended to a particular frame in the message area.

5.5   Protocol States

The TTP protocol distinguishes nine protocol states as depicted in Figure 5.11. The *init state* is the first state after the TTP controller has been switched on by the host. After the TTP controller has set its internal data structures to the initial values and the MEDL has been checked, it transits to the *listen state*. The listen state serves for the integration of the node when receiving a frame with explicit controller state from

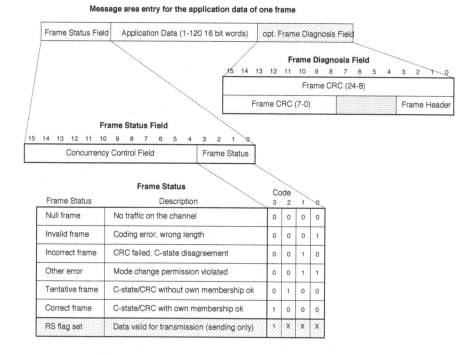

Message area entry for the application data of one frame

| Frame Status Field | Application Data (1-120 16 bit words) | opt. Frame Diagnosis Field |

Frame Diagnosis Field

| 15 | 14 | 13 | 12 | 11 | 10 | 9 | 8 | 7 | 6 | 5 | 4 | 3 | 2 | 1 | 0 |

| Frame CRC (24-8) |
| Frame CRC (7-0) | | Frame Header |

Frame Status Field

| 15 | 14 | 13 | 12 | 11 | 10 | 9 | 8 | 7 | 6 | 5 | 4 | 3 | 2 | 1 | 0 |

| Concurrency Control Field | Frame Status |

Frame Status

Frame Status	Description	Code 3	2	1	0
Null frame	No traffic on the channel	0	0	0	0
Invalid frame	Coding error, wrong length	0	0	0	1
Incorrect frame	CRC failed, C-state disagreement	0	0	1	0
Other error	Mode change permission violated	0	0	1	1
Tentative frame	C-state/CRC without own membership ok	0	1	0	0
Correct frame	C-state/CRC with own membership ok	1	0	0	0
RS flag set	Data valid for transmission (sending only)	1	X	X	X

FIGURE 5.10

Application Data and Status Fields in the Message Area [338]

another node. If the node receives such a frame, it enters the *passive state*. The *active state* is reached when the node sends a frame during its slot. The preconditions for becoming active are the setting of the host life-sign by the host and the reception of a minimum number of correct frames. The transition from passive to active state is also called the slot acquirement. After the slot acquirement, the node has sent successfully and is active in the membership.

If the node receives no frame in the listen state and it is allowed to send cold-start frames, it enters the *coldstart state*. In this state, the controller actively initiates the startup of the cluster by sending coldstart frames as described in Section 5.2.1.3. The coldstart state is replaced by the *active state* as soon as a correct frame is received from another node and thus two nodes are alive in the cluster. If the maximum number of coldstart frames is exceeded or if a frame is received during the startup timeout, the controller transits back to the listen state.

In order to configure the TTP cluster, the host can set the controller in the *await state*. After frames with download information arrive, the *download state* is entered and the configuration of the controller (i.e., MEDL) and the application code of the host are programmed into the node.

The *test state* serves for the execution of the built-in-self test by the TTP controller. The actual tests (e.g., memory test) depend on the controller implementation.

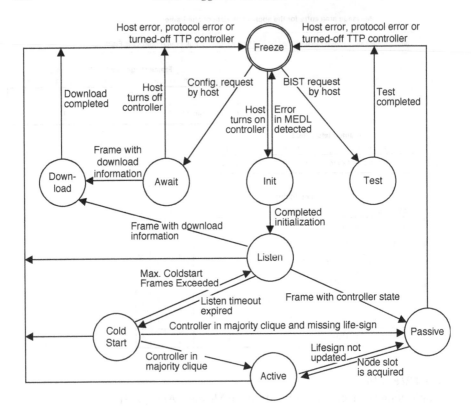

FIGURE 5.11
Protocol States

5.6   Validation and Verification Efforts

In the following, the verification of TTP and its algorithms by means of formal analysis and fault injection is described. The goal of these activities was to determine the correctness of the core algorithms of TTP and to show that an acceptable level of service is provided given the satisfaction of the fault hypothesis. For the users of TTP these results provide valuable assurance on the claims of TTP concerning reliability and suitability for safety-critical systems.

5.6.1   Formal Analysis of Clock Synchronization Algorithm

The clock synchronization algorithm of TTP is a specialization of the Welch-Lynch algorithm for tolerating a single fault by using four clock readings and no dedicated wires to communicate the clock readings. No dedicated wires are needed, because the deviations between the actual and the expected receive instants of frames at the TTP network provide the input to compute the clock differences. Also, the clock synchronization algorithm of TTP considers only nodes marked as having accurate oscillators.

Using the PVS theorem prover [243] the TTP clock synchronization algorithm was formally verified in [256] based on the verification of the Welch-Lynch algorithm [225].

The remaining challenges for the formal verification of the clock synchronization in TTP are pointed out in [285], e.g., the introduction of a hybrid fault model considering manifest faults and symmetric faults in order to model systems with fewer than four correct clocks. Although the requirements for handling arbitrary faults are no longer met with fewer than four clocks, insights into the behavior in these scenarios would be valuable as part of a never give up philosophy. In a safety-critical system, the computer system should never give up, even if the fault hypothesis is violated by reality. In a properly designed fault-tolerant system, chances are high that a violation of the fault hypothesis is caused by a correlated shower of external transient faults or by a Heisenbug and that a fast restart of the system will be successful [174].

5.6.2 Formal Analysis of Fault Isolation and Consistency

The time-triggered communication schedule temporally coordinates the frame transmissions of nodes in order to avoid collisions. The central and local guardians in TTP ensure fault isolation in case a node does not comply to the predefined send intervals defined in the communication schedule. Since clocks cannot be perfectly synchronized (e.g., due to drift rates), the permitted send intervals at the guardian take into account the inevitable difference of the local clocks bounded by the precision Π (cf. Section 5.2.5).

Based on the fault isolation of the guardians, TTP ensures the properties of agreement, validity and separation. *Agreement* requires that all correct nodes receive a frame if one correct node receives the frames. *Validity* is satisfied if all correct nodes receive a frame if the frame is sent by a correct sender node. *Separation* is satisfied when the temporal order of the frame receptions is equal to the order of the slots in the time-triggered communication schedule. This means that a frame sent by a correct node via a correct guardian arrives after the frame from the previous slot and before the frame from the next slot in the TDMA scheme.

The properties of agreement, validity and separation were formally verified using PVS in [283].

5.6.3 Formal Analysis of Membership Service and Clique Avoidance

The membership service presents to the host application in each node a consistent view of the operational state of all nodes in the cluster. This diagnostic information facilitates the construction of dependable systems and the implementation of application-level fault-tolerance.

In addition, the membership service is used by other algorithms in TTP for tolerating multiple subsequent node failures. For example, the clock synchronization uses the membership information for selecting the nodes that provide the clock values. Thereby, the clock synchronization becomes ready for tolerating a second node failure after the exclusion of the first failed node from the membership. The underlying fault assumption is that two subsequent node failures are at least two rounds apart.

The membership service of TTP is designed to provide the properties of agreement and validity [107]. The *agreement* property requires that the membership vector of all correct nodes be identical. *Validity* demands that all correct nodes are classified as such in the membership vector of correct nodes. In addition, at most one faulty node may be classified as correct in the membership vector at any given point in time.

These properties were verified in [255]. In this work, a set of disjunctively connected formulas is used, where each disjunct represents a particular configuration that the membership algorithm can reach. The correctness of the membership algorithm was proven by verifying that the system is always in one of these configurations.

The clique avoidance builds on top of the membership algorithm and tolerates asymmetric faults. In [43] the clique avoidance was verified by building a model for the algorithm in the language of ALV [45] and verifying the convergence to a single clique after precisely two rounds after the occurrence of a failure.

5.6.4 Fault Injection Experiments

Fault injection experiments were performed to validate TTP by testing its fault handling capabilities. Different configurations of TTP clusters were exposed to Software-Implemented Fault Injection (SWIFI) [132], pin-level fault injection [159], electromagnetic interference [159] and heavy-ion fault injection experiments [301].

In an experimental setup, which comprised a cluster with two redundant communication channels in a bus topology and local bus guardians, error propagation was observed due to the spatial proximity of the TTP communication controllers and the bus guardians. The central guardian, on the other hand, was effective in the containment of slightly-off-specification (SOS) failures, reintegration errors, asymmetric faults and babbling idiot failures [2].

In [253] fault injection was performed to judge whether the status information provided by the TTP controller is sufficient for the detection of connector faults. The derived results constitute an important input for maintenance-oriented online analysis mechanisms. The fault injection experiments have shown that the frame status information is a suitable indicator. Component-external faults and component-internal faults were simulated using electromagnetic interference, while a disturbance node was used for the simulation of connector faults.

5.7   Example Configurations and Implementations

TTP has been integrated into a number of commercial applications. In the railway domain, Thales Rail Signalling Solutions has used TTP for realizing the electronic interlocking system *LockTrac 6131 ELEKTRA* [157]. This system has been certified according to CENELEC standards with Safety Integrity Level 4 (SIL4). The ELEK-TRA system supports the basic interlocking functions, as well as additional features such as local and remote control, automatic train operation, blocking functionality and diagnosis capabilities.

In the aerospace domain, TTP was deployed for the FADEC (Full Authority Digital Engine Control) systems. Honeywell used TTP as the backplane bus in an electronic controller for General Electric's F110 jet engine on the Lockheed Martin F-16 fighter aircraft. In addition, Honeywell deployed TTP on Honeywell's F124 engine in the M-346 fighter-trainer of the aircraft manufacturer Aermacchi [232]. In the Airbus A380 TTP serves as the communication system for the cabin pressure control system. In the Boeing 787 Dreamliner, Hamilton Sundstrand Corporation and TTTech developed the communication system based on TTP for the electric and environmental control system [214]. Parker Aerospace selected TTP for a new generic fly-by-wire actuation platform, which will be initially used on the Bombardier CSeries and Embraer Legacy 450/500 aircraft programs [332].

Furthermore, prototypes of off-highway vehicles use TTP [91]. For example, Eaton Corp. introduced controllers for electrohydraulic systems that are designed for applications where control loop times are critical. These controllers meet SIL 2 and 3 requirements of the off-highway industry [138].

6

FlexRay

C. El Salloum

Vienna University of Technology

K. Bilic

Vienna University of Technology

CONTENTS

6.1 Protocol Overview

FlexRay is a deterministic, scalable and fault-tolerant digital serial bus system designed for automotive applications. It was specified and developed by the FlexRay consortium, which was a cooperation of automobile manufacturers and leading suppliers that existed from the year 2000 to the year 2009. The FlexRay consortium has concluded its work with the FlexRay specification Version 3.0.[1]

A major design driver of FlexRay was to keep costs low while delivering high performance in a rugged environment. Nodes are interconnected by using unshielded twisted pair cabling supporting either single-channel or dual-channel configurations. Dual-channel configurations can be used for increased (doubled) bandwidth or for fault-tolerance by sending redundant data over each channel. The effects of external noise on the network is reduced by employing differential signaling on each twisted pair [146].

FlexRay is a flexible protocol that provides both deterministic communication, where data is sent and received in predictable time-frames, and dynamic event-driven communication like in CAN [151] networks. This hybrid approach is accomplished by a communication cycle that provides pre-defined time intervals for dynamic and static data.

6.2 Protocol Services

This section describes the services of the FlexRay protocol.

6.2.1 Communication Services

As mentioned above, FlexRay supports both deterministic time-triggered communication and flexible event-triggered communication in a single communication protocol. In this section, we give a detailed overview of the protocol including communication modes, media access control, frame format, and coding and decoding on the physical layer.

In FlexRay, media access control is based on a recurring *communication cycle*

---

[1] www.flexray.com

which is the fundamental element of the media access scheme. The communication cycle is defined by a timing hierarchy encompassing four levels (see Figure 6.1 [64]).

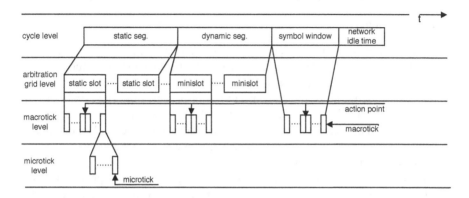

FIGURE 6.1
FlexRay Timing Hierarchy

- The highest level is the *communication cycle level*. This level defines the communication cycle and contains the *static segment*, the *dynamic segment*, the *symbol window* and the *network idle time*.

- The second level is called the *arbitration grid level* and forms the backbone of the arbitration in FlexRay. In the static segment the arbitration grid consists of the static slots, and in the dynamic segment of the minislots.

- The third level is the *macrotick level*. Macroticks are synchronized on a cluster-wide basis, and selected macroticks define *action points*, which are specific instants at which transmissions shall start or end.

- The lowest level is the *microtick level* which describes the time units that are directly derived from the communication controller's oscillator clock tick, optionally using a prescaler.

6.2.1.1 Temporal Structuring of Communication

The communication in FlexRay is organized in cycles, where each cycle consists of four segments: (1) the *static segment*, (2) the *dynamic segment*, (3) the *symbol window* and (4) the *network idle time*. A communication cycle always contains a static segment, while the dynamic segment is optional. The symbol window is also optional and contains a configurable number of macroticks. The network idle time is mandatory and contains the remaining number of macroticks within the communication cycle that are not allocated to the static segment, dynamic segment or symbol window. The network idle time is a communication-free time window used for clock corrections and to calculate and perform offset correction.

After the startup phase, the communication cycle is executed periodically. Communication cycles are numbered from 0 to 63. Each node shall maintain a *cycle counter* that represents the number of the current communication cycle.

Static Segment

Message transmission in the static segment is coordinated with the time division multiple access scheme. The static segment consists of a defined number of equally sized *static slots*, where each static slot is statically assigned to a unique frame via the *frame ID*. A frame is always sent in the slot of which the slot number is equal to the frame ID. On each channel, each frame ID is assigned to at most one node. A node is obliged to transmit a frame in all of its static slots. In case the node does not need the current slot because it has no new data, it has to send a *nullframe*, which is an ordinary frame with no data in the payload segment.

For a single node and a single slot, the following transmission patterns are possible (see Figure 6.2 [64]): (1) A node can send a frame on channel A and on channel B, (2) a node can send a frame only on channel A, (3) a node can send a frame only on channel B and (4) a node can send no frame at all during the slot. The major constraint is that in each slot on a given channel at most one node shall transmit a frame.

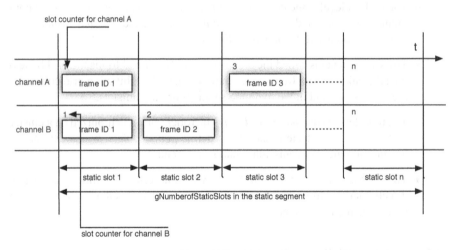

FIGURE 6.2
Structure of the Static Segment in FlexRay

All slots in the static segment consist of an equal number of macroticks, which is a global constant of the entire cluster. Within each slot there is an action point where the transmission of a frame should start. This action point is defined by a cluster-wide parameter which denotes the number of macroticks contained in the offset of the action point from the start of the slot [64]. At the action point, the channel is active and the transmission of a frame starts. After the transmission of a frame, there is an idle time before the next slot begins. The silence intervals within a slot are required to compensate for the clock drift of the nodes in order to establish a consistent view in the cluster. All nodes should observe the transmission of a given frame in the same slot.

Dynamic Segment

If configured, the dynamic segment starts after the static segment. It provides event-triggered communication based on a dynamic minisloting scheme used to arbitrate transmissions. Figure 6.3 depicts the structure of the dynamic segment within two FlexRay channels [64]. The dynamic segment consists of a definable number of equally sized *minislots*.

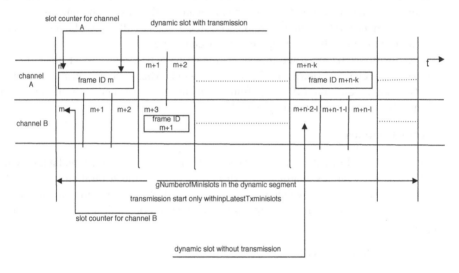

FIGURE 6.3
Structure of the Dynamic Segment in FlexRay

Based on the grid formed by the minislots, FlexRay defines so-called *dynamic slots*. In contrast to static slots, the duration of dynamic slots may vary to support frames of variable size. Each dynamic slot consists of one or more minislots. The dynamic slots in a communication cycle are enumerated in an ascending order starting with the ID of the last static slot + 1 (i.e., the numbering of slots continues in the dynamic segment). As in the static segment, a frame is statically assigned to a dynamic slot via the frame ID, and a frame ID is statically assigned to a unique node for each channel.

In contrast to the static segment, where each node is obliged to send a message in its static slot, a node will use its dynamic slot only when it has new data to send. A node is allowed to send a given frame when the actual number of the dynamic slot corresponds to the ID of that frame.

For arbitration, each node maintains two slot counters, one for each channel. A node's slot counter for a given channel is incremented by one at the end of every dynamic slot that has occurred on that channel.[2] If the value of a node's slot counter of a given channel equals the frame ID of a frame that has been assigned to that node for that channel, the node has the opportunity to send the corresponding frame. If the node does not start to transmit the frame (i.e., the channel is idle throughout the entire minislot), the duration of the dynamic slot is only one minislot, and the next dynamic slot starts at the next minislot. If the node starts the transmission of the frame, the length of the dynamic slot will be prolonged by additional minislots until the transmission of the frame is finished.

Frames with a lower frame ID are sent earlier in the dynamic segment, and thus have a higher priority. It might happen that a frame with a low priority (i.e., a high frame ID) cannot be sent before the end of the dynamic segment has been reached. In this case, the corresponding node has to wait for the next communication cycle for a new chance to transmit the frame. The slot counters will always be reset to zero when the end of the dynamic segment has been reached.

As for the static slots, the minislots contain an action point offset defined by a global parameter. In contrast to the static segment, where frame transmission only starts at the action point offset, the frame transmission starts and stops at the action point offset of the corresponding minislots.

Symbol Window

The symbol window can be used to transmit specific symbols (e.g., a media test symbol to test the local bus guardian). FlexRay does not provide arbitration for multiple nodes within the symbol window. If this is required, it has to be implemented on a higher level. The number of macroticks within the symbol windows is a cluster-wide parameter.

Network Idle Time (NIT)

The network idle time is comprised of the remaining number of macroticks of the communication cycle that are not allocated to the static segment, the dynamic segment or the symbol window. During this time interval, each node calculates and applies clock correction terms.

6.2.1.2   Frame Format

A FlexRay frame is divided into three segments, namely the *header segment*, the *payload segment* and the *trailer segment* (see Figure 6.4 [64]).

FlexRay Header Segment

The FlexRay header segment is the first segment of the FlexRay frame and consists of five bytes. It includes a *reserved bit*, the *payload preamble indicator*, the *null frame*

---

[2] In the dynamic segment, the slot counters of both channels do not necessarily have the same values.

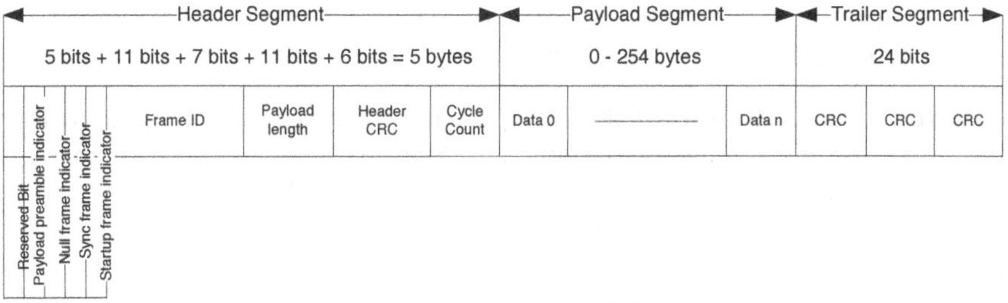

| Reserved Bit | Payload preamble indicator | Null frame indicator | Sync frame indicator | Startup frame indicator | Frame ID | Payload length | Header CRC | Cycle Count | Data 0 | ———— | Data n | CRC | CRC | CRC |

Header Segment: 5 bits + 11 bits + 7 bits + 11 bits + 6 bits = 5 bytes

Payload Segment: 0 - 254 bytes

Trailer Segment: 24 bits

FIGURE 6.4
FlexRay Frame Format

indicator, the *sync frame indicator*, the *startup frame indicator*, the *frame ID*, the *pay-load length*, the *header CRC* and the *cycle count*.

Reserved bit: The reserved bit is not used by the protocol and shall not be used by the application. It is reserved for future protocol use. There are two rules given by the FlexRay specification which define this bit in the node environment:

- The transmitting node shall set the reserved bit to logical 0.
- The receiving node shall ignore the reserved bit.

Payload preamble indicator: If the payload preamble indicator is set to logical 1, it indicates that an optional vector is contained within the payload segment of the frame which is transmitted. If the payload preamble indicator is set to logical 0, then no optional vector is transmitted in the frame. There are two rules for the payload preamble indicator which distinguishes between the payload preamble indicator in the static FlexRay segment and the payload preamble indicator in the dynamic FlexRay segment.

- If the frame is transmitted in the static segment, the payload preamble indicator indicates the presence of a network management vector at the beginning of the payload.
- If the frame is transmitted in the dynamic segment, the payload preamble indicator indicates the presence of a message ID at the beginning of the payload.

Null frame indicator: The null frame indicator indicates whether the frame is a *null frame*. A null frame is a frame which contains no useable data in the payload segment of the frame. If the null frame indictor is set to zero, then the payload segment contains data.

Sync frame indicator: If the sync frame indicator is set to logical 1, it indicates that the frame is a sync frame. A sync frame is a frame which is used for system wide synchronization of the communication between nodes in a FlexRay network. If the sync frame indicator is set to logical 1, every receiving node shall utilize the frame for synchronization. The clock synchronization mechanism makes use of the sync frame indicator.

Startup frame indicator: If the startup frame indicator is set to logical 1, it indicates that the frame is a startup frame. Startup frames are special frames which have an important role in the startup mechanism. The startup frame indicator shall only be set to 1 in the sync frames of *coldstart* nodes. Therefore, a frame with the startup frame indicator set to 1 shall also have the sync frame indicator set to one. Since the startup frame indicator can only be set to 1 in sync frames, every coldstart node can transmit exactly one frame per communication cycle and channel with the startup frame indicator set to 1.

Frame ID: The frame ID is an 11 bit-field which defines the slot in which the frame should be transmitted to a receiver. With these 11 bits it is possible to have an ID range between 1 and 2047. The ID 0 is an invalid frame ID in FlexRay. Every frame ID is only assigned to one frame and every frame has its own frame ID.

Payload length: The payload length is a 7 bit-field which is used to indicate the size of the payload segment. It defines the number of 16-bit words in the payload segment. The upper limit of the payload length in the FlexRay protocol is 127 two-byte words.

The payload length shall be fixed and identical for all frames sent in the static segment of the communication cycle. In the dynamic segment, the payload length may be different for different frames of a communications cycle. It is also possible that the payload length of a specific frame changes from cycle to cycle and may also be different on each configured channel.

Header CRC: The header CRC is an 11 bit-field that contain a cyclic redundancy check (CRC) code that is computed over the sync frame indicator, the startup frame indicator, the frame ID and the payload length. The header CRC of transmitted frames is not calculated online by the transmitting communication controller, but offline and is provided to the communication controller as a configuration input. The header CRC of every frame which is received must be calculated by the communication controller in order to check the header for correctness.

For all configured channels, the CRC is computed in the same manner. The initialization vector is 0x01A and the CRC polynomial is:

$$x^{11}+x^9+x^8+x^7+x^2+1 = (x+1)*(x^5+x^3+1)*(x^5+x^4+x^3+x+1) \quad (6.1)$$

To compute the header CRC, the sync frame indicator must be shifted in first, followed by the startup frame indicator, followed by the most significant bit of the frame ID, followed by subsequent bits of the frame ID, followed by the most significant bit of the payload length and followed by subsequent bits of the payload length [64].

Cycle count: The cycle count is a 6 bit-field and represents the value of the node's cycle counter at the time of frame transmission. Valid values for the cycle counter are between 0 and 63.

FlexRay Payload Segment

The FlexRay payload segment contains 0 to 254 bytes (0 to 127 two-byte words) of data which can be transmitted between nodes in the FlexRay network.

In frames transmitted in the dynamic segment, the first two bytes of the payload segment may optionally be used as a message ID field. With this message ID, the nodes in the FlexRay network can filter data with a specific message ID.

For frames transmitted in the static segment, the first 0 to 12 bytes of the payload segment may optionally be used as a network management vector. The length of the network management vector could be between 0 and 12 bytes. The usage of the network management vector is application-specific and not defined by the FlexRay protocol.

FlexRay Trailer Segment

The FlexRay trailer segment is a 24 bit-field which contains a cyclic redundancy check code (CRC) computed over the header segment and the payload segment of the frame.

The same CRC polynomial is used on both channels but the initialization vector is not the same for both channels. For channel A, the initialization vector is 0xFEDCBA and for channel B, the initialization vector is 0xABCDEF. The polynomial used for CRC calculation is given as follows:

$$x^{24} + x^{22} + x^{20} + x^{19} + x^{18} + x^{16} + x^{14} + x^{13} +$$
$$+ x^{11} + x^{10} + x^{8} + x^{7} + x^{6} + x^{3} + x + 1$$
$$= (x+1)^2 * (x^{11} + x^9 + x^8 + x^7 + x^5 + x^3 + x^2 + x + 1)$$
$$(x^{11} + x^9 + x^8 + x^7 + x^6 + x^3 + 1)$$

6.2.1.3 Coding and Decoding

FlexRay employs *Non Return to Zero* (NRZ) coding for transmitting frames over the communication channel. Every transmitted byte is pre-fixed with a high-bit and a low-bit (see Figure 6.5 [64]). Therefore, the transmission of each byte requires 10 bits on the physical medium. These preceding bytes are called the *Byte Start Sequence* (BSS), and are used for bit synchronization between the sender and the receiver.

In addition, every frame is preceded with a *Transmission Start Sequence* (TSS), which can be configured from 3 to 15 bits. The TSS is required for star couplers which require a given activation time before they can forward the frames. Star couplers can compensate this time by omitting some bits of the TSS.

The TSS is followed by a high-bit called the *Frame Start Sequence* (FSS). After

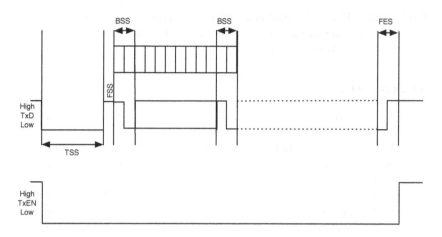

FIGURE 6.5
FlexRay Coding

the FSS, the actual bytes of the frame are transmitted, starting with the header. The last byte of the frame is followed by the *Frame End Sequence* (FES), which consists of a low-bit and a high-bit.

6.2.2 Protocol Operation Control

The major state machine of the FlexRay protocol is called *Protocol Operation Control* (POC). The circles in the state diagram depicted in Figure 6.6 represent states or superstates, which are collections of states [274]. The names of all states and superstates in this state machine are prefixed with the sequence *POC:* (e.g., *POC:ready*). The arrows represent transitions between these states.

After power-on, a communication controller is in the state *POC:default config*. From this state, the host can trigger the transition to the state *POC:config*, where the host is able to configure the communication controller. After the host has finished the configuration of the communication controller, it can trigger the transition to *POC:ready*. In this state, the controller cannot send or receive any messages and waits for further commands from the host.

In the *POC:wakeup* state, the communication controller can wake up other nodes, via transmitting a *wakeup pattern* over the network. The transition to the *POC:wakeup* state is also triggered by the host. After the wakeup procedure is finished, the communication controller returns to the *POC:ready* state.

The *POC:startup* state is a superstate that consists of multiple states that are relevant for startup. The transition from the *POC:ready* state to the *POC:startup* state is also triggered by the host. If the startup was successful, the communication controller enters the state *POC:normal active*.

POC:normal active is the major operational state, where a node sends and re-

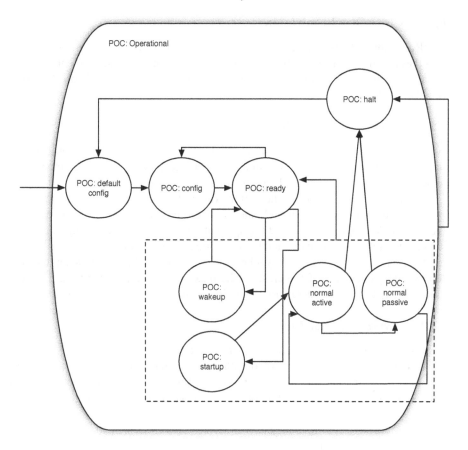

FIGURE 6.6
FlexRay Protocol Operation Control

ceives frames. If a failure occurs, the communication controller enters—depending on its configuration—either the state *POC:normal passive* or *POC:halt*. In the state *POC:normal passive*, a communication controller can still receive frames, but is not allowed to send any frames. In the state *POC:halt*, all processes within the controller are stopped. The only possible transition from the *POC:halt* state is to *POC:default config*.

6.2.3   Clock Synchronization

FlexRay employs a distributed clock synchronization mechanism in which each node observes the timing of transmitted *sync frames* from other nodes and synchronizes itself to the cluster. The clock synchronization is based on the *Fault Tolerant Midpoint* (FTM) algorithm.

6.2.3.1   Global and Local Time

According to the FlexRay specification, the term *global time* denotes a uniform notion of time within a cluster. An absolute time or reference time is not defined. Each node has its own local view of the global time called *local time*.

The *local time* of a node is the current value of its local clock. It is represented by the cycle counter, the macrotick and the microtick, where the cycle counter and the macrotick have to be visible at the application level. At the beginning of a cycle, the update of the cycle counter and the macrotick is always atomic.

Every node executes a clock synchronization algorithm in order to adapt its local time to the global time of the cluster. The maximum difference between the local clocks of any two synchronized nodes within the cluster is the *precision* of the cluster.

6.2.3.2   Synchronization Process

In FlexRay, each node performs *offset correction* and *rate correction* of its local clock [274]. Offset correction is concerned with adjusting the value of the local clock to the value of the global time at a given point in time. Rate correction is concerned with adjusting the frequency of the local clock to the frequency of the global time.

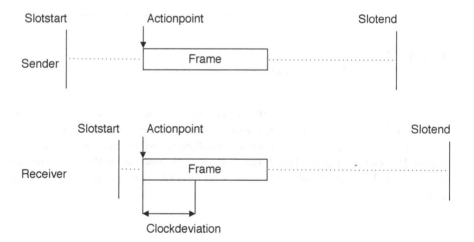

FIGURE 6.7
Deviation of Local Clocks

For *offset correction*, every communication controller uses an FTM algorithm to calculate a correction term in order to adjust its local clock. For this purpose, it measures the difference between the expected and the actual arrival times of sync frames from other nodes (see Figure 6.7). A node's offset correction term is derived by the following steps:

1. Depending on the number of correctly received sync frames, a value k is derived according to table 1.

Number of values	k
1 - 2	0
3 - 7	1
>7	2

2. The k minimum and maximum measured values are removed (see Figure 6.8 [274]).

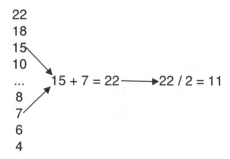

FIGURE 6.8
Fault Tolerant Midpoint Algorithm Used in FlexRay

3. The smallest and the largest of the remaining values are averaged, and the resulting value is the offset correction term for a node.

In order to derive a node's clock frequency deviation from the frequency of the global time, the offset correction terms of two successive communication cycles are used. The difference of two successive offset correction terms is the rate correction term, which indicates by how many microticks a node's cycle length should be changed.

As depicted in Figure 6.9 [64], the offset correction term is determined every communication cycle while the rate correction term is determined every second communication cycle. After the rate correction term has been derived, it is applied for the next two communication cycles, by adding or removing microticks distributed over the entire double cycle. Offset correction is only applied every second communication cycle in order not to disturb the calculation of the rate correction term. It is performed during the *Network Idle Time* (NIT), by either shortening or enlarging the NIT.

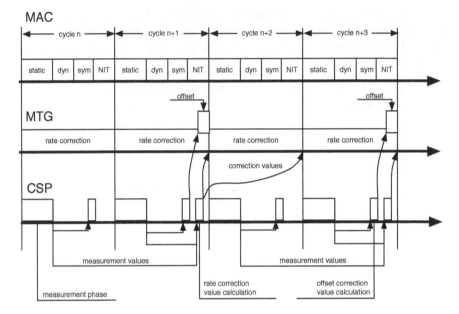

MAC = media acces schedule

MTG = clock sync correction schedule

CSP = clock sync calculation schedule MTG

FIGURE 6.9
Clock Synchronization in FlexRay

6.2.4   Wakeup and Startup

This section describes the wakeup and startup procedure of a FlexRay cluster.

6.2.4.1   Wakeup

There are different possibilities to switch the nodes of an embedded system into an operational state. The simplest way would be to power-on all nodes of a system at once via a central switch for the power supply of all ECUs. Such a solution is simple from a technological perspective but involves a significant cabling overhead. FlexRay provides a more advanced alternative, which is waking up the nodes via the communication medium, thus evading the need for additional cabling [274].

Wakeup Pattern

The host of a node can initiate the wakeup procedure by commanding the communication controller to transmit a *Wakeup Pattern* which consists of two or more *wakeup symbols*. A wakeup symbol consists of a low phase, where the communication medium is set to LOW for 6 μs, followed by an idle phase of 18 μs (see Figure 6.10 [274]).

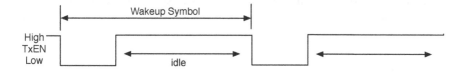

FIGURE 6.10
FlexRay Wakeup Pattern

Wakeup Process

The entire wakeup procedure is application-specific and, to a significant part, controlled by the hosts. It cannot be performed by the communication controllers and the FlexRay protocol alone. In the following, we describe a possible wakeup scenario:

1. A node is woken up by some external event (e.g., the key of a car has been turned), performs the configuration of its communication controller and makes the transition to the *POC:ready* state.

2. After the communication controller has been initialized, the host enables its bus drivers.

3. The host enters the *POC:wakeup* state. In this state, it will listen to the bus for two consecutive communication cycles and, if no communication is ongoing, it transmits the wakeup pattern on one channel. According to the specification, a communication controller shall not send a wakeup pattern on both channels at the same time, in order to prevent an erroneous communication controller from causing a global communication failure.

4. The bus drivers of all other nodes that are attached to the channel on which the wakeup pattern has been transmitted will wake up themselves and then wake up their hosts.

5. The awakened hosts enable their communication controllers.

6.2.4.2   Startup

The purpose of the startup procedure is to establish an initial common notion of time among the nodes of the FlexRay cluster which is a prerequisite for time-triggered communication. In order to avoid single-points of failure, the startup procedure is distributed and does not rely on a single master.

With respect to startup, FlexRay differentiates between two types of nodes, cold-start nodes and non-coldstart nodes [274]. The coldstart nodes may actively partici-pate in the startup process while non-coldstart nodes may integrate themselves only in a running cluster (i.e., after there are at least two synchronized coldstart nodes). A *leading* coldstart node is the first node that sends a *startup frame* during the startup process.

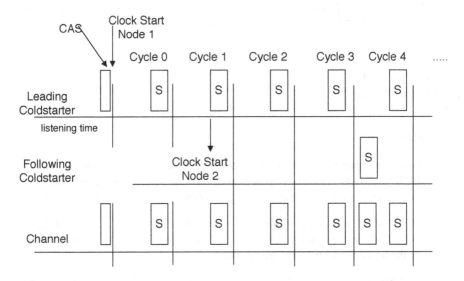

FIGURE 6.11
FlexRay Startup

After a coldstart node is initialized, it can begin with the startup procedure which is initiated by the host via a dedicated command. The startup procedure consists of the following steps (see Figure 6.11 [274]):

1. A coldstart node listens to the bus for at least two communication cycles. If there is no ongoing communication, it starts its own clock and sends a so-called *collision avoidance symbol* (CAS) which is a defined number of zero-bits. By sending the CAS, the node becomes a leading coldstart node. The CAS indicates to all other listening nodes that there is activity on the bus, and that they have to wait again for at least two communication cycles before they can become a leading coldstart node.

2. The leading coldstart node starts executing its time-triggered schedule, but dur-ing the startup phase it will send only the frame that has the *startup frame in-dicator* bit set to 1 in its header. Since every node can have at most one frame configured as a startup frame, the leading coldstart node will send only one frame per communication cycle during the startup phase.

3. The startup frames are received by the other coldstart nodes. Due to the frame

ID, which corresponds to the slot ID and the value of the cycle counter which is also part of the frame, the listening coldstart nodes know the time of the leading coldstart node.

4. The receiving nodes wait until they have received two successive frames, measure the time between the two frames and synchronize their clocks in the following way: (1) After the reception of the second frame, the node initializes its local clock based on the slot ID and the cycle counter of the frame. (2) The time difference between the two successively received frames should be equal to the cycle time, and any deviation will be applied to perform rate correction of the own local clock.

   This way, the startup procedure accomplishes both *state correction* and *rate correction*.

5. The clocks of the receiving coldstart nodes are now synchronized. Nevertheless, they have to wait for two more cycles, where they check whether they stay in sync before they are allowed to send frames on their own.

6. Non-coldstart nodes may communicate after they have observed at least two coldstart nodes that are in sync. Therefore, there have to be at least two coldstart nodes in the system, but for fault-tolerance reasons it is recommended to have more than two.

6.3 Diagnostic Services and Fault Isolation

FlexRay employs several services and mechanisms for diagnostics and fault isolation, like redundant communication channels, local and central bus guardians and validity checking of received frames. In this section, we will introduce these services.

6.3.1 Redundant Communication Channels

FlexRay provides scalable fault-tolerance by supporting either *single-channel* or *dual-channel* configurations. In safety-critical systems, dual-channel configurations will be employed and safety-critical data will be redundantly transfered on both channels. If redundancy is not required, a node can either be connected to only one channel, or it can use both channels to transfer non-redundant data in order to increase the bandwidth [146].

6.3.2 Bus Guardians

FlexRay supports two general types of bus guardians, the *local bus guardian* and the *central bus guardian*.

6.3.2.1   Local Bus Guardian

In an architecture with local bus guardians, every FlexRay node has its own bus guardian that guards the sending behavior of the attached node. In order to do so, it requires knowledge about the global time and the node's sending schedule (i.e., allocated slots). Furthermore, it requires the ability to prevent a node's access to the communication medium.

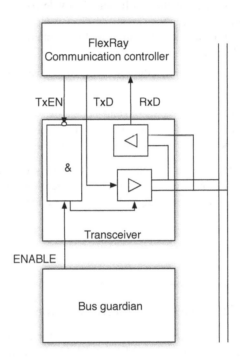

FIGURE 6.12
Local Bus Guardian

Figure 6.12 [274] depicts a possible realization of a local bus guardian. Both the transmit enable signal and the enable signal of the bus guardian are required to enable the transmission on the medium. In [274] it is recommended to use different logical levels for the two enable signals (e.g., LOW for the communication controller and HIGH for the bus guardian) to prevent sending in the case of common mode failures like short circuits.

The coupling of a bus guardian to the communication controller is a trade off between cost and the level of protection. If the bus guardian has no own local clock, but relies on the clock of the attached communication controller, it cannot detect any failure of the node's clock (rate or offset error). The only property that can be checked is whether the node sends in its assigned slots with respect to its local clock.

If the bus guardian is able to receive frames and to perform clock synchronization on its own, it can establish its own view of the global time. Such a bus guardian is able also to detect faults in a node's local clock.

6.3.2.2 Central Bus Guardian

A central bus guardian is based on an active star coupler and is able to perform clock synchronization on its own. As depicted in Figure 6.13 [274], the bus guardian is directly connected to all nodes.

Based on its view of the global time, it switches the transmission lines according to the schedule of all attached nodes. If the schedule of the nodes in the network is changed, the internal configuration of the bus guardian also has to be adapted. A significant advantage of a central bus guardian compared to a local bus guardian is that it can also protect the startup procedure and that it can prevent the formation of cliques [274].

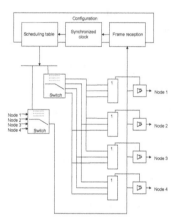

FIGURE 6.13
Inner Structure of a Central Bus Guardian

6.3.3 Checks on the Reception of a Frame

The FlexRay protocol performs several syntactic and semantic checks upon the reception of a frame [274]. The syntactic checks include verifying the *Byte Start Sequence* (BBS) of all bytes, the *Transmission Start Sequence* (TSS), the *Frame Start Sequence* (FSS), the *Frame End Sequence* (FES), the *Header-CRC* and the *Frame-CRC*. A frame that is syntactically incorrect is dropped.

The semantic checks include verifying the *frame length*, the *frame ID* and the *cycle count*. The frame ID and cycle count in the frame header should match the current slot ID and the actual cycle count. In addition, it is verified whether the transmission of a frame did obey the slot borders.

6.4 Protocol Parameterization

This section introduces the parameters of the FlexRay protocol. The FlexRay specification uses a uniform naming convention for all parameters that is based on two prefixes [64]. The first prefix is one of six alphabetic characters, which are *a, c, v, g, p, z* and have the following meaning:

- **a: (Auxiliary Parameter)** Auxiliary parameters are used for the definition or derivation of other parameters or for the derivation of constraints.

- **c: (Protocol Constants)** They define characteristics of the protocol. These values are fixed and cannot be changed.

- **v: (Node Variable)** They describe values which vary due to the progression of time or events.

- **g: (Cluster Parameter)** Cluster parameters must have the same value in all nodes of a cluster. These values can only be changed in the *POC:config* state of the communication controller.

- **p: (Node Parameter)** Node parameters may have different values in different nodes of a cluster.

- **z: (Local SDL Process Variable)** These are variables used in the *Specification and Description Language* (SDL) in order to facilitate the accurate representation of the required algorithmic behavior. Their scope is local to the process where they are declared and their existence in any particular implementation is not mandatory [64].

The second prefix is one of the two alphabetic characters *d* and *s*. This prefix is optional and has the following meaning:

- **d: (Time duration)** The time duration parameter describes a duration between two points in time.

- **s: (Set)** The set parameter describes a set of values such as variables, parameters etc.

6.4.1 Cluster Parameters

The cluster parameters are used to describe properties of the entire cluster. Therefore, they must have the same values in all nodes of the cluster. In this section, we introduce the most relevant cluster parameters of the FlexRay protocol:

- **Bus Speed:** Currently, FlexRay only supports a bus with a speed of 10 Mbit/s. Support for other bit rates is planned.

- **gColdStartAttempts:** Maximum number of attempts a node is allowed to try to start the cluster by initiating schedule synchronization (2-31).

- **gSyncNodeMax:** Number of nodes which are allowed to send synchronization frames (2-15).

- **gdCycle:** Is the length of one cycle (10-16,000 μs).

- **gdMacrotick:** Duration of the macrotick (MT) given in μs (1-6 μs).

- **gChannels:** The channels which are used by the cluster (A, B, A&B).

- **gdStaticSlot:** Duration of the static slot (4-661 MT).

- **gdActionPointOffset:** Number of macroticks the action point is offset from the beginning of a static slot (1-63 MT).

- **gNumberOfStaticSlots:** Number of static slots in the static segment. The number of static slots is defined between 2 and 1023. The minimum size of two slots comes from the fact that two nodes are required for startup and synchronization.

- **gPayloadLengthStatic:** Payload size for static frames in two-byte words (0-127).

- **gdMinislot:** Duration of the minislot used in the dynamic segment (2-63 MT).

- **gdMinislotActionPointOffset:** Number of macroticks the action point is offset from the beginning of a minislot (1-31 MT).

- **gNumberOfMiniSlots:** Number of minislots used in the dynamic segment (0-7986).

- **gdSymbolWindow:** Duration of the symbol window (0-142 MT).

- **gdNIT:** Duration of the Network Idle Time (2-805 MT).

- **gdTSSTransmitter:** Number of bits in the transmission start sequence (TSS) (3-15 gdBit).[3]

6.4.2 Node Parameters

The node parameters can be different for each node in the cluster. In this section, we list the most relevant parameters:

- **pChannels:** The channels to which the node is attached (A, B, A&B).

- **pDelayCompensation[A], pDelayCompensation[B]:** Value for the compensation of the reception delay on the individual channels (0-200 microticks).

---

[3]Nominal bit time

- **pKeySlotId:** ID of the slot used to transmit the startup or sync frame (1-1023).

- **pOffsetCorrectionOut:** Maximum permissible offset correction value (13-15,567 microticks).

- **pRateCorrectionOut:** Maximum permissible rate correction value (2-1923 microticks).

- **pMicroPerCycle:** Number of microticks per communication cycle (640-640,000).

- **pWakeupChannel:** Channel used to transmit the wakeup pattern (A, B).

6.5   Controller Host Interface

The *controller host interface* (CHI) is the interface between the host processor and the protocol engine. The CHI depends on the actual implementation of the employed communication controller. In this section, we give an overview of the CHI of the *E-Ray IP Module* [112] which is a widely used communication controller developed by the Robert Bosch GmbH.

6.5.1   Overview of the E-Ray IP Module

The E-Ray IP module is a FlexRay v2.1 communication controller developed by the Robert Bosch GmbH. It can be realized as part of an ASIC, within an FPGA or as a stand-alone device. It can store up to 128 message buffers and supports a payload with up to 254 data bytes. Message buffers can be configured for static and dynamic messages as well as part of a FIFO for incoming messages. The controller performs all the required synchronization tasks for accesses of the host or the protocol unit to the message RAM. The E-Ray IP module provides a generic 8, 16 or 32 bit CPU interface and supports data rates of up to 10 Mbit/s on each channel.

Figure 6.14 [112] depicts a block diagram of the E-Ray IP module consisting of the following components:

Generic Interface (GIF) The GIF is adaptable to different customer-specific CPUs and supports 8, 16 or 32 bit interfaces.

Input Buffer (IBF) and Output Buffer (OBF) These buffers are used to synchronize the read and write operations of the host to the message RAM.

FlexRay Protocol Units (PRT A and PRT B) The two protocol units execute the FlexRay protocol and are connected to transient buffer RAMs for intermediate message storage.

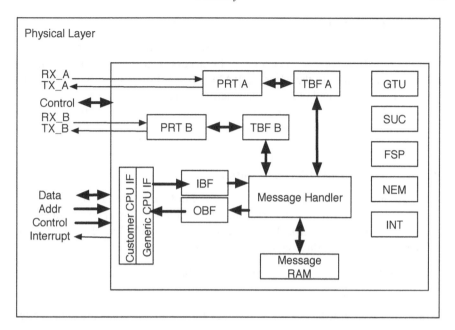

FIGURE 6.14
Block Diagram of the E-Ray Controller

Transient Buffer RAM (TBF A and TBF B) The TBFs are realized as double buffers and function as intermediate storage for frame transmission and reception.

Message RAM (MRAM) The message RAM stores the message buffers and the corresponding configuration data. It is realized as a single-ported RAM and is directly accessed only by Message Handler.

Global Time Unit (GTU) The global time unit performs fault-tolerant clock synchronization and timing control of the static and dynamic segment.

System Universal Control (SUC) The SUC controls the wakeup, startup and the reintegration of a node in a cluster.

Frame and Symbol Processing (FSP) The FSP tests the syntactical and semantical correctness of the received frames as well their correct timing.

Network Management Vector (NEM) The NEM handles the network management vector.

Interrupt Control (INT) The INT controls the generation of interrupts.

6.5.2 Programmers Model

The E-Ray IP Module allocates and addresses space of 2 KB. The registers are organized as 32-bit registers, but 8 or 16 bit access is also supported. The host can access the messages in the Message RAM (MRAM) exclusively via the Input and Output Buffer (IBF and OBF), as depicted in Figure 6.14. Write and read access to the MRAM is exclusively controlled by the Message Handler (MHD) in order to avoid conflicts between the accesses of the host, and receptions or transmission of frames.

6.5.2.1 Assignment of Message Buffers

The assignment of message buffers is depicted in Figure 6.15 [112]. The maximum number of message buffers is 128, while the available number of message buffers depends on the configured payload length of the message buffers. The set of message buffers is divided into three consecutive groups:

Static Buffers This group starts with message buffer 0 and contains only static message buffers.

Static and Dynamic Buffers The message buffers in this group are either assigned to the static or to the dynamic segment. These buffers can be reconfigured during runtime from static to dynamic and vice versa.

FIFO The buffers in the third group form a single FIFO for incoming messages.

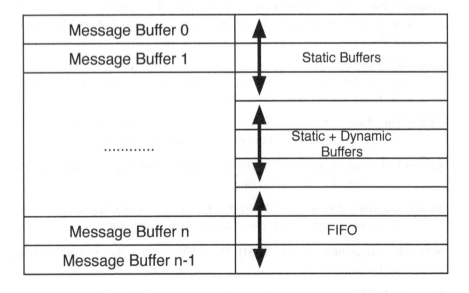

FIGURE 6.15
Assignment of Message Buffers in an E-Ray Controller

The separation of the message buffers into these three groups is configurable, but can be configured exclusively in the POC:DEFAULT_CONFIG or POC:CONFIG state.

6.5.2.2 Structure of the Message RAM

The Message RAM is organized in 2048 32 bit words, where each 32 bit word is protected by one parity bit. Thus, the size of the Message RAM is 2048*33 bit. In order to efficiently support different payload sizes, the Message RAM is structured in a *Header Partition* and a *Data Partition* (Figure 6.16 [112]).

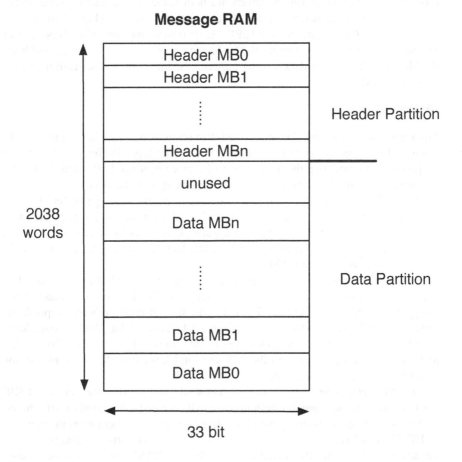

FIGURE 6.16
The Message RAM of an E-Ray Controller

The header partition consists only of the headers of the message buffers, where each header has the same size of four 32 bit words (plus one parity bit per word), and holds an 11 bit data pointer that points to the corresponding payload in the data partition. The data partition stores the data sections of the message buffers configured in the header section. The number of data bytes for each message buffer can vary from 0 to 254. The beginning and the end of a message buffers payload is determined by the data pointer and the payload length, both configured in the message buffer's header.

6.5.2.3 Message Handling

The data transfer between the host and the Message RAM is controlled by the Message Handler. The host can access the contents of the Message RAM only via the Input Buffer and Output Buffer, which are both realized as a double buffer structure. One side of the double buffer is accessible by the host (Input Buffer Host (IBF Host) / Output Buffer Host (OBF Host)) while the other side (Input Buffer Shadow (IBF Shadow) / Output Buffer Shadow (OBF Shadow)) is accessible by the Message Handler for data transfer between the Input and Output Buffer and the Message RAM.

Writing Messages

The Input Buffer contains the header and data sections that should be transfered to a selected message buffer in the Message RAM. The Input Buffer is used for two purposes, (i) for updating the data section of a given message buffer and (ii) for the configuration of the message buffers by writing the specific headers.

When the host wants to configure or update a specific message buffer, it writes the header section or the data section or both sections in the host side of the Input Buffer (IBF Host). By using the Input Buffer Command Mask Register (IBCM), the host can set whether the header section, the data section or both sections should be updated in the Message RAM.

When the host writes the number of the selected message buffer to the IBF Request Host register (IBRH[6:0]), the contents of IBF Host and IBF Shadow are swapped as depicted in Figure 6.17 [112]. After the buffers have been swapped, the Message Handler starts to transfer the selected data from the IBF Shadow to the Message RAM. The controller indicates the ongoing transfer by setting the bit IBSYS to 1. After the transfer between the IBS and the Message RAM is completed, the controller sets IBSYS back to 0.

If a write access to IBRH[6:0] occurs while IBSYS is 1, the content of IBF Host and IBF Shadow are not swapped immediately and the controller sets the bit IBSYH to 1. After the ongoing transfer is finished, the controller swaps the contents of IBF Host and IBF Shadow, sets IBSYS back to 0 and starts the next transfer of the new content of the IBF Shadow to the Message RAM. IBSYS remains 1 until the transfer is completed.

A write access to the IBF while IBSYH is 1 has no effect on the contents of the IBF and is recorded by an error flag.

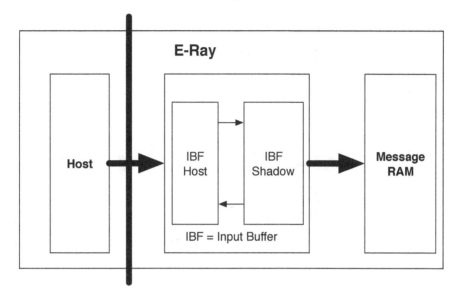

FIGURE 6.17
E-Ray Controller: Input Double Buffer

Reading Messages

The Output Buffer contains the header and data sections that should be transfered from a selected message buffer in the Message RAM to the host.

If the host wants to read the contents of a specific message buffer, it writes the number of that message buffer to the register OBRS[6:0]. By setting the bit REQ to 1, the host triggers the transfer of the message buffer from the Message RAM to the OBF Shadow. The bit OBSYS is automatically set to 1 while the transfer is ongoing, and set to 0 after the transfer is completed. By setting the bit VIEW to 1, the contents of OBF Host and OBF Shadow are swapped, and the host can read the contents of the specified message buffer by accessing OBF Host (see Figure 6.18 [112]). REQ and VIEW can only be set to 1 while OBSYS has the value 0.

FIFO Functionality

As mentioned above, a consecutive set of message buffers can be assigned to a FIFO for incoming messages. Every valid message that is not matching any configured static or dynamic message buffer and that passes the programmable FIFO Rejection Filter (FRF) will be stored in the FIFO. If the FIFO is full, the new message will overwrite the oldest message in the FIFO. The FRF can filter incoming messages based on the channel through which the message was received, the Frame ID and the cycle counter. The messages in the FIFO are read by writing the number of the first message buffer that is assigned to the FIFO to the OBRS[6:0] register. Any read request to that message buffer will read and consume the oldest message in the FIFO.

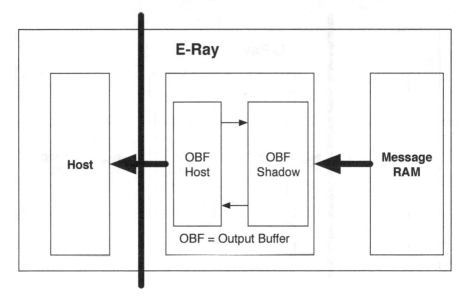

FIGURE 6.18
E-Ray Controller: Output Double Buffer

6.6   Example Configurations and Implementations

6.6.1   Topology and Layout of a FlexRay Network

This section gives an overview of the different network topologies and layouts that are supported by the FlexRay protocol. With respect to topologies, FlexRay is very flexible. It supports single-channel or dual-channel bus networks, single-channel or dual-channel star networks or various hybrid combinations of bus and star topologies.

A FlexRay cluster consists of at most two channels, called *channel A* and *channel B*. Each node in a FlexRay cluster can be connected to a single channel or to both of them. A node that is connected to a given channel can communicate with every other node that is connected to the same channel. A communication controller can be connected to at most one cluster, and if a node should be connected to more than one cluster, it requires a dedicated communication controller for each cluster.

In the following subsections, we will describe the different topologies supported by FlexRay.

6.6.1.1   Passive Bus Topology

The example in Figure 6.19 [64] shows a dual-channel bus topology, with nodes connected to both channels, nodes connected only to channel A and nodes connected only to channel B. Bus topologies can also be realized with a single channel, in which case all nodes in the network would be connected to this bus.

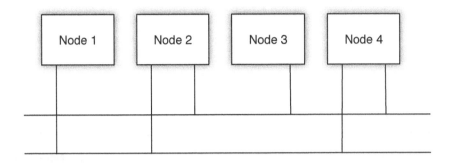

FIGURE 6.19
FlexRay Topologies: Passive Bus Topology

6.6.1.2   Active Star Topology

Another possible topology in FlexRay is the multiple star topology. As for the bus topology, the multiple-star topology supports redundant communication channels. The restriction is that there can be no more than two star couplers per network channel. Furthermore, FlexRay supports also cascaded stars (see the example in Figure 6.20 [64]).

6.6.1.3   Hybrid Network

Star and bus topologies can be combined to form a hybrid network. This approach allows us to combine the ease-of-use and cost advantages of bus topologies with the performance and the reliability of star topologies. The FlexRay Specification describes two representative topologies among the large number of topologies that are possible. Figure 6.21 [64] shows an example, where some nodes are connected using a point-to-point connection to a star coupler, while other nodes are connected via a bus, which is also connected to the star coupler.

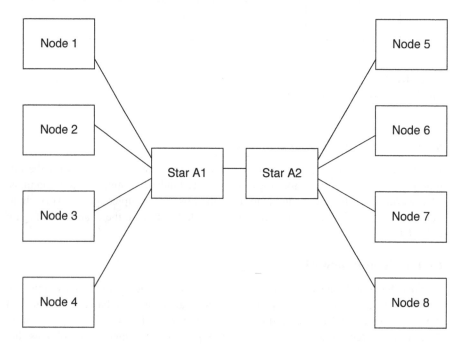

FIGURE 6.20
FlexRay Topologies: Cascaded Stars

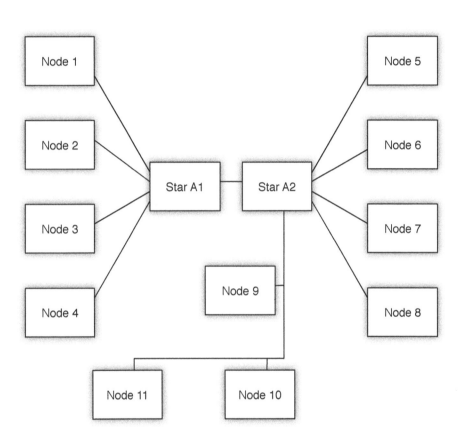

FIGURE 6.21
FlexRay Topologies: Single Channel Hybrid Example

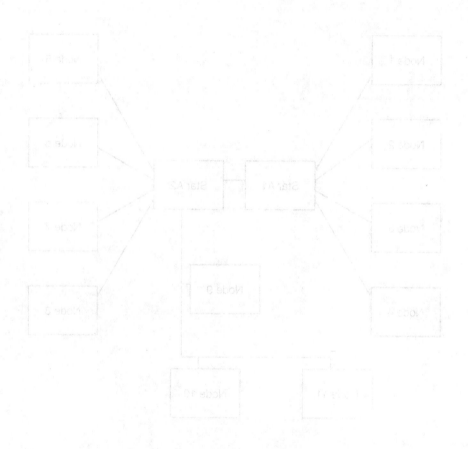

7

SAFEbus

M. Paulitsch

EADS

K. Driscoll

Honeywell

CONTENTS

7.1   SAFEbus

SAFEbus[1] is the only backplane or local area network standard (ARINC 659) that provides fail-op / fail-safe fault tolerance with near unity coverage for all of its components — signal lines, terminations, interface electronics, clock sources and power supplies. This coverage includes tolerating a Byzantine fault. SAFEbus provides a time-based protocol that delivers messages with a precision on the order of 100 nanoseconds over a backplane network. The SAFEbus protocol can be implemented in an integrated circuit or FPGA with just 70,000 gates.

7.1.1   Background

In the late 1980s, there was a push toward integration of multiple functions on a common computing and I/O platform, also referred to as integrated modular avionics (IMA). This push was due to the advantages of IMA systems over then prevalent federated architectures. The advantages are decreased size, cost and weight, increased reliability, less-frequent maintenance and more flexibility. The success of an IMA system hinges on a backplane bus connecting Line Replaceable Modules (LRMs). The backplane bus must be designed to support the requirements of *space and time partitioning*. These requirements are derived from the concept of "robust partitioning" that prevents functions on a common platform from adversely influencing each other, even when some functions may be faulty. Honeywell designed SAFEbus as a backplane for the Aircraft Information Management System (AIMS), which is the IMA part of the avionics for the Boeing B-777 airplane.

Boeing provided some additional design requirements. One requirement was for the number of days the Boeing 777 could be dispatched without maintenance following a failure. The goal was to allow a plane with a failure in an AIMS component to follow its normal schedule, which will eventually bring it to a maintenance base. This requirement meant that individual components of the AIMS system had to be reliable and that the system as a whole had to be fault-tolerant. A second design requirement was that the backplane-bus interface not force complexity on the functions in an LRM. Some LRMs might be high-performance processors, but others might be simple hardwired logic. A third requirement, implicit in the notion of an integrated cabinet, was that the design support a multiprocessor architecture. In particular, the backplane had to provide adequate net throughput for the initial set of functions together with 50 percent extra capacity to allow for growth. Fourth, the integrity requirements for the avionics system as a whole meant that the backplane bus had to exhibit total fault containment. There had to be less than one chance in a billion per hour of operation that an error occurring within the backplane system would be passed undetected from or to application software. Finally, the design had to be one that would support the certification of the system and the re-certification of modified functions. In particular, the design could not be one that would force

---

[1] SAFEbus is a registered trademark of Honeywell International, Inc.

the re-certification of all functions when only one function was modified. Honeywell designed SAFEbus because no existing backplane bus met these requirements. SAFEbus builds on the Multi-Processor Flight Control System (M^{2}FCS) [80] research done for the United States Air Force.

7.2 Protocol Overview

The SAFEbus protocol is heavily dependent on its hardware. In order to understand the SAFEbus protocol, one needs to know the basic building blocks of the hardware architecture. The SAFEbus interface logic within each LRM uses paired hardware; each half of the pair consists of a Bus Interface Unit (BIU) ASIC, a Table Memory, an Intermodule Memory (IMM) and Backplane Transceivers. This logic is paired to provide immediate fault detection and containment. The backplane bus lines are configured in a unique fault-tolerance topology that lies somewhere between quad redundancy and dual-dual redundancy [82]. This topology simultaneously provides high integrity and availability (see Figure 7.1). SAFEbus consists of two Self-Checking Buses (SCBs), A and B, called bus pairs. Each SCB is itself composed of two buses, x and y. Figure 7.2 presents a nomenclature of the different bus-related items. One of the BIUs in an LRM transmits data on one of the buses in each SCB, and its partner BIU transmits on the other bus within each SCB. The data on any two buses which come from different BIUs are compared at each receiving LRM. Only bit-for-bit identical data are written into the Intermodule memories. A transmitting LRM checks its transmission using a local loopback. That is, the receiving circuitry in the transmitting LRM also checks what is actually put on the bus for errors. Such self-

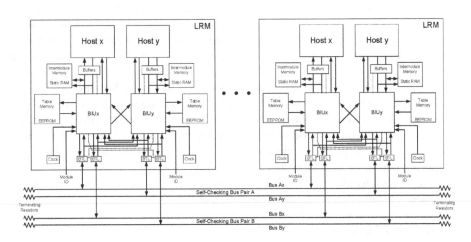

FIGURE 7.1
SAFEbus Interface Logic

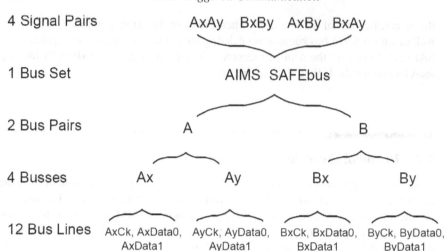

FIGURE 7.2
SAFEbus Nomenclature

checking ensures a babbling LRM will be self-detected and will remove itself from SAFEbus. This removal is enforced within an LRM by having each BIU control the other BIU's drivers. If either BIU thinks it should not be transmitting, neither BIU can transmit. Each bus consists of three wires, two for data and one for clock. Thus, the entire SAFEbus set of buses uses a total of 12 wires. The data is transmitted synchronously, two bits at a time, at 30 MHz (for throughput of 60 Mb/s). A "bit time" is the time it takes to send one bit on one wire. Using multiple wires, two bits and one dock cycle are sent during one bit time. For time, the term "bit time" is preferred to "clock period" because there are multiple clocks within each BIU. SAFEbus uses the "wired OR" capable Backplane Transceiver Logic (BTL) defined in the IEEE 1194 standard. The wired-OR drivers and the use of extended clock pulse-widths allows SAFEbus to do some out-of-band signaling that supports some features of the protocol.

Table 7.1 defines some SAFEbus terminology that is further described in this paragraph. The bus time is divided into "windows" of design-time configurable sizes. Each window is sized to contain either one message (of from 1 to 256 32-bit words), or one synchronization pulse or some fixed idle time. A set of static cyclic schedules defines the sequence of windows, the size of each window, which LRM(s) may transmit during each window and which LRM(s) may receive a message from the window. Each window ends with a small, fixed intermessage gap time. A typical intermessage gap is 2 to 4 clock periods (i.e., about 50 ns). Fixed idle time may be inserted to adjust the duration of the full cycle or the time between messages, to a precision of 33 ns.

Messages that are to be transmitted or have been received over the backplane are placed in buffers in Intermodule memories, which are pseudo dual-port memories

TABLE 7.1

SAFEbus terminology.

Term	Explanation
BIU	Bus Interface Unit
Command	Defines what a BIU does during a Window (TX, RX, skip, sync, send interrupt to host)
Frame	A cyclical repeating sequence of windows (NB: in SAFEbus and avionics generally, "frame" refers to a repetition of an execution cycle; whereas, in non avionics data communication, "frame" is often roughly synonymous with "message"
IMM	Intermodule Memory (shared message buffer memory between the BIU and its host)
Intermessage Gap	The constant reserved minimum idle time between transmissions on the bus
LRM	Line Replaceable Module (a node on SAFE bus)
Message	A single, unique transmission of data on SAFEbus; has a length (in words) fixed at design time
Table	One or more sequences of commands controlling one or more frames plus their resync jump points, and a BIU Configuration area
Table Memory	The nonvolatile memory that stores one or more tables
Window	The bus time reserved for a message, sync pulse or idle plus the trailing intermessage gap; has a duration (in bit times) fixed at design time

shared by the host(s) and the BIUs. This organization permits a simple host interface, because all the hosts can view SAFEbus as a shared multi-port memory.

7.3 Protocol Services

7.3.1 Communication Services

SAFEbus provides data communication with very low jitter (on the order of 100 ns) that is fail-op / failsafe with near unity coverage, even with a Byzantine failure.

The data for a message that is to be transmitted over SAFEbus is calculated typically by independent self-checking pair Hosts. As with all self-checking pairs, this data is bit-for-bit identical in the fault-free case. The Hosts write this data into buffers in the Intermodule Memories. Associated with each buffer is a Buffer Control Word (BCW). Whenever a host completes assembling a message buffer in the Intermodule Memory, it sets a bit in the BCW to say that the buffer is ready for transmission.

The SAFEbus protocol is driven by sequences of commands stored in the BIUs' Table Memories (see Section 7.3.7.1). Each command corresponds to a single window on the bus. The command indicates whether the BIU should transmit, receive or ignore the message in that window. The BIUs in every LRM on SAFEbus are synchronized to equivalent points in their respective tables and mechanisms are provided to quickly attain synchronization if it is ever lost. The tables also contain the local address (in the IMM) of the data to be transmitted or received. The commands in each BIU's Table are organized into multiple frames. Each frame controls a repetitive sequence of windows, and each frame has a fixed total period.

According to a schedule stored in its Table Memory, each BIU checks its Intermodule Memory for the buffer assigned to the next time window. If that buffer's BCW says the buffer is ready for transmission, the BIU begins pre-fetching the message from the Intermodule Memory. An LRM broadcasts its messages on the four buses when it is scheduled to do so. The two BIUs in an LRM are sync'ed to within two bit times of each other via the SAFEbus protocol on the buses (no "backdoor" sync between them). Because a BIU doesn't know if it is faster or slower than its partner, it turns on its partner's drivers two bit times before its first bit transmission and off two bit times after its last bit transmission for each message. Off-line scheduling ensures that messages from different sources never collide, taking into account worst-case clock drifts, metastability behavior, resync intervals, read timing errors, LRM positions along the bus (the speed of light does make a difference), etc. Time windows for messages can be set up such that up to four LRMs share a time window in a Master/Shadow arrangement, using a mini-slotting scheme to arbitrate for that window.

Receiving BIUs write validated input data into buffers within each of their Intermodule Memories. At the completion of a message, the BCWs associated with each buffer are updated with status that includes whether the buffer has valid data and what time it arrived. The fact that the BIUs are synchronized to protocol time allows each of them to independently write identical BCW timestamps into each one of their Intermodule Memories without doing an exchange of the received time between the BIUs. Protocol time is measured in bit times from the first window transmitted. The time of each SAFEbus event (e.g., message reception) is the protocol time of the window for which it occurred.

One of the benefits of the table-driven protocol is extremely high efficiency. Control applications typically generate short messages, and most serial protocols perform poorly when messages are short. Efficiencies of between 10% and 30% are typical. In contrast, the SAFEbus protocol is over 89% efficient for a continuous stream of 32-bit messages. Ethernet messages with the same payload would be less than 5% efficient. Because buffer addresses are kept in the tables, they do not need to be transmitted on the bus. The use of transmit and receive commands in the individual tables eliminates the need to send source or destination LRM addresses. Within a message, all clock periods contain data (zero overhead). And, because transmissions are scheduled, no transmission time is consumed arbitrating between contending BIUs (with the rare exception of the optional use of Master/Shadow that typically consume about 9 bit times per arbitration).

7.3.1.1 Determinism and Partitioning

SAFEbus' determinism and support for robust partitioning warrants more detailed examination, since no other protocol provides these features to this extent. When a system has functions with different levels of criticality, the functions must be partitioned both in space and in time. Space partitioning means that no function can prevent another from obtaining adequate memory space and that the memory space assigned to one function cannot be corrupted by the behavior of another function. Memory-management units (MMUs) are usually adequate for simple uniprocessor main memory. But, problems arise with multi-port memory (including network interfaces) and "memory" that must be shared (CPU registers, cache, I/O registers). Time partitioning means that one function's demand for shared hardware resources will never prevent another function from obtaining a specified level of service and, more importantly, that the timing of a function's access to these resources will not be adversely affected by variable demand or by failure of another function.

Any protocol that includes a destination memory address in a message is a space-partitioning problem. It is extremely difficult to verify correct address usage in a partitioned multiprocessor. To ensure correct usage, the BIU would have to duplicate the typical processor MMU function. Then, a difficult protocol would have to be implemented to ensure all BIUs used the same MMU information.

Protocols that use contention arbitration cannot be made strictly time-deterministic. Such arbitration is meant to ensure that when two modules contend for the bus, the one with the highest priority request is granted access. But minor jitter in the execution of functions can change which modules contend for the bus on any given bus cycle. As a result, the order in which the modules obtain access can vary from one arbitration to another in ways that cannot be predicted at design time.

SAFEbus achieves both time and space partitioning by placing all message location (IMM) and bus-timing information in its Table Memories that are frozen at design time. This Table information is held in the BIUs' Table Memories where it cannot be corrupted by any errant software or communications errors. The contents of these memories can be changed only by a very well guarded interface (see Section 7.3.6).

To extend SAFEbus' time determinism to include the functions' software, the software execution can be synchronized with the execution of the commands in the bus Table. Thus, the software is at the same point during the same bus transmission window in every frame. One benefit is that message latencies can be reduced to insignificance. Results can be scheduled to be transmitted just after they are generated and input data can be delivered just before it is needed (software never has to ask for input data to be transferred over SAFEbus as data is sent autonomously). A second benefit is that there is less latency jitter on cabinet outputs, which means that a SAFEbus IMA can be used in tighter control loops. A third benefit is that double buffering is rarely necessary because it is possible to schedule the transmission of a data block for a time when it is known that software will not be reading it or modifying it. The elimination of double buffers means the Intermodule memories can be smaller and memory access faster. While the use of software synchronized to the bus

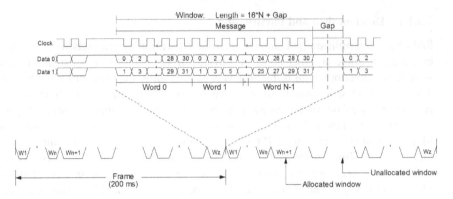

FIGURE 7.3
Basic Message Structure

and single buffers is the preferred operation, SAFEbus does allow for asynchronous software and double buffering.

7.3.1.2 Data-Message Structure

There are two data-message types: Basic and Master/Shadow. The Basic message structure has been chosen to maximize the efficiency of data transmissions. The Master/Shadow structure supports data transfers by redundant or aperiodic functions.

Basic and Master/Shadow Message Structures

Basic messages have a simple structure (see Figure 7.3). Each message consists of a string of 1 to 256 32-bit data words followed by a programmable intermessage gap of two to nine bit times. The Master/Shadow mechanism allows LRMs or applications to be reconfigured or spared without disturbing the traffic pattern on the bus. Master/Shadow windows are identified by a field in the associated Table command. As many as four transmitters can be assigned to one Master/Shadow window. Time-slot arbitration determines which of the transmitters actually gets control of the window. If the Master is alive and has fresh data to send, it starts transmitting at the beginning of the window. The first Shadow begins transmitting "delta" bit times into the window, only if the Master did not use its opportunity to transmit. The second Shadow begins transmitting two delta bit times into the window, only if the Master and the first Shadow did not use their opportunities to transmit. Finally, the third Shadow begins transmitting three delta bit times into the window, only if none of the other candidate transmitters use their opportunities to transmit. Delta is a programmable value that is typically set at one bit time larger than the selected intermessage gap (values from three to ten bit times may be selected). The selected value depends on the propagation characteristics of the backplane. Examples of the transmission over SAFEbus when the Master or third Shadow transmits are shown in Figure 7.4.

Time-slot arbitration could re-introduce non-determinism, but strict measures

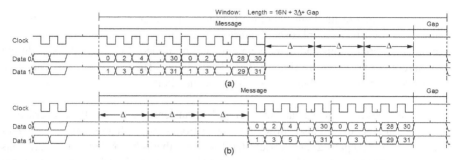

FIGURE 7.4
Master/Shadow Message Structure: (a) Master transmits; (b) Shadow 3 transmits

have been taken to eliminate this danger. First, the total size of a Master/Shadow window is always the size of its message's data plus three deltas, no matter what happens during arbitration. Thus, the time window remains the same size no matter which transmitter "wins" the arbitration. Second, recipients of a Master/Shadow message always place the data in the same memory location, no matter which transmitter wins the arbitration. Third, delta can be made large enough to guarantee that the candidate transmitters will never mistake a busy bus for an idle one and begin transmitting in error. Fourth, recipients of a message from this window will be alerted to the presence of this message at exactly the same time after the end of the window, regardless of which LRM wins the arbitration. This allows completely transparent redundancy among the Master and the Shadows.

The Master/Shadow mechanism also can be used for sharing bandwidth among asynchronous functions. Of course, partitioning is not maintained in such Windows.

7.3.1.3 Bus Encoding

To improve error-detection coverage, data on the four SAFEbus serial lines are encoded in four different ways. Data on Bus Ax have normal polarity. Data on bus Bx are inverted. On bus Ay, every other bit is toggled, starting with the second bit. Bus By is the inverse of bus Ay. This is illustrated in Figure 7.5. This encoding can be seen as a form of "encryption" in which the four buses Ax, Ay, Bx, and By are XORed with the running keys "0000...", "0101...", "1111...", and "1010..." respectively [81].

One type of failure that this encoding catches is bus-to-bus shorts. With identical data on all four buses, a short between the buses could not be detected until another failure occurred such that the failure propagated from the bus with the second fault to the bus with which it was shorted. Having a first fault lay dormant for indefinitely long periods of time and then to make its appearance exactly when a second failure occurs, cannot be tolerated. This problem is largely ignored by homogeneous fault-tolerant architectures. One way to mitigate this problem is to do periodic scrubbing of the buses by putting different data on the buses that would ordinarily be the same and see that they actually are received as different. There are a number of problems with this approach. The scrubbing has to disrupt normal communication and it has to disable some of the fault tolerance features. The latter is dangerous and must be invoked only with a complex set of interlocks. Because of these complications, scrubbing is not continuous and there is an engineering trade-off between the scheduling of scrubbing vs. exposure time to these latent faults. The use of this bus encoding detects these shorts within two bit times of the short onset and without requiring scrubbing.

This encoding scheme can detect unipolar transient upsets that affect several data lines simultaneously. It also allows quick detection of bus collisions caused by a malfunctioning BIU pair (a specific dual-fault scenario that is beyond the basic SAFEbus fault hypothesis). Because bus lines are "wired OR," if a faulty LRM tries to transmit at the same time as another LRM, illegal encodings appear within two bit times of the LRMs starting to transmit differing data.

An additional virtue of this encoding scheme is that power consumption is independent of the data being transmitted. Two bus lines are always high and two are always low (constant average DC power). When the data change, two of the buses change state and two do not (constant average AC power). Because power consumption is constant, the power supply does not have to be designed for a worst-case data pattern. The fact that bus pair A and B are inverses of each other provides some of the characteristics of differential signaling, without an additional doubling of the number of signal lines required.

7.3.1.4 Out-of-Band Signaling Pulses

Because SAFEbus messages are pure data (no message-framing overhead) and all data values are used, the only way that protocol specific information can be conveyed is through the use of out-of-band signaling. One method SAFEbus has for out-of-band signaling uses the clock lines. For data transfer, the clock lines are alternating low for one half bit time and high for one half the time (as depicted in Figure 7.3). For out-of-band signaling, the clock lines are driven low for four bit times, creating a uniquely identifiable pulse, called the Sync Pulse. There are four possible variants of this signal depending on the values of the Data0 and Data1 lines during this pulse. Two of these variants are used for synchronization, one of them is used for a debugging mechanism (see Section 7.3.4.1), and one is not used.

The other out-of-band signaling method that the SAFEbus uses drives data lines low while the clock lines remain high. Of the three variants that are possible with this method, SAFEbus only uses one. It drives Data0 low to indicate Initial Resync (see Section 7.3.3).

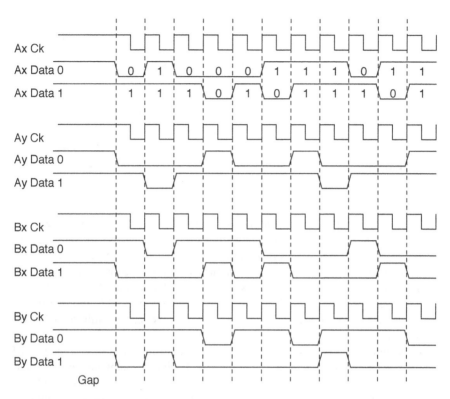

FIGURE 7.5
Bus Encoding Example

7.3.2 Clock Synchronization

SAFEbus' Long Resync and the Short Resync messages both perform precision (sub bit time level) clock synchronization. The purpose of this synchronization mechanism is to maintain separation of adjacent messages in the presence of oscillator drift and keep the two BIUs in an LRM within two bit times of each other. Since the mechanism is the same for both and the Short Resync message is simpler, only the Short Resync message will be discussed here. The additional functionality of the Long Resync message is described in Section 7.3.3. The Short Resync message is shown in Figure 7.6. It consists of the Sync Pulse on the clock lines and both data lines being high. To provide availability, all LRMs transmit the Sync Pulse. The multiple drives are combined into a single pulse by the "wired OR" action of the open-collector BTL drivers. While all LRMs are scheduled to transmit this pulse at the same time, only one needs to succeed. Because of clock drift, each of the LRMs might turn on its driver at slightly different times. However, these resync pulses happen often enough such that the drift can never be more than the width of the pulse. Thus, this pulse can appear to be from four bit times up to eight times wide. When a BIU sees this pulse,

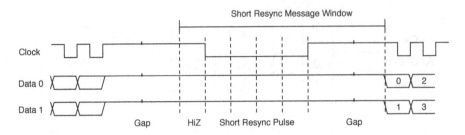

FIGURE 7.6
Short Resync Message

it freezes its internal time, and at the end of the pulse it releases the freeze. It uses an internal effective 4x clock to sample the clock lines. This is shown in Figure 7.7. Because each BIU's time was frozen, its time will have fallen behind real-time. The BIUs then enter a catch-up phase where the internal time counters are incremented at a 2x rate until the internal time has caught up with real-time (Figure 7.8). The duration of this catch-up phase is equal to the freeze duration. This allows all BIUs to have identical times after each resync event without ever causing time to go backward in any BIU.

The aggregate effect of this resync mechanism is that the slower of the two BIUs in the quickest LRM pace the system timebase. Each BIU maintains a counter (called SAFEbus Time) driven by its synchronization-corrected oscillator. The synchronization mechanisms make the values in these counters identical in all BIUs with respect to the time that protocol events happen (e.g., the receipt of a message). The Time value may be used to time stamp data.

7.3.3   Restart, Re-Integration, Integration

Synchronization States

Figure 7.9 shows the major synchronization states of a BIU and their transitions. The major states are:

Initializing While in this state, the BIU performs such operations as Table Memory CRC checks, BIU Configuration Area loading and IMM tests. The Full-resolution SAFEbus Time register value is not valid in this state. IMM access is disabled during much of the initialization process, while the BIU performs IMM pattern testing.

Out_of_Sync While in this state, the BIU hunts for resynchronization messages transmitted over SAFEbus, and attempts to synchronize with them if they are present. If enough time elapses without a synchronization message being seen, the BIU issues an Initial Sync pulse to start up the backplane. The Full-resolution SAFEbus Time register value is not valid in this state.

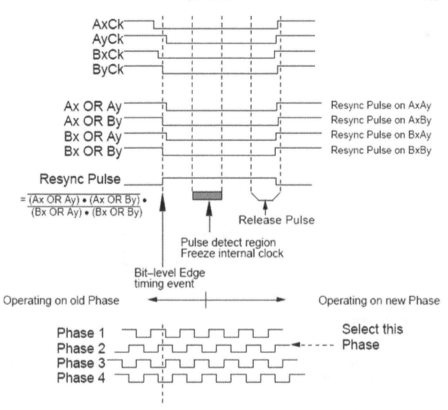

FIGURE 7.7
Resynchronization Pulse Timing

In_Sync While in this state, the BIU executes command sequences out of a Table Command Sequence Area in the Table Memory. It transmits when executing an appropriate transmit command (and data is fresh), and receives data when executing a receive command. Synchronization is maintained via the transmission and reception of programmed synchronization messages. The Full-resolution SAFEbus Time register value is valid in this state.

Halted This is a sub-state of the In_Sync state. The BIU enters the Halted state when it is executing in debug mode, and encounters a breakpoint, or completes a single step operation.

Disconnected While in this state, the BIU suspends all transmission and reception activity on SAFEbus. The Host may still read and write memory and BIU registers, but no backplane data will get written into memory. The BIU will enter this state if commanded to by the host, or if initialization fails or if it receives a Long Resync message with mismatching Version while in the Out_of_Sync state. It leaves this state if commanded by the host or if an Initial Sync Pulse

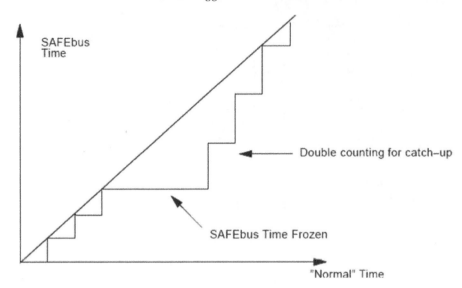

FIGURE 7.8
Time Adjustment

is detected on SAFEbus. The Full-resolution SAFEbus Time register value is
not valid in this state.

Synchronization Messages

Three uniquely identifiable transmission patterns are provided to support bit-level
and frame-level synchronization of the SAFEbus backplane. The Initial Sync Message is used to initialize the SAFEbus after a power-up or in the pathological case
of a cabinet-wide loss of synchronization. The Short Resync Message is provided
to maintain bit-level synchronization between all BIUs in the cabinet by correcting for oscillator drift between BIUs. The Long Resync Messages are provided to
allow lost modules to regain synchronization with an active bus. Long Resync Messages come in two variants. The Entry Resync variant is provided simply to allow
lost modules to resync to the current frame. The Frame Change variant is provided
to switch between different frame programs in the current Table. Both Versioned
and Unversioned forms of the Long Resync Messages exist. Long Resync Messages
also implement the bit-level resynchronization operation (as provided by the Short
Resync message).

Long Resync

The structure of the Long Resync Message (Figure 7.10) is separated into two distinct
sub-windows. The Long Resync Pulse sub-window starts with a unique Long Resync
pulse identified by a low level on the clock lines with all associated Data0 lines
low. The pulse is nominally four bit-times long followed by a Maximum Gap. The

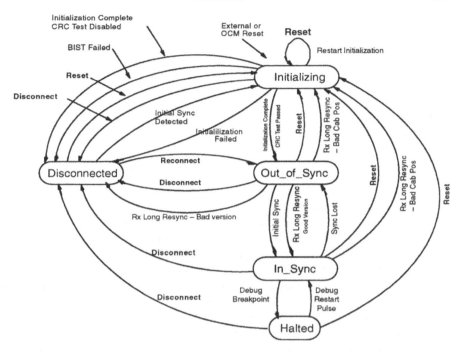

FIGURE 7.9
Synchronization State Diagram

Long Resync Pulse is transmitted by all BIUs that are executing either a Transmit or Receive Long Resync command.

The second part of the Long Resync Message is the Long Resync Information sub-window. A single, unique BIU must transmit in the Long Resync Information sub-window. This consists of an 8-bit Resync Code (a value from 0 to 255), a 1-bit Versioned Frame indicator, 1 bit of reserved space, a 4-bit Cabinet position, 7 bits of reserved space, 43 bits containing SAFEbus Time and a 32-bit Table Version. The Resync Code allows the BIU to determine which of 256 locations in the Table it should jump to in order to align itself with the other BIUs. In software, this would be called an indirect jump. The code indexes into a Resync Jump Table (see Figure 7.13) which contain addresses that point to the main Table Command Sequence Area command that should follow this Long Resync message in the normal sequence of command execution. The Version Frame indicator, Cabinet position and Table Version are used as part of versioning enforcement as described in Section 7.3.7.3. The SAFEbus Time field is used to supply this information to LRMs receiving this message while not in sync.

Long Resync messages can be sent Master/Shadow, which combines the properties of both these message types. This benefits the multiway trade-off among fault tolerance, bandwidth used and resynchronization latency.

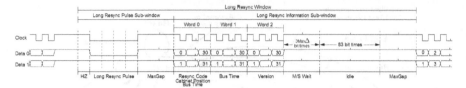

FIGURE 7.10
Long Resync Message

Frame Change

A variant of the Long Resync message is the Frame Change message. The message form is identical. The differences are in the way that it is used. In the Long Resync message, the Resync Code points to the next command in the normal sequence. In the Frame Change message, the Code points elsewhere, usually to another frame. The Frame Change is a conditional jump. An LRM which is scheduled to transmit a Frame Change cannot do so unless its host writes a code into the BIU which matches the Frame's code (a "lock and key" mechanism). A tightly controlled mechanism is provided to switch between frames based on explicit commands generated by Level-A operating system software using the "lock and key." Normally, "lock and key" fault-containment mechanisms are very weak. However, in the case of SAFEbus, this mechanism is protected against hardware failures by the use of self-checking pair hardware and it is protected against software failures through use of an MMU which limits the lock's access only to the Level A software.

If the Frame Change message is not transmitted, the "jump" is not taken and the command/window sequence continues with the command which is next after the Frame Change in the table. If the Frame Change message is transmitted, the "jump" is taken and all LRMs fetch their next command from the location pointed to by the Frame Change's Resync Code. Use of this feature is described in Section 7.3.6.1.

Initial Sync

Initial Sync messages (Figure 7.11) are transmitted by an Out_of_Sync BIU that waits longer than the Initial Sync Wait Limit without seeing any Resync Pulses on the backplane. All BIUs which are out of sync at this point will use the Initial Sync message to synchronize into an Unversioned Initial Frame. The Initial Sync message starts with the Initial Sync Pulse. It has a unique pattern identified by a signal pair of buses with a low level on the Data0 line and a high level on the clock line for at least two bit times. In addition, for a bus to be considered part of an Initial Sync Pulse signal pair, the Data0 line must have been high at one point during the time while the BIU waited for the Initial Sync Wait Limit. The pulse is nominally four bit-times long.

To more accurately resync the clocks, and to allow the BIU time to fetch the command for the first message in the Initial Frame, the Initial Sync Pulse is followed by a Long Resync message with no BIU transmitting the Information sub-window.

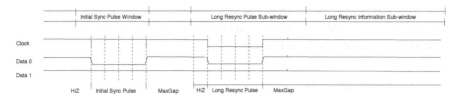

FIGURE 7.11
Initial Sync Message

The missing information is assumed to be zero. In particular, the Resync Code is zero and the SAFEbus Time is zero. The BIU will enter the In-Sync state at the first bit time of the Gap which separates the idle data portion of the Long Resync message from the first window of the Initial Frame. The first bit time of the Initial Frame occurs $(106 + 6\text{Max}\Delta + 2\text{MaxGap})$ bit times after the leading edge of the Long Resync pulse which follows the Initial Sync Pulse.

7.3.4 Diagnostic Services

SAFEbus uses masking fault tolerance, which does not need the complicated diagnostic services required by other protocols, e.g., it does not need to keep track of membership. However, it does supply a rich set of diagnostic mechanisms and information for maintenance purposes. This includes full BIST capability via dual (x and y) IEEE 1149.1 (JTAG) test buses. The x JTAG bus connects only to the x BIU and the y JTAG bus connects only to the y BIU in order to prevent fault propagation from the two sides of the self-checking pair. For commanded BIST, the BIU goes off line and scans a set of pseudo-randomly generated test vectors into the logic, clocks it through the logic and accumulates the result of the clocking into the BIST Signature Result register. The BIST also includes scrubbing of all the critical fault-tolerant circuitry within the BIUs. For scrubbing protection mechanisms, hardware is invoked to test if the protection mechanism is really invoked when a failure occurs.

7.3.4.1 Debugging Mechanisms

Another unique feature of SAFEbus is its ability to breakpoint and single step an entire system, including the processors connected to SAFEbus. This is the same as the breakpoint and single step functionality commonly seen in software debuggers, but on a systemwide basis. The reason this is possible is because SAFEbus acts as the central clock for the system, providing the timing ticks that the operating systems of the processors attached to it use for dispatching tasks. If the SAFEbus freezes, so do all of the processors.

A breakpoint can be set for any time within the global SAFEbus timeline. A breakpoint can also be initiated by driving the Ck, Data0 and Data1 lines low simultaneously for at least four bit times. To resume from a breakpoint, some LRM transmits a Short Resync pulse.

Because setting a breakpoint can be dangerous during normal operation, breakpoints are only enabled for special test table versions. This is indicated by the upper two bits of the Table Version being "11."

7.3.5   Fault Isolation

SAFEbus, with its self-checking approach, provides near perfect coverage. The checking at the receiving end provides near perfect error detection coverage for many faults, including Byzantine faults [78]. It provides better coverage than signature-based error detection techniques (such as CRCs) [250] while simultaneously not incurring the overhead of these schemes.

Recapping Figure 7.1: SAFEbus consists of four buses (Ax, Ay, Bx, and By) that connect several self-checking pair Line Replaceable Modules (LRMs). Each of the four buses and each half of an LRM are independent fault containment zones. If there are N LRMs, there are 2*N + 4 fault containment zones. Each of the BTL bus interface parts is in its associated bus' fault containment zone and gets its power supply from that bus. Thus, the fault containment zone boundaries are between the BTL parts and the BIU parts. Names ending in x and y denote redundant parts for integrity. Bus names beginning with A and B denote redundant bus pairs for availability. The BTL drivers are connected such that BIUx transmits only on Ax and Bx; and BIUy transmits only on Ay and By. Thus, a faulty BIUx can contaminate only buses Ax and Bx; and a faulty BIUy can contaminate only buses Ay and By.

The dashed lines in this diagram are control signals. In particular, BIUx enables BIUy's bus drivers and BIUy enables BIUx's bus drivers. Thus, an LRM can't transmit unless both BIUx and BIUy agree to do so. The set of buses is fail-op, fail-stop. Each LRM is fail-stop. N redundant LRMs are N-1 fail-op, fail-stop. The major failure scenario that SAFEbus does not cover is two simultaneous active faults in the same LRM that are somehow complementary to escape the bit-for-bit checking and cross-coupled driver enables. This has a probability that has been calculated to be much less than 10^{-10}.

7.3.5.1   Babble Protection

To detect errors, the transmitting LRM checks what it actually puts on the bus. If a BIU sees a miscompare, it stops transmitting and it disables its partner's drivers. The dual nature of this comparison ensures that a babbling module cannot stay on the bus. This is an availability feature rather than an integrity feature. Integrity fault containment is done by the receivers. This availability feature is only applicable to Master/Shadow Windows. In order to prevent loss of resources due to transient failures, LRMs are allowed to restart for a limited time by implementation of a strike counter. This strike counter is incremented for each fault found in a specific interval (say a maximum of three failures are allowed within ten minutes). This ensures that transient faults are dealt with in a constructive manner and permanent or intermittent faults are isolated after the LRM has hit the strike counter limit.

7.3.5.2 Byzantine Protection

SAFEbus has a unique way of tolerating Byzantine faults. Because the transfer of a message from one LRM to another LRM uses four fault zones, it is possible for it to tolerate one Byzantine fault. The BTL receivers are cross-linked to the two BIUs such that each receiving BIU gets a copy of the message from all four buses. This can be seen as the first round of the classical Byzantine exchange. Each BIU creates two 4-bit status vectors, collectively called the "syndrome," for each 16 bits received within a message.[2] The first vector has a bit for each bus saying whether anything came in from that bus. The second vector is the result of the comparisons: Ax = Ay, Bx = By, Ax = By, Ay = Bx. The BIUs exchange their syndromes. From these eight bits, the two BIUs can determine which (if any) of the data bus inputs have arrived error free. If there is such an error-free source, both BIUs select it as the source data. This can be seen as the second round of the classical Byzantine exchange. This prevents Byzantine failures arriving from outside a pair from confusing a pair into thinking that one of the halves of the pair is faulty. If a message arrives with uncorrectable errors, the BCW associated with its buffer is updated to that status. A self-checking pair host reading a BCW status that indicates an error is not allowed to read the data buffer because differing data in the two Intermodule Memories could cause the host pair to split.

While the syndrome exchange prevents a Byzantine fault from splitting a pair, an additional mechanism is needed for Byzantine agreement among pairs. Before SAFEbus, Byzantine algorithms either did a full exchange of an entire message's content or used signatures. The problem with the former method is the large amount of bandwidth it requires. The problem with the latter method is that it does not provide full coverage. SAFEbus introduced a new method: hierarchical Byzantine agreement. In this method, a lower-level agreement prevents Byzantine faults from affecting a pair, as described above. An upper-level agreement only needs to send one bit of information from every receiving LRM of the message. This bit would indicate whether the LRM rejected the message as being faulty or not. Because the SAFEbus granularity is 32 bits, it can send 32 bits just as well as one bit. Using 32 bits allows this upper level of exchange to be implemented with no additional hardware and with no software. All that is done is to have the buffer for this upper-level round of exchange be placed into the IMM such that the data word of this buffer overlaps with the BCW for the original message. Thus, what is exchanged on this upper-level is the BCW of the original message. If the original message was rejected, the BCW is zero.

7.3.5.3 Availability vs. Integrity Trade-Off

The syndrome exchange mechanism includes an option to select a preference for availability or integrity, for those cases where there is not a generally applicable best choice. For example, if Ax = Ay and Bx = By; but Ax ≠ By and Bx ≠ Ay. This can only happen if there are at least two identical bit errors (after decode). A receiver

---

[2]The granularity of 16 bits was chosen as an engineering trade-off between the desire for a smaller size to increase availability and the desire for larger size to minimize metastability errors and to make it easier to meet timing constraints on the exchange between the BIUs.

cannot tell if pair A or pair B is correct. For integrity, both have to be thrown out because neither can be trusted. For availability, either A or B could be arbitrarily chosen.

7.3.5.4   Zombie Module Protection

Each LRM that plugs into SAFEbus must contain a Table of commands that is compatible with all other LRMs on that SAFEbus. One mechanism for preventing LRMs with an incompatible Table would be to have the LRMs exchange Table version information at startup. However, such a scheme would not cover the scenario where an LRM is dead or comatose at startup and then "wakes up" in the middle of a critical operations with an incompatible Table. This is called the "zombie module" problem. To solve this, SAFEbus requires all LRMs joining a SAFEbus that has active traffic to compare the Table version information in a Long Resync message to its own. It is allowed to join the network only if the versions are compatible. See Section 7.3.7.3 for details on version enforcement.

7.3.6   Configuration Services

The SAFEbus Tables are loaded into each BIU's Table Memory using the same dual IEEE 1149.1 (JTAG) test buses that are used to support BIST. These Table Memory images can contain multiple tables so that one LRM can play several different roles depending on which slot and in which cabinet the LRM is located. This mechanism can reduce the number of spares required to be held in maintenance facilities. For example, the Boeing 777 AIMS cabinets had one I/O module (IOM) design that was used in eight places. This reduced the number of types of spares required from eight down to one, and reduced the total number of spares needed to be held in the logistics pipeline. In theory, this could also allow for on-board sparing and manual reconfiguration. Each slot in every SAFEbus cabinet has five slot ID pins that can allow up to 32 slots per cabinet. To prevent wrong-ID masquerade faults, these pins are protected by parity and cross compare between the two halves of a pair. Each cabinet also has a Cabinet ID. To eliminate the need for Cabinet ID pins on every slot, SAFEbus allows some subset of LRMs within the cabinet to have these pins and then these LRMs broadcast this cabinet identification in their Long Resync messages' Cabinet Position field. With the existing four bits of Cabinet Position and five bits four slot ID, SAFEbus can currently accommodate up to 512 slots systemwide. The Long Resync message has seven spare bits adjacent to this field that could be used to expand this to a total of 65,536 systemwide slots.

After the tables have been loaded, SAFEbus can be used to download software and other data. To optimize bandwidth usage, special frames can be used for these downloads (as described in Section 7.3.6.1).

7.3.6.1   Frame Changes

This section provides an example for frames to do hardware initialization, software initialization, data loading, built-in self test (BIT) and normal application communication; plus, the Frame Change transitions between them. The following is a brief textual description of each frame.

1. The Hardware Initialization (HW Init) frame follows Arinc 659's definition for the Initialization frame and, as such, is an Unversioned frame. The only data traffic during the HW Init frame is the transfer of Version Messages from all LRMs. The HW Init frame is approximately 117 microseconds in duration.

2. The Software Initialization (SW Init) frame provides time for the processors' environment to be setup before the functional partitions are dispatched in the Flight frame.

3. The Built in Test (BIT) frame is a Versioned frame that provides time to perform power-up BITE.

4. The Flight frame is a Versioned frame that provides the data transfers for running software partitions.

5. The Dataload frame is a Versioned frame that provides the Dataload function, using the majority of the SAFEbus bandwidth. While dataloading may be performed in the Flight frame, a dataloading session using the Dataload frame is much quicker.

Figure 7.12 shows these frames and possible transitions. Note that the Frame Changes are unidirectional. This means, for example, that once the Flight frame starts executing, it cannot be Frame Changed to any other frame. In order to go from the Flight frame to another frame, the SAFEbus must be reset. This is an additional safety precaution against accidental transitions out of flight frame.

7.3.7   Protocol Parameterization

7.3.7.1   Table Memory

As shown in Figure 7.13, a SAFEbus Table is divided into three areas: Resynchronization Jump Table, Table Command Sequence Area and BIU Configuration Area. The Resynchronization Jump Table is used by the BIU to rapidly locate the address in the Table Memory corresponding to the Resync Code (see Section 7.3.3). The majority of the Table Memory is used for the cyclic command sequences that control frames. More than one frame schedule can be present for different system "modes" (such as system initialization, ground check-out, flight operation, software loading, etc.), with each mode having a different schedule. The different frames for different modes are contained within a single Table Command Sequence Area. Therefore, this selection is in addition to the selection of different tables depending on location (slot and cabinet) roles. The BIU Configuration Area contains information for BIU customization options such as memory speeds, host interface characteristics, selection of availability or integrity as preferred for those cases where one is not universally preferred over the other, intermessage gap, Master/Shadow delta and SAFEbus Time increment rate. The contents of the Table Memories which are associated with the

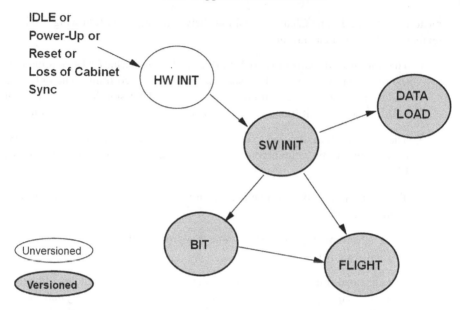

FIGURE 7.12
Example Frames and Their Frame Change Transitions

two BIUs on a single module are bit-for-bit identical. However, the Table Memory contents are different for each module on SAFEbus. This is because the local IMM addresses can be different and the commands are different for TX versus RX, Master versus Shadow, etc.

7.3.7.2   Frame Description Language

SAFEbus uses a Frame Description Language to define the contents of each frame. This is an intermediary language that can be produced by off-line schedulers. A back-end tool specific to each BIU design translates this FDL into a bit image that can be loaded into Table Memories. This language has been standardized by ARINC 659 to allow decoupling between vendors of scheduling tools and vendors of BIU hardware. For AIMS, a software tool parses a database of ICD information and generates a schedule that has much more freedom than schedulers for most time-triggered protocols.

7.3.7.3   Table Versioning

The SAFEbus protocol includes a mechanism to ensure that only LRMs with compatible tables are allowed to transmit on the bus. To support this mechanism, SAFEbus uses two types of frames: Unversioned and Versioned. Unversioned frames allow LRMs of any SAFEbus version to communicate. These frames are used for LRMs to exchange their Table version information with each other. Such frames always op-

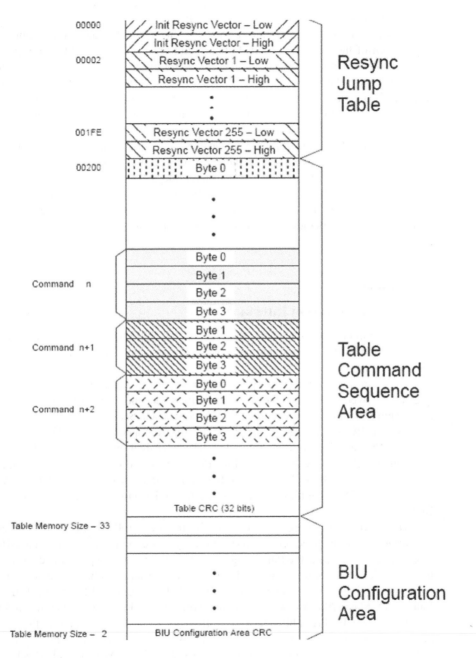

FIGURE 7.13
Table Memory Structure

erate with the maximum programmable gap size between messages in order to be compatible with even the slowest LRM implementations.

After power on, all LRMs enter a standard Unversioned Initialization Frame in which each of the possible 32 LRMs that could exist on one SAFEbus are given a window in which to transmit their Table version information. Up to eight of the LRMs are given the authority to do a Frame Change from the Initialization Frame to either a Versioned frame or another Unversioned frame. The decision of which Frame Change to do is an application dependent function which uses the gathered Table Version information as input.

The Version field in the Frame Change command informs all BIUs of the destination frame's version. BIUs with a Table Version that doesn't match the destination frame's version cannot follow the Frame Change and will drop out of sync. BIUs with a Table Version not matching a currently running frame are prevented from joining the bus traffic because they can't sync to the Versioned Long Resync messages. Most normal application operations are done in a versioned frame.

7.4   Communication Interface

All data communications between modules on the SAFEbus backplane occur via buffers stored in the Intermodule Memory address space. Two buffer formats are provided: Controlled and Non-controlled.

The Controlled buffer memory format consists of a single buffer control word (BCW) followed by one or two data sub-buffers (depending on whether the data item requires double buffering) of up to 256 words each. See Figure 7.14.

When a BIU executes a command to transmit from a controlled buffer, it determines the buffer type from the command. The BIU then reads the Buffer Control Word and checks the TX Fresh bit and the Ping/Pong bit (if the buffer type is double buffered). If the TX Fresh bit is set, the BIU transmits the data and writes the BCW back with the TX Fresh bit cleared to 0. If a transmission failure occurs, the entire BCW is set to 0.

When a BIU executes a command to receive into a controlled buffer, it determines the buffer type from the command. If the buffer type is double sub-buffered, the BIU reads the BCW and tests the Ping/Pong bit to determine where to place the data. The BIU then places the received and validated data into the stale sub-buffer (always Ping for single sub-buffers) and then forms a BCW with the current SAFEbus time, Buffer Valid is set, Master/Shadow winner bits set appropriately, Ping/Pong set to the buffer just written and TX Fresh bit cleared to 0. If errors occur in reception after the BIU modifies a location in the buffer, the entire BCW is set to 0. When the host reads a controlled receive buffer, it must first read the BCW. If the BCW is 0, a self-checking pair host cannot read any other words in the buffer because erroneous receive data that causes the BCW to be set to zero is not guaranteed to be identical between the x and y sides of the pair. If it is non-zero, the host can read the sub-buffer indicated

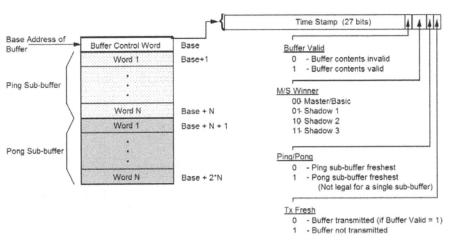

FIGURE 7.14

IMM Buffer Structure

as freshest by the Ping/Pong bit. This bit is valid for both buffer types, since single sub-buffered controlled buffers will always have the Ping/Pong bit set to Ping (0). When the host writes a controlled transmit buffer, it must first read the BCW. If it is non-zero, the host can write the sub-buffer indicated as not being the freshest by the Ping/Pong bit. The host then sets the TX Fresh bit by requesting the BIU to do a Set_Ping or Set_Pong operation depending upon the sub-buffer to be transmitted.

The Non-controlled format consists of 1 to 256 words without a BCW. No freshness indication or validity information is provided in such a buffer. If such indications are required, they must be provided by the user in the data itself. The BIU considers a Non-controlled buffer to be fresh whenever it attempts to transmit the contents of the buffer.

The BIUs contain special "coincidence" circuitry to detect the case that the Host pair is updating the BCW at the same time that the BIUs are trying to read it. Even though these memories are only pseudo-dual port instead of true dual port, the independence of the clocks between the BIUs, and between the BIUs and the Host pair means that read-write order can be different between the x and the y sides of the pairs. The coincidence circuitry ensures that the BIUs see the same order. If this were not done, it is possible that the two BIUs could disagree on whether a buffer was ready to transmit. A coincidence mechanism similar to that used for the BCW on transmit is also used on receive to prevent the Hosts from getting inconsistent data.

Synchronization of the bus schedule and the application-software's execution is done by embedding interrupt commands in the SAFEbus Tables. On receiving this interrupt, the processor's operating system releases an application task (or task set) for execution. This interrupt takes the place of the clock or hardware timer other operating systems (OS) or real-time executives employ. That is, this SAFEbus interrupt becomes the OS' real time clock tick. As part of the time and space partitioning,

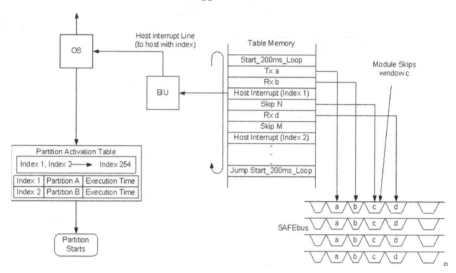

FIGURE 7.15
Data Stream Time Partitioning

the tasks are grouped into sets of "partitions." Tasks in one partition are guaranteed not to interfere with tasks of another partition in a mechanism similar to processor virtualization. See Figure 7.15.

7.5   Validation and Verification Efforts

SAFEbus was part of the Boeing 777 airplane certification and passed rigorous testing for design requirements to support Level A applications, the highest level of safety requirements. Newer versions of the Boeing 777 use upgrades to the original SAFEbus BIU design, which also have been certified. Derivatives of SAFEbus have been certified on several other aircraft.

7.6   Example Configurations and Implementations

SAFEbus is deployed in the various members of the Boeing 777 aircraft family, in two generations of the Information Management System (AIMS). The AIMS system consists of two redundant cabinets implementing seven primary airplane functions. The functions integrated are a mix of hard real-time functions (data-conversion gateway, display system, flight data acquisition, flight management and thrust management) and non-real time (central maintenance and data communication management) with repetition rates ranging from 1 Hz to 80 Hz [51]. A derivative of SAFEbus is used in Versatile Integrated Avionics (VIA) boxes on some Boeing 737 variants and other aircraft.

Standardization

SAFEbus is standardized as ARINC 659 "Backplane Data Bus" [8]. This is in a family of associated standards including: ARINC 653 Avionics Application Software Standard Interface, ARINC 651 Design Guide for Integrated Modular Avionics, and ARINC 650 Integrated Modular Avionics Packaging and Interfaces.

8

Time-Triggered Ethernet

W. Steiner

TTTech

G. Bauer

TTTech

B. Hall

Honeywell

M. Paulitsch

EADS

CONTENTS

8.1   Protocol Overview

Ethernet is the dominant network standard for local area networks (LAN). While originally designed for classic office applications, the growing communication demands in real-time systems led to adapting Ethernet for time-critical applications. Today, we can find Ethernet variants everywhere: In industrial applications (EtherCat, Ethernet Powerlink, ProfiNet, Ethernet IP), in aerospace applications (ARINC 664-p7), in military naval applications (Gigabit Ethernet Data Multiplex System), in consumer audio/video systems (AVB), as well as in datacenters and cloud computing (DCB). All of these Ethernet variants aim to achieve a certain degree of Quality of Service (QoS) such that end-to-end transmission guarantees can be ensured. In this multitude of Ethernet-variants, *TTEthernet* introduces the deterministic time-triggered communication paradigm in an Ethernet flavor, which allows the use of standard Ethernet in safety-critical systems and systems with applications of mixed-criticality.

TTEthernet is the industrial further development of the academic TT-Ethernet research [179], conducted within a joint research project[1] between the Vienna University of Technology and TTTech Computertechnik AG. The main objective of the academic project has been the integration of time-triggered with event-triggered messages on a single physical Ethernet network: As event-triggered messages are not

---

[1] The FIT-IT project TT-Ethernet has been funded by the Austrian Ministry for Transport, Innovation, and Technology (BM-VIT) under contract No 808197.

synchronized, they typically result in conflicts with time-triggered messages at the outgoing ports in a network switch. The solution proposed within the academic TT-Ethernet project has been a preemptive switch. This TT-Ethernet switch [315] identifies the reception of a time-triggered message based on an identifier within the message and preempts all event-triggered messages under transmission to free the outgoing ports for the time-triggered message. The merit of this solution is twofold: first the switch latency for time-triggered messages is constant with negligible error and, secondly, the switch can be kept almost free from additional configuration data.

This academic TT-Ethernet research has been continued within a joint industrial development between TTTech and Honeywell. Here, the objectives have been extended toward scalable fault-tolerance and a finer classification of event-triggered messages into rate-constrained and best-effort traffic classes. The main driver for these additional objectives has been a shift in the target application domain toward safety-critical applications, in particular civil aerospace. Fault tolerance is a standard requirement in these applications and the rate-constrained traffic class has been intended as a compatible communication mode to ARINC 664 part 7, which specifies an unsynchronized real-time Ethernet variant for airborne applications. Another key differentiator to the academic version is a non-preemptive integration mode of event-triggered and time-triggered messages to minimize the number of damaged Ethernet frames on the network, thereby easing diagnosis. This industrial development has led to Time-Triggered Ethernet in its current form which is called *TTEthernet* and will be discussed in more detail in this chapter.

TTEthernet is intended as cross-industry communication infrastructure: originally designed according to aerospace standards, it is scalable in several directions. As different industrial areas define and use different variants of real-time Ethernet already, the nature of the *TTEthernet* technology is one of a set of services added to existing standards rather than their replacement. The *TTEthernet* services are currently being standardized by SAE, and define realizations of the synchronization concepts discussed earlier in this book, as well as how to communicate according to the time-triggered paradigm. The SAE AS6802 standard is being developed at the time of this writing, and is expected to be ready for balloting by the end of 2010. At a minimum, a device has to implement the functionality as required by SAE AS6802 to be called a *TTEthernet* device.

In the next section, Section 8.2, we introduce the *TTEthernet* technology in detail: Communication Services (Section 8.2.1) discusses the time-triggered, rate-constrained and best-effort traffic classes and options on how to integrate them onto a single physical network. We discuss the frame format of these traffic classes and introduce *protocol control frames* (PCF) which are used for synchronization. As a result of the non-preemptive traffic integration policy, PCFs are subject to network jitter. As the quality of synchronization is directly proportional to the network jitter, we present the permanence function as a means to transform most of this network jitter into network latency. Section 8.2.2 discusses the two-step clock synchronization approach realized in *TTEthernet*, and focuses on the *compression function* used as convergence function in the first one of these steps. In Section 8.2.3, we introduce the startup and restart algorithm and discuss its variant for dual-fault tolerance. Diagnosis mechanisms beyond clique detection algorithms are considered in Section 8.2.4. In Section 8.2.5, the concept of high-integrity design is discussed in the form of a COM/MON pair. Here, we also show the role of a central guardian in *TTEthernet* for time-triggered and event-triggered messages. Finally, Section 8.2.6 briefly discusses configuration and re-configuration options for *TTEthernet* devices. In Section 8.3, we continue with the discussion of protocol parameterization with respect to the different traffic classes. We introduce the concept of "porosity" as a property of a communication schedule for time-triggered traffic, which directly influences the dataflow performance of event-triggered messages. Section 8.4 discusses the communication interface of *TTEthernet* in a generic way and the particular realization in current *TTEthernet* products. The correctness of time-triggered protocols for safety-critical and mixed-criticality systems is essential for successful mission operation. The *TTEthernet* algorithms have therefore been subject to formal studies. Section 8.5 summarizes the results of the formal analysis. We conclude this chapter with some example configurations in Section 8.6, and discuss *TTEthernet* implementations in software and FPGA.

8.2　Protocol Services

A *TTEthernet* network consists of end systems and switches and bi-directional communication links to connect these devices to each other. Furthermore, an end system and a switch may be integrated into a single device. Figure 8.1 depicts an example *TTEthernet* network consisting of five end systems (SM1..SM5) and two redundant communication channels. Channel 1 consists of three switches (SC1,SC2,CM1), while channel 2 uses one switch only (CM2). Current *TTEthernet* implementations allow up to three channels.

The *TTEthernet* protocol defines three different roles for synchronization: the *Synchronization Master* (SM), the *Synchronization Client* (SC) and the *Compression Master* (CM). As depicted in Figure 8.1, the SM role is typically assigned to end systems (SM1..SM5), while the switches act as CMs (CM1,CM2). As a minimum, one component per channel is configured as CM. Both, switches and end systems, can be configured as SCs.

Synchronization demands the exchange of information between the devices. In the case of *TTEthernet*, this information is transported in standard Ethernet frames that are called protocol control frames (PCF). From a top-level view, the *TTEthernet* synchronization algorithm operates in two steps. In the first step, all SMs concurrently send PCFs to all CMs. From the received PCFs, each CM produces a new PCF called the "compressed" PCF. In the second step, all CMs send their compressed PCFs back to the SMs and SCs. In the SMs and SCs, the compressed PCFs are consolidated to derive a new reference point in time for synchronization. This two-step clock synchronization process is periodically executed throughout the mission time.

The *TTEthernet* synchronization algorithms closely synchronize the local clocks

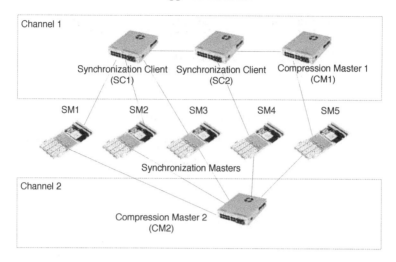

FIGURE 8.1

An Example *TTEthernet* Network Consisting of Two Channels and Five End Systems

in the end systems and switches, which enables time-triggered communication. We collectively refer to the *TTEthernet* synchronization algorithms and the time-triggered communication as "*TTEthernet* services." The *TTEthernet* services are intended to be realized on top of Ethernet as standardized in IEEE STD 802.3-2005.

8.2.1 Communication Services

8.2.1.1 Communication Modes

TTEthernet defines the term "traffic class" to differentiate communication modes. The prime communication mode of *TTEthernet* is, of course, the time-triggered traffic class (TT). Besides this mandatory traffic class, *TTEthernet* also names the optional rate-constrained traffic class (RC) and the optional best-effort traffic class (BE). *TTEthernet* is an integrative communication protocol capable of communicating frames of these three traffic classes on the same physical network. Figure 8.2 gives an overview of the different traffic classes and their relation to the common internet protocols. As depicted in Figure 8.2, the *TTEthernet* services (TT Services) complement layer two of the Open Systems Interconnection model (OSI model).

An end system that implements the *TTEthernet* services is able to synchronize its local clock with the local clocks of other end systems and switches in the system. The end system can then send messages at off-line planned points in this synchronized global time. These messages are said to be time-triggered.

Time-Triggered (TT) messages are used when tight latency, jitter and determinism are required. All TT messages are sent at predefined times. In cases where an end system decides not to use one of its assigned timed slots, for example if there is no new data to be sent, the switch recognizes the inactivity of the sender and frees the

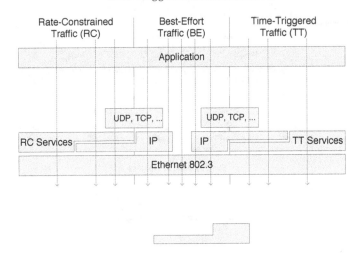

FIGURE 8.2
Interaction of Standards (© 2009 IEEE [309])

bandwidth for the other traffic classes. TT messages are optimally suited for communication in distributed real-time systems.

Rate-Constrained (RC) messages realize a communication paradigm that aims at establishing well-shaped dataflows: successive messages belonging to the same rate-constrained dataflow are guaranteed to be offset by a minimum duration as configured. RC messages are used when determinism and real-time requirements are less strict than provided by time-triggered communication. For RC messages, sufficient bandwidth must be allocated such that delays and temporal deviations have defined limits. In contrast to TT messages, RC messages are not sent with respect to a system-wide synchronized time base. Hence, different communication controllers may send RC messages at the same point in time to the same receiver. Consequently, the RC messages may queue up in the network switches, leading to increased transmission jitter and requiring increased buffer space. As the transmission rate of the RC messages is bound a priori and controlled in the network switches, an upper bound on the transmission latency can be calculated off-line and message loss is avoided. The rate-constrained communication paradigm is used in ARINC 664 part 7.

Best-Effort (BE) messages implement the classic Ethernet approach. There is no guarantee whether or when these messages can be transmitted, what delays occur and if BE messages arrive at the recipient. BE messages use the remaining bandwidth of the network and have lower priority than TT and RC messages. However, BE traffic may be attractive, for example, during maintenance and configuration phases: As during such phases no critical traffic in the form of TT or RC may be present, the whole network bandwidth is available for BE traffic without explicitly changing the network mode. RC messages will not be sent and bandwidth reserved for TT messages is automatically reclaimed by the switches.

8.2.1.2 Frame Formats

TTEthernet is fully compliant with the Ethernet frame format as standardized in IEEE 802.3. An overview of the fields in an Ethernet frame is given in Figure 8.3: The first row lists the number of octets for a given field, the second row the field's name. An octet is an entity having exactly eight bits.

7	1	6	6	2	46 - 1500	4	12
Preamble	SOF	Destination Address	Source Address	Type/ Length	Payload	CRC	IFG

FIGURE 8.3
Ethernet Frame Format

An Ethernet frame starts with a preamble of seven octets followed by a start of frame delimiter (SOF) of one octet. For addressing of the frame, Ethernet specifies the Destination Address and the Source Address, each six octets long. Following the address fields, Ethernet IEEE 802.3 specifies two octets that are either used as a type (EtherType) or as a length field. By convention, a value from 64 to 1522 (decimal) of this field indicates its usage as length field; a value of 1536 (decimal) and higher means that the field is used to reflect an EtherType. An example EtherType is 0x0800, which defines the Ethernet frame to carry an Internet Protocol (IPv4) packet. The address fields together with the EtherType/Length field is commonly referred to as the MAC Header. The actual payload of an Ethernet frame has a size of 46 to 1500 octets. Ethernet specifies a 32-bit CRC which follows the payload field. The minimum interframe gap (IFG) is 12 octets. As a result, the overall length of an Ethernet frame, as a sum of the fields discussed above, is between 84 and 1538 octets.

+	0-15			16-31
0	Integration_Cycle			
32	Membership_New			
64	Reserved			
96	Sync_Priority	Sync Domain	Type	Reserved
128	Reserved			
160	Transparent Clock			
192				

FIGURE 8.4
Contents of a *TTEthernet* Protocol Control Frame

TTEthernet encapsulates all protocol-related information in PCFs. A PCF is a standard Ethernet frame with minimum payload (46 octets) whose Ethernet Type field is set to $0x891d$. Figure 8.4 depicts the contents of a PCF as carried in the Ethernet payload field. We discuss these fields briefly below and in more detail in the context of startup/restart and clock synchronization.

- Integration_Cycle: In *TTEthernet*, time is represented cyclically with a period called the cluster cycle. In order to provide integration on multiple points throughout the cluster cycle, it is divided into integration cycles which are numbered from 0 to *max_integration_cycle* − 1. The Integration_Cycle field carries the number of the current integration cycle. An integrating component uses this information to set up its synchronized time. The timing hierarchy is further discussed in Section 8.2.2.1.

- Membership_New: This field is used to keep track of the *TTEthernet* devices that provide synchronization messages for the clock synchronization process.

- Sync_Priority: *TTEthernet* provides hooks for systems-of-systems synchronization by means of priorities. The synchronization priority mechanism is one way to deterministically synchronize multiple *TTEthernet* subnetworks to each other.

- Sync_Domain: Within a *TTEthernet* network, two or more synchronized timebases can coexist. In this context, we call one synchronized timebase a synchronization domain. A *TTEthernet* device may belong to one or many synchronization domains. Two *TTEthernet* devices belonging to different synchronization domains will never synchronize to each other. The Sync_Domain field is used to differentiate PCFs of different synchronization domains from each other.

- Type: The synchronization algorithms use different types of PCFs. Coldstart frames (CS) and Coldstart Acknowledgment frames (CA) are used for startup and restart, Integration frames (IN) are used for clock synchronization and clique detection.

- Transparent_Clock: This field is used to store the accumulated delay of a PCF from its generator to the consumer. Time is represented as multiples of picoseconds. The content of the Transparent_Clock field is defined to be the accumulated transmission delay of the frame through the network in units of 2^{16} nanoseconds. For example, one nanosecond is represented by $0x10000$; 2.5 nanoseconds is represented by $0x28000$. The use of the transparent clock is further discussed in Section 8.2.1.5.

A particular Ethernet variant for avionics applications has been standardized by ARINC as ARINC 664 part 7 (for short, ARINC 664). One particular aspect in ARINC 664 is the specification of "virtual links." Figure 8.5 depicts the Ethernet Destination Address field as used by ARINC 664. The address field is subdivided into a constant 32-bit field and a 16-bit field called the Virtual Link Identifier (VL ID). As a rule, in the constant field the two least significant bits of the first octet must be 1. All other bits may be selected by the user, but have to be fixed for all ARINC 664 frames in a given system. The Virtual Link Identifier is used as the actual addressing scheme in an ARINC 664 network. Each VL ID relates a single sender to, potentially, multiple receivers. ARINC 664 switches store these VL IDs in the form of statically configured tables. A switch that receives an ARINC 664 frame on a given incoming port knows from this statically configured table to which outgoing ports the frame has to be relayed. Another specific of ARINC 664 is the implementation of a sequence number that uses the last octet in the Ethernet payload. A sender uses the sequence number to cyclically number the frames of a given virtual link. A receiver of a virtual link uses this sequence number for redundancy management and integrity checking.

Destination Address 48 bits	
Constant Field 32 bits	VL ID 16 bits
xxxx xx11 xxxx xxxx xxxx xxxx xxxx xxxx xxxx xxxx	xxxx xxxx xxxx xxxx

FIGURE 8.5
Ethernet Destination Address is Used to Specify the Virtual Link Identifier (VL ID) in ARINC 664 Part 7

The first realization of *TTEthernet* is implemented in the context of ARINC 664, and we use the term *TTEthernet/ARINC664* to refer to this realization. Succeeding realizations of *TTEthernet* may be implemented on top of other industry-specific Ethernet variants or just on plain Ethernet. *TTEthernet/ARINC664* uses the ARINC 664 addressing mechanism for the time-triggered and rate-constrained traffic classes. In the case of *TTEthernet/ARINC664*, the information of the PCF is also encoded in the VL ID of the frame. Hence, a *TTEthernet/ARINC664* device derives the PCF payload from a locally stored lookup table indexed by the VL ID rather than from the explicit fields carried in its payload.

As Ethernet itself keeps evolving, it is subject to extensions. We discuss one next: the VLAN tags. The IEEE 802.1*Q* (IEEE Standard for Local and Metropolitan Area Networks - Virtual Bridged Local Area Networks) defines "Virtual LANs (VLANs)." A VLAN is identified by a VLAN identifier (VID), and characterizes a group of end systems. One end system can be a member of several VLANs. Standard Ethernet messages can be VLAN-tagged, via a four octet field before the EtherType/Length field. The VLAN-tag holds a priority and the VLAN ID (VID). This priority is used to derive the outgoing queue (and hence, Quality-of-Service) used for forwarding the frame in switches implementing IEEE 802.1*Q*. The VLAN extension to Ethernet is of particular interest to Audio/Video Bridging (AVB) activity which aims to enhance Ethernet with Quality of Service (QoS). AVB and *TTEthernet*, both being standard Ethernet compliant, can seamlessly co-exist in a single physical network from a pure communication point of view. However, compatibility modes between *TTEthernet* and AVB that maintain the tight temporal guarantees across protocols are subject to further development.

8.2.1.3   Coding and Decoding

The *TTEthernet* services operate on the MAC layer (layer two of the OSI reference model). As such, they are largely independent of the underlying PHY layer. To support the independence between the MAC and PHY layers, the IEEE developed a set of interface standards. The media-independent interface (MII) is used in a 100 Mbit/s Ethernet network to connect the MAC layer to the PHY layer. For 10 Mbit/s, 10 Gbit/s and 100 Gbit/s, the respective interface specifications are the Attachment Unit Interface (AUI), the Gigabit Media Independent Interface (GMII) and the 10 Gigabit Media Independent Interface (XGMII). Although not explicitly stated in the SAE AS6802 standard, it is expected that a *TTEthernet* device will also adhere to these interface standards.

8.2.1.4   Media Access Control

The synchronized local clocks and the pre-configured schedule specify the temporal position of the TT frames on the timeline. Intervals that are not assigned to TT frames are free for RC or BE communication. Figures 8.6(a) and 8.6(b) present an example *TTEthernet* network and an associated schedule for TT frames.

Figure 8.6(a) depicts an example topology of 20 *TTEthernet* end systems connected to each other via bidirectional communication links. Each bidirectional communication link can be decomposed into unidirectional dataflow links. In the figure, these dataflow links are numbered from 1 to 50 (dataflow links 1..6 are shown). Figure 8.6(b) gives an example schedule for TT frames for the topology on the top. In the schedule, the numbered dataflow links are listed on the x-axis. The y-axis presents time in the granularity of slots. In this example, one cluster cycle is split into 1200 slots. An appropriate integration cycle in this example can be 300 slots. Hence, within one cluster cycle, an integrating component may integrate at any one of four different integration points.

Time progresses from bottom to top, along the y-axis, and is cyclically repeated afterward. A dot in the plot means that the respective slot on the given dataflow link is uniquely assigned to a TT frame. The remaining bandwidth (positions not marked for TT traffic) is free for RC and BE frames. The presented schedule is specifically designed to accommodate unsynchronized traffic, like RC and BE, by concentrating the TT traffic to several segments on the timeline. Schedule synthesis for TT traffic in multi-hop networks is formally specified in [307], while static scheduling for mixed-criticality systems is discussed in [308].

The integration of TT with unsynchronized traffic, like RC or BE, leads to access conflicts on the dataflow links. For example, assume a scenario in which dataflow link 5 is transmitting an RC frame received from dataflow link 1. Now, a TT frame received from dataflow link 3 becomes ready that also has to traverse dataflow link 5. In this case, the switch connecting these dataflow links exhibits a conflict at its outgoing port to dataflow link 5.

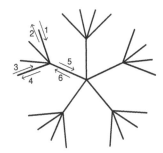

(a) *TTEthernet* network consisting of twenty end systems connected in "snowflake" topology

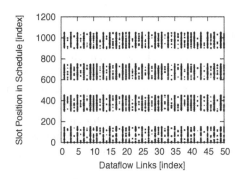

(b) Schedule for TT frames designed with blank intervals for RC and BE traffic. Time is depicted on the y-axis in granularity of slots.

FIGURE 8.6
Example Network Topology and Assigned Schedule for the Different Traffic Classes (TT, RC and BE)

TTEthernet addresses the collision problem associated with the integration of synchronized and unsynchronized traffic by assigning each frame a priority. While the SAE AS6802 standard does not require a particular priority scheme, it recommends the following in decreasing order: PCF, TT, RC, BE. Within a traffic class, *TTEthernet* allows multiple priorities.

If the conflict occurs between messages with equal priority, these messages will be served in FIFO. When a high-priority message (**H**) is being served and a low-priority message (**L**) becomes ready, **L** will be queued. The third case, though, is of particular interest: there are three integration methods to resolve conflicts when a low-priority (**L**) message is already in transmission and a new high-priority (**H**) message becomes ready for transmission [309]. The three methods are preemption, timely block and shuffling, and are depicted in Figure 8.7.

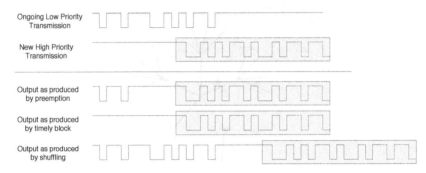

FIGURE 8.7

Integration Methods for High-Priority (H) and Low-Priority (L) Traffic

Preemption [179]

If an **L** message is being relayed by a switch when an **H** message arrives, the relay process of the **L** is stopped. The switch establishes the minimum time of silence on the channel and relays the **H** message an a priori specified duration later.

High Real-Time Quality: Preemption guarantees that the switch introduces an almost constant and a priori known latency for an **H** message.

Generation of False Messages: As truncated messages are now systematically generated by this mechanism, it has to be guaranteed that a truncated message does not appear to a receiver as a correct message. This can be ensured by the preemption mechanism to generate a signal pattern that violates the line encoding rules when a message is truncated or to include the original message length within the message. Hence, all receivers of this truncated message will identify a syntactically faulty message.

Resource Inefficiency: If standard Ethernet components are to be supported, each truncation action results in a loss of bandwidth, as the already transmitted fraction of a message is discarded by the receiver and the whole message has to be retransmitted. Add-on functionality to standard Ethernet that allows reconstruction of fragmented Ethernet messages at a receiver [223] supports reclaiming of the lost bandwidth, at least to some extent.

Timely Block

If the **H** message is a TT message, the switch in the network knows a priori when this **H** message will arrive on which port, and to which ports (or an internal buffer) this message has to be forwarded. Timely Block means that the switch will not forward messages at those times when a TT message is expected.

High Real-Time Quality: Since the outgoing ports for an **H** message are scheduled to be free when the **H** message arrives, the integration-imposed delay is almost constant.

Resource Inefficiency: When the message lengths of the **L** messages are not known, the timely block has to have a duration of at least the maximum possible **L** message. For Ethernet, this means either $123.040\mu s$ for 100 Mbit/s or $12.304\mu s$ for 1 Gbit/s. As the minimum-sized Ethernet frames are $6.72\mu s$ for 100 Mbit/s and $0.67\mu s$ for 1 Gbit/s, a single timely block may prevent up to 19 Ethernet frames from being delivered. To overcome this resource inefficiency, the switch/end system can act more intelligently, if the message lengths of the **L** messages are known or transported in the messages themselves: At any point in time when an **L** message is ready to be relayed, the switch will only relay this **L** message if it is guaranteed that the **L** message is completely relayed before the **H** message has to be relayed.

Shuffling

If an **L** message is being relayed by a switch when an **H** message arrives, the **H** message is delayed until the relay process of the **L** message is finished. Hence, in the worst case, the **H** message is delayed for a maximum-sized **L** message. This delay will also impact **H** messages that follow until the bandwidth required for the **L** message is compensated for the sum of the inter-frame gaps between succeeding **H** messages.

Resource Efficiency: In contrast to preemption and timely block, shuffling will not truncate a message, nor block the outgoing ports for **L** messages. Hence, from a utilization point of view, shuffling is an optimal solution.

Low Real-Time Quality: If the **H** message is a TT message, the good real-time quality of time-triggered traffic is degraded. However, as time-triggered traffic is dispatched according to a synchronized timebase, this dispatch point in time is known to the receivers a priori. Messages of other traffic classes can include a global timestamp within the message to inform the receivers of their dispatch point in time. Hence, the synchronized timebase mitigates the increased transmission jitter. Network latency, on the other hand, cannot be mitigated: An **H** message will potentially be delayed by at most one **L** message at each communication step. However, in a 100 Mbit/s network, or 1 Gbit/s network, shuffling still gives sufficient real-time quality for a broad range of applications, such as avionics or automotive applications.

In a standard Ethernet network with 100 Mbit/s, the frame lengths are between $6.72\mu sec$ and $123.36\mu sec$; in a 1 Gbit/s network frame lengths are between $0.672\mu sec$ and $12.336\mu sec$. Hence, shuffling may impose an additional communication delay of $123.36\mu sec$ or $12.336\mu sec$, depending on the network speed. This communication delay is also directly added to the transport jitter through the network. For PCFs which are used to establish the synchronized timebase, we mitigate this additional jitter by means of the permanence function which we discuss in the following section. For other **H** messages (e.g., TT or RC frames), the actual transmission jitter can then be mitigated by the synchronized time itself. TT frames have an a priori specified dispatch point in time and RC frames may include a timestamp of the synchronized timebase in their payload.

Interference with BE Traffic: As an **H** message may be delayed by an **L** message in general, this is also true in particular for those cases where **L** is a BE message. Though the influence of BE messages is bounded and guaranteed by *TTEthernet*, shuffling may not be acceptable in applications where BE messages are sent from uncontrolled sources, e.g., a flight passenger connecting their laptop to a cabin network, where alarm data is also communicated.

Increased Complexity in Composability/Scalability for TT: Shuffling requires constraints on the scheduler of TT messages: The increased latency and jitter affect the communication schedule. Incremental scheduling is only successful if changes of the TT timing (due to RC and BE messages) are taken into account when creating the initial communication schedule. However, there will be no issue for the application and task-level scheduler if these constraints are taken into account right from the beginning.

The *TTEthernet/ARINC664* products realize the timely block and shuffling integration method. The preemption integration method is available in a prototype state. For the integration strategy we define a priority scheme in decreasing order: PCFs, TT dataflows, k priorities for RC dataflows and BE dataflows. Figure 8.8 presents an example *TTEthernet* network consisting of three end systems and a single switch.

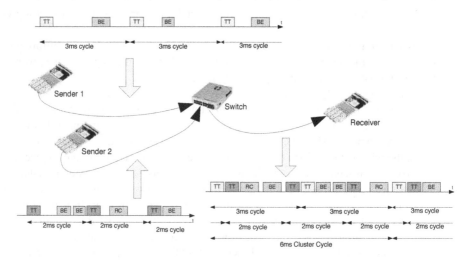

FIGURE 8.8
TTEthernet – Example of Dataflow Integration (© 2009 IEEE [309])

The example consists of two end systems that send frames, a switch that integrates the frames from the two senders and a receiver that receives the integrated dataflow from the switch. As depicted, Sender 1 sends a time-triggered frame (TT) with a period of 3 milliseconds and best-effort frames (BE). Sender 2 sends a time-triggered frame (TT) with a period of 2 milliseconds, best-effort frames (BE) and rate-constrained frames (RC). The resulting integrated dataflow is depicted on the right.

8.2.1.5 Permanence Function

TTEthernet allows several applications of different criticality to share a common network. Conceptually, the synchronization strategy itself can be interpreted as an application of highest criticality and the timely transport of the PCFs is of highest importance. The synchronization strategy cannot rely on the synchronized timebase to resolve conflicts in the shared network. PCFs can be generated at points in time when a *TTEthernet* device is already sending, receiving or relaying an unsynchronized application's message. When shuffling is used as a traffic integration method (e.g., in *TTEthernet/ARINC664*), the conflicts of PCFs with application messages are resolved by queuing, which introduces significant queuing delays. For the synchronization of the local clocks in a system, it is essential that these delays are mitigated and this is exactly the purpose of the permanence function.

Figure 8.9 gives an overview of the operation of the permanence function. In this case the transmission of a PCF from an SM to a CM is depicted. The synchronization module is the part of a *TTEthernet* device that executes the synchronization algorithms. This part may be realized, e.g., in the form of a module on an FPGA or as software running on an embedded CPU. At some point in time, the logic in the synchronization module requires the dispatch of a PCF. This point in time is called the *dispatch point in time*. As discussed above, there are several reasons that cause a delay from this dispatch point in time until the first bit of the PCF is transmitted on the wire (e.g., propagation of the signal through the software/hardware stack and conflicts with unsynchronized traffic). We call this delay the *send delay*. Similar delays occur in switches and the receiver of the PCF, and we call these delays the *relay delay* and the *receive delay*, respectively.

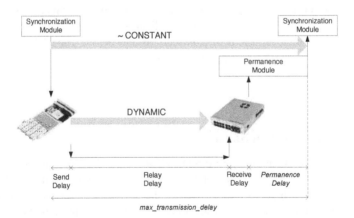

FIGURE 8.9
The Permanence Function Transforms Network Jitter into Network Latency

TTEthernet uses the *transparent clock* mechanism to measure the actual end-to-end transmission time of each PCF. This mechanism requires all devices to measure and to add their delay (send delay, relay delay, receive delay) imposed on a PCF into its payload field `pcf_transparent_clock`. Static delays, such as wire delays, can be compensated for by static configuration within the *TTEthernet* devices. The quality of the delay measurements is a key factor for the clock synchronization process. In an FPGA-based realization, for example, it is possible to measure these delays with the accuracy of one hardware-clock tick. Software-based solutions running on low-cost embedded controllers may introduce much larger measurement errors.

Hence, within bounded measurement and digitalization errors, the receiver of a PCF knows by the value of `pcf_transparent_clock` the actual transmission delay of the received PCF. After reception of the PCF, the receiver makes the PCF *permanent* by delaying it for the remaining difference to the maximum delay (depicted by the permanence module in Figure 8.9):

$$permanence\_delay \quad = \quad max\_transmission\_delay -$$
$$\text{pcf\_transparent\_clock} \qquad (8.1)$$

max_transmission_delay is an off-line calculated parameter that defines the maximum one-way transmission delay for any PCF in the system (from any SM to any CM or vice versa from any CM to any SM or SC).

In an ideal system, the permanence point in time of a PCF is *max_transmission_delay* after its original dispatch point in time. Hence, the transparent clock and the permanence function transform the actual dynamic network delay into the constant maximum delay. In the real world, measurement and digitalization errors occur, such that the permanence point in time will occur within an interval around the nominal permanence point in time. This interval is implementation-dependent (e.g., jitter through the physical layer, clock speeds of different clock domains, etc.). In typical realizations of *TTEthernet*, this interval can be on the order of several tens of nanoseconds and below. The *TTEthernet* services do not correct for this uncertainty and account it as part of the precision in the system.

8.2.2   Clock Synchronization

8.2.2.1   Clock Synchronization Overview

The clock synchronization algorithm in *TTEthernet* is illustrated in Figure 8.10. The "cluster cycle" is the time frame for *TTEthernet*. It denotes one full cycle through the schedule of time-triggered frames. Hence, the length of the cluster cycle is defined by the least-common multiple of all time-triggered frame periods. For long cluster cycles, it is necessary to re-synchronize, which means to execute the clock synchronization algorithm more than once in order to keep the precision in the system small. The period of clock synchronization is denoted by "integration cycle" and depicted in Figure 8.10 numbered from 0 to $n-1$. Multiple integration cycles per cluster cycle contribute to high-quality precision, and allow fast re-integration of a component that has lost synchronization.

A *TTEthernet* device uses the `local_clock` variable to cyclically count within an integration cycle and the `local_integration_cycle` to cyclically count the number of integration cycles. Together, `local_integration_cycle` and `local_clock` represent the local view of a *TTEthernet* component of the synchronized global timebase. *TTEthernet* devices typically also provide programmable counters for applications that demand a longer time frame than one cluster cycle.

The clock synchronization algorithm is executed at the beginning of each integration cycle and can be separated into two steps. In the first step, the SMs send PCFs to the CMs. The CMs execute a first convergence function, the compression function. The compression function collects the PCFs and generates a compressed PCF which is sent back to the SMs and SCs in the system. In the second step, the SMs and SCs collect the compressed PCFs from different CMs and execute a second convergence function. We discuss the compression function and the clock correction in more detail next.

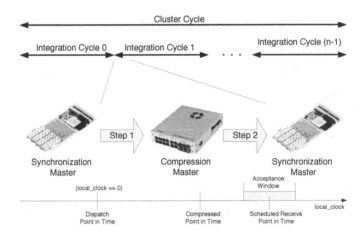

FIGURE 8.10
The Timing Hierarchy in *TTEthernet*

8.2.2.2 First Step Convergence: Compression Master

Once synchronized, the SMs dispatch their PCFs at the same points in time (according their local clocks) to the CMs. Due to drifts in the oscillators, the actual dispatch points in the SMs and the resulting permanence points in time in the CMs will deviate. Therefore, the CMs implement the compression function that runs unsynchronized to the synchronized global time. The compression function collects the PCFs from different SMs and produces a new PCF which it sends to the SMs as a response. The dispatch point in time of this new PCF is calculated as a function of the relative permanence points in time of the PCFs from the SMs. The dispatch point in time from the CM is called the *compressed point in time*.

The compression function runs unsynchronized to the synchronized timebase; hence, it is started upon the reception of a PCF, rather than upon the synchronized local clock in the CM reaching a particular point in time. Therefore, it has to be guaranteed that faulty SMs that may send early or late will not cause the compression function to recognize only a subset of PCFs from correct SMs in the generation of the new PCF.

An example scenario of the compression function is depicted in Figure 8.11. In this example, three end systems that are configured as SMs dispatch PCFs (in this case, a special type called Integration Frame [IN]) to a switch that is configured as CM. Note that the depicted deviations of the dispatch points in time stem from the relative differences in the oscillators of the end systems; in a perfect world, these dispatch points in time would be perfectly aligned.

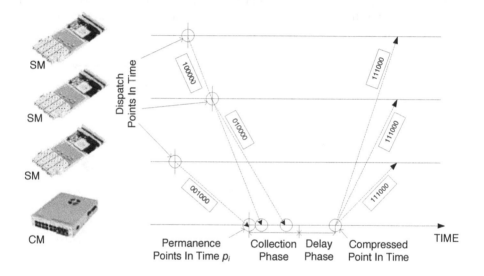

FIGURE 8.11

Compression Function Overview: Three End Systems Configured as Synchronization Masters Provide Their Local Clock Readings to a Switch Configured as Compression Master; in the Real World, the Network Jitter is Compensated for by the Permanence Function (In Formal Verification Studies [310], We Take Credit for the Permanence Function by Abstracting from the Network Delays)

The CM will use the permanence function discussed previously to derive the permanence points in time of the PCFs. The first permanence point in time (p_1) will cause the compression function to start the collection phase. As the successive PCFs with the same value in the pcf_integration_cycle field become permanent, the CM records their relative offsets to the first permanence point in time ($p_i - p_1, i > 1$) and stores these offsets in a local data structure that we call the *clock synchronization stack*. Should PCFs with differing values in their pcf_integration_cycle field become permanent in the CM, a parallel compression function is started for each new value in pcf_integration_cycle.

The duration of the collection phase is given by the following rules, where "observation window" specifies the maximum deviation of two correct local clocks in the system as measurable by a clock within the network:

- The first permanence point in time will cause the compression function to collect the following permanent PCFs for one observation window.

- When the compression function collects at least a second permanent PCF during the first observation window, the collection phase is prolonged for a second observation window.

- The collection phase will end when the number of permanent PCFs collected during observation window i is equal to the number of permanent PCFs collected during observation window $i - 1$ (i.e., when no new PCF becomes permanent for the duration of one observation window). Otherwise collection will be continued for another observation window.

- The collection phase will stop at the latest after the $(k+1)^{th}$ observation window, where k is the configured number of faulty SMs to be tolerated.

After the collection phase the relative permanence points in time of the collected PCFs are used to determine a correction value for the following delay phase. In order to minimize the impact of the faulty SMs we use a variant of the fault-tolerant median (where $p_i, i \geq 1$ represents the permanence points in time):

- One permanence point in time: *correction_value* $= 0$

- Two permanence points in time: *correction_value* $= \frac{p_2 - p_1}{2}$

- Three permanence points in time: *correction_value* $= p_2 - p_1$

- Four permanence points in time:
  correction_value $= \frac{((p_2 - p_1) + (p_3 - p_1))}{2}$

- Five permanence points in time: *correction_value* $= p_3 - p_1$

- More than five permanence points in time: Take the average of the $(k+1)^{th}$ largest and $(k+1)^{th}$ smallest inputs, where k is the number of faulty SMs that have to be tolerated.

In the delay phase, the compression function will wait for

$$
\begin{aligned}
delay\_phase\_duration \quad = \quad & correction\_value + \\
& (k+1) * observation\_window - \\
& collection\_phase\_duration \quad (8.2)
\end{aligned}
$$

where *collection_phase_duration* is the length of the preceding collection phase.

The overall time from the first permanence point in time until the CM dispatches the resulting compressed PCF is denoted by the *compression_master_delay* and calculated as follows:

$$
\begin{aligned}
compression\_master\_delay &= collection\_phase\_duration \\
&\quad +delay\_phase\_duration \\
&= (k+1) \times observation\_window \\
&\quad +correction\_value
\end{aligned}
\tag{8.3}
$$

This equation assumes negligible overhead for the calculation of the fault-tolerant median.

At the compressed point in time, the CM generates the compressed PCF with `pcf_integration_cycle` set to the same value as in the original PCFs from the SMs. Furthermore, in the `pcf_membership_new` field of the compressed PCF, the CM sets those bits to one that indicate the SMs from which the CM received PCFs in the collection phase. All other bits in `pcf_membership_new` are set to zero.

The compressed PCF may be delayed by a configured duration before dispatch. This is necessary, for example, if the CM executes a central guardian function. This duration can be added in the transparent clock of the compressed PCF upon dispatch to the SMs. Otherwise, *compression_master_delay* has to be extended by this duration.

8.2.2.3   Second Step Convergence: Synchronization Master

The SMs off-line configure the *scheduled_receive_pit* as the nominal point in time when they expect the permanence of the compressed PCFs received from the CMs.

$$
\begin{aligned}
scheduled\_receive\_pit &= dispatch\_pit \\
&\quad +2 \times max\_transmission\_delay \\
&\quad +compression\_master\_delay
\end{aligned}
\tag{8.4}
$$

Starting from an SM's dispatch point in time (`local_clock==0`), it takes *max_transmission_delay* until a PCF from an SM becomes permanent in the CM. The delay through the CM is given by *compression_master_delay* (Equation 8.3 setting *correction_value* = 0). Then it takes another *max_transmission_delay* until the compressed PCF becomes permanent in the SMs.

The *scheduled_receive_pit* is the nominal point in time of permanence of the compressed PCF. In the real world, oscillator drifts as well as digitalization errors and timing errors imposed by crossing clock domains as well as faulty components require the specification of an interval around the *scheduled_receive_pit*. This interval is called the *acceptance window*. Compressed PCFs that become permanent within the acceptance window are called "in-schedule," while all other PCFs are called "out-of-schedule."

At the end of the acceptance window, the SMs evaluate the in-schedule compressed PCFs. As there is at least one CM per channel, an SM will receive at least as many compressed PCFs as channels it is connected to. Current *TTEthernet* implementations allow one, two or three channels. In cases where more than one CM is configured per channel and in certain failure scenarios, an SM will receive more than one compressed PCF per channel. The evaluation of the PCF is done as follows:

1. Per-Channel Selection:

   (a) From the possibly many in-schedule compressed PCFs on a channel, the SM selects the PCF with the highest number of bits set in the pcf_membership_new field.

   (b) In case of equal maximum number, the PCF that has become permanent latest with maximum number of bits set in pcf_membership_new is selected.

   (c) The output of this selection process is at most one PCF per channel and we call these PCFs "best-channel" PCFs.

2. Low-Membership Exclusion:

   (a) From the best-channel PCFs the SM removes those PCFs with a relatively low number of bits set in the pcf_membership_new field.

      • The relatively low number is determined by the maximum number of bits set in the pcf_membership_new of the best-channel PCF minus a configurable parameter.

      • This parameter is typically set to the number of faulty SMs to be tolerated.

3. Clock-Correction Calculation

   (a) If only one best-channel PCF remains, then the clock correction value is determined by the scheduled receive point in time minus the permanence point in time of this PCF.

   (b) If only two best-channel PCFs remain, then the clock correction value is determined by the arithmetic mean of the two time differences between the scheduled receive point in time and the permanence points in time.

   (c) If three best-channel PCFs remain, then the clock correction value can be configured to either be the arithmetic mean of the two extreme time differences or the middle time difference between the scheduled receive point in time and the permanence points in time.

8.2.3   Startup and Restart

This section discusses the *TTEthernet* startup/restart on an informal level. The startup/restart algorithm executed in the SMs is depicted in pseudo-code using standard IEEE formalism in Figure 8.12. The CMs realize a similar algorithm which we do not discuss in this chapter. SCs implement a much simpler state machine, as SCs only passively integrate to an already established synchronized time. We refer the interested reader to the SAE AS6802 standard for a complete discussion of

Time-Triggered Communication

the state machines and the formal description of the algorithm's parameters. The startup/restart algorithm has been designed to tolerate multiple failures. Downgraded versions for less demanding fault-tolerance applications are available and similar to the algorithm discussed in this section.

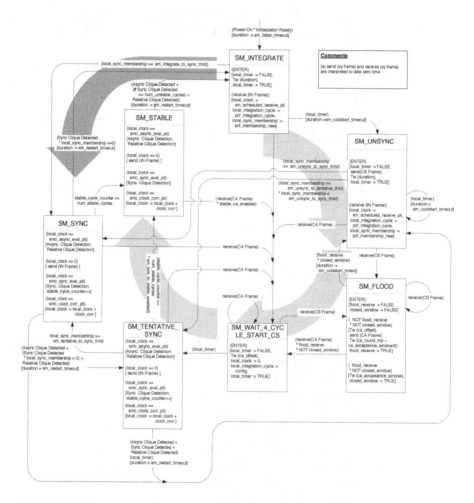

FIGURE 8.12

Protocol State Machine for the SMs: Coldstart Proceeds Clockwise, Regular Integration Counter-Clockwise

In Figure 8.12, each box represents a state in the protocol state machine. Within each box, the state name is depicted, e.g., SM_UNSYNC, followed by guarded commands. Guarded commands are of the form (list of conditions), e.g., (ENTER), followed by a list of commands to be executed when a guard evaluates to true, e.g., [local_timer:=FALSE]. Tw(duration) is used to describe a timeout; * and + denote logical AND and OR, respectively. Boxes are connected to each other by transitions, which are labeled by the conditions to be met and commands to be executed. The *TTEthernet* synchronization protocol uses the following variables:

- `local_timer`: An unsynchronized timer used to measure timeouts, e.g., the duration for which an SM tries to integrate before coldstart. In order for conformity to the IEEE standard, we use `local_timer` as Boolean in the state machine above.

- `local_clock`: The synchronized timer used to measure the current point in time relative to the current integration cycle.

- `local_integration_cycle`: A synchronized counter that cyclically counts the integration cycles.

- `local_sync_membership`: A membership vector with a one-to-one mapping of bit to SM; used in the synchronous and relative clique detection function.

- `local_async_membership`: A membership vector with a one-to-one mapping of bit to SM; used in the asynchronous and relative clique detection function.

In a multiple-failure tolerant configuration of *TTEthernet*, the startup/restart protocol has the character of an end-to-end protocol between the SMs. Hence the CMs have only minor impact on the protocol execution: the CM will not react to CS and CA frames, but will only passively integrate to IN frames. Once a CM perceives that sufficient SMs are operational, it will stop forwarding CS frames (note: IN frames will always be forwarded).

The startup/restart protocol realizes four functions: integration, coldstart, clique detection and restart. We will discuss these functions next with a focus on their fault-tolerance capabilities and with references to the state machine depicted above.

8.2.3.1 Integration

A component that is powered-on or reset will start in an integration state. In particular, the SM will start in the SM_INTEGRATE state, where it tries to integrate to an already established global synchronization. The regular integration is depicted by the counter-clockwise background arrow in the state machine. As discussed in Section 8.2.2.1, during normal operation mode synchronization messages, the IN frames, are periodically exchanged with a period of the integration cycle. An integrating component will thus wait for two integration cycles in which it is guaranteed that it will receive an IN frame from the CM should the global synchronization already be established and maintained by a sufficient set of components in the system. Once a component receives an IN frame, it checks the number of bits set in the `membership_new` payload-field of the IN frame. If this number is sufficiently high, the component enters normal operation mode and starts executing the clock synchronization service. If this number is too low, the IN frame is ignored and the component continues waiting for an IN frame with sufficient weight. If the two integration cycles time out, this means that the SM did not integrate. Hence, it concludes that there is no global synchronization available and executes the coldstart procedure, by entering SM_UNSYNC state.

8.2.3.2 Coldstart

The coldstart is depicted by the clockwise background arrow in the state machine; it proceeds in four steps that implement the classic two-round end-to-end message exchange required to overcome inconsistent message transmissions (this is executed in states SM_UNSYNC and SM_FLOOD in the state machine). The first round is initiated by an SM with the transmission of a Coldstart Frame (CS) to all CMs (Step 1). The CMs will distribute the CS to all SMs in the network (Step 2). In the two fault-tolerant configurations, all SMs except the original CS sender will acknowledge the CS by sending a Coldstart Acknowledgment Frame (CA) a configurable duration after the CS became permanent (Step 3). Finally, the CMs will again relay the CA to all SMs in the network (Step 4). The reception of the CA in an SM concludes the coldstart procedure and the SM transits to the SM_WAIT_4_CYCLE_START_CS state.

A defined offset (the CA_offset) after the received CA frame becomes permanent, the SMs will test whether normal operation mode can be entered (SM_TENTATIVE_SYNC state). This is done by sending out an IN frame. The CM will then generate and dispatch a new compressed IN frame as described in Section 8.2.2.1. When the received compressed IN frame becomes permanent, the SMs will check the number of membership bits set and enter normal operation mode if this number is sufficiently high.

The four step startup/restart is necessary because of the two-fault hypothesis: In the worst case an SM and a CM may be faulty, as depicted in Figure 8.13 (SM1 and CM1 are faulty). The scenario starts with the faulty SM (SM1) dispatching its CS frame only to the faulty CM (CM1), which in turn relays the faulty CS frame only to an arbitrary subset of SM (SM3). To establish consistency, the SM (SM3) answers the CS frame with a CA frame dispatched to both CMs. The correct CM (CM2) will now distribute the CA frame consistently. Note: All SMs (even the faulty one, see Section 8.2.5.2) will answer only CS frames that they did not send themselves. In a faulty SM plus faulty CM scenario, it is, therefore, guaranteed that either the original CS frame or the CA as response to a CS frame will be dispatched by a correct SM to all CMs in the network.

In order to generate a CA frame, three components have to be involved, two SMs and one CM: (a) a first SM sends a CS frame, (b) a CM relays the CS frame and (c) a second SM acknowledges the CS frame by sending a CA frame. Hence, under a two-failure hypothesis it is impossible that two faulty components can maliciously cooperate in order to produce the startup sequence of frames. Either two SMs are faulty, in which case no CM would relay the initial CS frame, or an SM and a CM are faulty, in which case no SM will acknowledge a faulty CS frame.

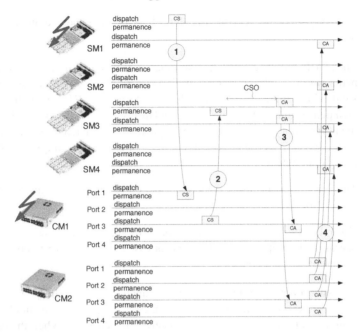

FIGURE 8.13
Startup/Restart Example in Presence of a Faulty SM and a Faulty CM: the CA Frame
Resolves an Inconsistent Startup Attempt

8.2.3.3 Restart

Once a component detects the loss of synchronization (by using the clique detection
mechanisms discussed in the following section), it will try to regain synchronization.
Depending on the current state in the protocol state machine, the component will either try to re-integrate or re-coldstart. Re-integration will be executed when a clique
is detected in SM_STABLE state. Re-coldstart will be executed when the component is in SM_TENTATIVE_SYNC state. When a component is in the SM_SYNC
state, it will re-integrate only if the synchronous clique detection function detected
a clique and the component did not receive any synchronization message confirm-
ing its current schedule; in all other cases the component will try to re-coldstart.
This subtle differentiation is a result of the quality of synchronization a component has already reached: SM_STABLE is an indication that there were already a
number of rounds of successful synchronization exchange, while a component in
SM_TENTATIVE_SYNC state is still testing if a newly generated global synchronization has been established successfully.

8.2.3.4 Clique Detection

In the synchronized states (SM_TENTATIVE_SYNC, SM_SYNC, and SM_STABLE
state), *TTEthernet* uses clique detection algorithms to reliably detect all clique sce-

narios. Three types of clique detection algorithms are implemented that can be active in parallel:

- The synchronous clique detection function is essential to move faulty components out of synchronization, e.g., in the case of a faulty high-integrity SM, a valid failure mode of an SM is that it only accepts IN frames that match its own perception of synchronization and drops all other IN frames. Hence, both the asynchronous and the relative clique detection function will not cover this clique scenario as frames belonging to other cliques are simply ignored by the faulty component.

- The asynchronous clique detection function is the usual means by which correct components will detect cliques.

- The relative clique detection function is of significance when the number of operational components is low and is essential to resolve temporary clique formations during coldstart.

The synchronous clique detection function uses the `local_sync_membership` variable to store the `pcf_membership_new` field with the highest number of bits set of IN frames received in-schedule. When the number of bits set in `local_sync_membership` is below an a priori defined threshold, a synchronous clique is detected. This check is executed at the end of the acceptance window (see Section 8.2.2.1).

The asynchronous clique detection function uses the `local_async_membership` variable to record the `pcf_memberhip_new` fields of IN frames received out-of-schedule. When the number of bits set in the `local_async_membership` variable goes above a configured threshold, an asynchronous clique has been detected. This check is executed at the beginning of each integration cycle before the transmission of the IN frame.

The relative clique detection function detects a clique if `local_sync_membership` is equal to or less than `local_async_membership`. This check is executed at the beginning of each Integration Cycle before the transmission of the IN frame.

8.2.4 Diagnostic Services

The clique detection algorithms discussed in the previous section are low-level diagnostic services. They identify situations in which *TTEthernet* end systems and/or switches have lost synchronization. In addition to the clique detection algorithms, more sophisticated diagnostic routines can be implemented, for example, to identify unexpected behavior. Although such diagnostic algorithms are outside the SAE AS6802 standard that defines the *TTEthernet* services, diagnosis is generically supported by the different traffic classes. For example, time-triggered frames can be used to periodically distribute detailed information of the internal synchronization state of a *TTEthernet* device (e.g., maximum and minimum clock correction values or even a history of these values). On the other hand, best-effort messages allow the realization of arbitrary on-demand diagnosis protocols in the network.

8.2.5 Fault Isolation

As *TTEthernet* is designed for ultra-highly dependable systems, we have to assume that the failure mode of any chip IP in the *TTEthernet* network may become arbitrarily faulty. This means that the faulty chip IP may assign arbitrary signals on an arbitrary selection of its pins. In *TTEthernet*, we find essentially two types of chip IP: The end system and the switch (we discuss the impact of their integration into a single chip IP toward the end of this section).

In *TTEthernet*, we assume each chip IP to be a fault-containment unit. This means that a fault will not propagate directly from one device to another one. However, a fault in one device may manifest in an error state and ultimately result in a failure of a *TTEthernet* device. This failure may then become visible as faulty or missing Ethernet frames on the interface from the faulty device to the *TTEthernet* network. To tolerate faulty Ethernet frames, *TTEthernet* specifies two ways to construct error-containment units: The central guardian and the high-integrity design. A third type of error-containment is based on triple-modular redundancy and is currently under development.

An error-containment unit consists of at least two independent fault-containment units. In the case of the central guardian, the error-containment unit is constructed from one end system and one switch chip IP. The high-integrity design constructs the error-containment unit from either two end system chip IPs or from two switch chip IPs. In this case, we also speak of a high-integrity end system or a high-integrity switch. Both, central guardian and high-integrity design aim at transforming the failure model of a faulty device from an arbitrary failure model to a more benign one. In the case of *TTEthernet*, arbitrary failure modes are transformed into inconsistent-omission failures.

8.2.5.1 Central Guardian

The central guardian functionality is implemented in the *TTEthernet* switch and protects the *TTEthernet* network against arbitrarily faulty end systems. Figure 8.14 depicts the central guardian control and enforcement actions for time-triggered and rate-constrained traffic.

A *TTEthernet* end system transmits a time-triggered frame according to the synchronized global time when the scheduled dispatch point in time (*dispatch_pit*) of the frame is reached. As discussed in Section 8.2.1.4, depending on the integration method of unsynchronized and synchronized traffic, the actual instant when the frame is transmitted on the dataflow link, the send point in time *send_pit*, may be delayed. However this delay is bounded by *max(send_delay)*.

$$send\_pit \in [dispatch\_pit, dispatch\_pit + max(send\_delay)] \qquad (8.5)$$

As a consequence of the latency of the dataflow link (*link_latency*), the *TTEth-*

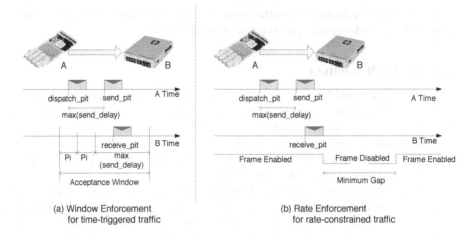

(a) Window Enforcement
for time-triggered traffic

(b) Rate Enforcement
for rate-constrained traffic

FIGURE 8.14

Central Guardian Enforcement Actions for Synchronized Time-Triggered Traffic and
Unsynchronized Rate-Constrained Traffic

ernet switch to which the end system is connected receives the first bit of the time-
triggered frame at *receive_pit*:

$$receive\_pit = send\_pit + link\_latency \qquad (8.6)$$

A *TTEthernet* switch that is configured to execute the window enforcement
checks whether the receive point in time of a time-triggered frame happens within an
acceptance window. The start of this acceptance window has to consider the earliest
possible *receive_pit*. This directly follows from the *link_latency* and the maximum
offset of the local clocks in the end system and switch, the precision Π.

$$acceptance\_window\_start = dispatch\_pit + link\_latency - \Pi \qquad (8.7)$$

Likewise, the end of the acceptance window has to consider the latest possible
receive_pit for the time-triggered frame.

$$
\begin{aligned}
acceptance\_window\_end \;=\; & acceptance\_window\_start \\
& +2 \times \Pi \\
& +max(send\_delay) \qquad (8.8)
\end{aligned}
$$

When the window enforcement algorithm for time-triggered traffic is enabled in
the *TTEthernet* switch, the switch will forward a time-triggered frame only when
receive_pit falls within the acceptance window as specified by Equations 8.7 and 8.8.

Rate-constrained traffic is not dispatched according to the synchronized global
time and, therefore, the window enforcement is not directly applicable. Instead a

TTEthernet switch supporting the rate-constrained traffic class implements a rate-enforcement algorithm as depicted in Figure 8.14 on the right. A rate-enforcement algorithm controls and enforces the temporal distance between two succeeding frames with the same frame identifier. A frame is only forwarded when this temporal distance exceeds a minimum gap as specified by a configurable parameter.

8.2.5.2 High-Integrity Design

The high-integrity design of a device aims for error-containment already within the device rather than via a remote instance as in the case of the central guardian. A particular high-integrity design method is the commander/monitor (COM/MON) design method depicted in Figure 8.15.

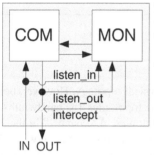

Core COM/MON Assumptions:
- COM and MON fail independently
- MON can intercept a faulty message produced by the COM
- COM cannot produce a valid message such that this message appears as two different messages on listen_out and OUT; though it may be valid on listen_out but detectably faulty on OUT or vice versa
- MON cannot itself generate a faulty message, neither by inverting listen_out to an output, nor by toggling the intercept signal

FIGURE 8.15
Realization of a High-Integrity Component as a Commander/Monitor (COM/MON) Architecture

A COM/MON pair is constructed out of two fault-containment units, where one unit operates as the commander (COM) and the other as monitor (MON). In *TTEthernet*, both end system and switch may be realized in a COM/MON design. Current realizations use a single oscillator for COM and MON and a dedicated clock monitor to prevent common-mode failures of the clock. The input (IN) received by the device is forwarded to both, COM (IN) and MON (listen_in). COM and MON execute an agreement protocol on the information received before further usage. This agreement protocol is also called the congruency exchange. When COM and MON agree on the received information, they process the input in their state machines. In *TTEthernet*, COM and MON execute the same state machine and as they agree on the input and derive their timing from the same clock source, they will produce the same output at about the same point in time. This means that COM and MON operate in a replica-determinant manner [258]. The output from the COM (OUT) is forwarded to the MON, which cross-compares OUT with its own output. When the output from the COM differs from the output of the MON, then the MON terminates COM's transmission on OUT.

8.2.6   Configuration Services

In standard Ethernet networks, the simple network management protocol (SNMP) is widely accepted. However, in case of a network for mixed-criticality systems we also have to consider faulty devices. *TTEthernet* realizes a fault-tolerant reconfiguration protocol. Here, a switch may only change its configuration when a configurable set of end systems send "unlock" frames to the switch within a small temporal duration. End systems can be re-configured through the host interface. If necessary, the host or a middleware layer can implement an interactive consistency protocol before the dataflow is adapted. Future *TTEthernet* products may harmonize the configuration and maintenance routines with existing state-of-the-art approaches like SNMP.

8.3   Protocol Parameterization

8.3.1   Physical Topology

TTEthernet is largely agnostic to the underlying physical topology. However, being fully compatible to standard Ethernet, the preferred topologies for bandwidths of 100 Mb/s and above are redundant star and tree topologies. A "channel" in the star topology consists of exactly one switch and the communication links between this switch and the attached end systems. For a redundant star topology, the switches and communication links are replicated such that an end system is connected to as many switches as there are redundant instances. Each redundant switch and its attached communication link form one channel. The tree topology extends the concept of a channel by allowing multiple switches per channel. Hence, in a tree topology different end systems may be connected to different switches and these switches are then connected directly or indirectly to each other via communication links. The communication links between switches of the same channel are commonly called multi-hop links. It is not required that the redundant channels in the *TTEthernet* network have the same topology. The example depicted in the beginning of this chapter in Figure 8.1, for example, shows two redundant channels in which Channel 1 is configured in tree topology with three switches, and Channel 2 is configured in star topology.

For availability reasons, additional communication links to the ones present in the star/tree topology can interconnect the *TTEthernet* devices, thereby forming a mesh topology. *TTEthernet* as specified in the AS6802 standard does not specify redundancy management using these additional links directly. However, adaptive routing protocols can be implemented in addition to *TTEthernet*. For example, when a switch becomes fail-silent, the network can reconfigure the respective channel and the additional links compensate for the faulty switch.

For the correct execution of the synchronization algorithms, it is required that all non-faulty CMs receive the PCFs from all non-faulty SMs and vice versa. Devices configured as SCs may be attached to the network such that they receive only some of the PCFs from non-faulty CMs, which means that in the failure case these

SCs may lose synchronization to the network. However, as SCs do not source PCFs, connecting them only with partial redundancy to the network impacts only the SCs themselves.

The topology freedom comes with the implementation of the permanence function (Section 8.2.1.5) which masks the network-imposed jitter on the PCFs. The *max_transmission_delay* parameter gives the upper bound in time for the transmission of a PCF between an SM or SC and the CM. The parameter *max_transmission_delay* is therefore also the restricting parameter of the number of devices in the transmission path of a PCF.

8.3.2 Protocol-Control Flow Parameterization

In a *TTEthernet* network, the precision is the main synchronization quality parameter from which most other parameters are derived. The precision itself is a function of the drift rates in the device's oscillators, the integration cycle, the network jitter (not covered by the permanence function), as well as the failure hypothesis. The failure hypothesis and the network topology are often given by the system requirements. Hence, the precision in the system may either be improved by increasing the quality of the oscillators and/or by decreasing the integration cycle.

TTEthernet provides scalable fault-tolerance. The key parameters determining the degree of fault-tolerance are the integrity level of the devices (high-integrity or standard-integrity), the number of SMs and CMs, as well as the number of independent channels between the SMs and CMs. The number of SMs in the *TTEthernet* network is reflected in the threshold parameters in the protocol state machines in SMs, SCs, and CMs. For example, the threshold that gives the number of bits to be set in a PCF such that an SM may integrate is determined by the overall number of SMs in the network.

8.3.3 Dataflow Parameterization

All messages communicated in a *TTEthernet* network are standard Ethernet frames. In a particular implementation of *TTEthernet*, on top of ARINC 664-p7 (*TTEthernet/ARINC664*) time-triggered and rate-constrained frames are specified in more detail, to carry a particular frame id, called the Virtual Link Identifier, and a sequence number (see Section 8.2.1.2). The number of Virtual Link Identifiers that a device supports turned out to be a critical parameter. The first ARINC 664-p7 devices supported a few hundred of these identifiers. *TTEthernet/ARINC664* devices increase this number up to a few thousand.

TTEthernet identifies the traffic class of a particular Ethernet frame by combinations of bits in the Ethernet frame. In *TTEthernet/ARINC664*, a device classifies a frame as time-triggered or rate-constrained when the Constant Field in the Destination Address is set to a configured value as depicted in Figure 8.5. All other frames are classified as best-effort traffic. The Virtual Link Identifier is then used to differentiate between time-triggered and rate-constrained traffic, by a table-lookup in a device. It is, therefore, possible that different *TTEthernet/ARINC664* devices within the same *TTEthernet* network change the traffic class of a frame. For example, a frame may be sent as a time-triggered frame to the first switch, but the switch forwards the frame as a rate-constrained frame.

8.3.3.1 Time-Triggered Parameters

A time-triggered frame f_i on a dataflow link $[v_k, v_l]$ (that is the unidirectional connection between two *TTEthernet* devices v_k and v_l), $f_i^{[v_k,v_l]}$, is fully temporally specified by the following triple (as defined by the Time-Triggered Architecture [177]):

$$f_i^{[v_k,v_l]} = \{f_i.period, f_i^{[v_k,v_l]}.offset, f_i.length\} \tag{8.9}$$

The time-triggered frame period and frame length are given a priori; it is the task of the scheduler to assign values to $F^L.offset$ for all frames F on all dataflow links L in the network.

A crucial aspect in time-triggered protocols is the number and kind of different frame periods they support. Some protocols, for example, define a base period and allow only harmonic multiples of the base period as valid frame periods. *TTEthernet* is more flexible in this respect. In particular, *TTEthernet/ARINC664* supports eight non-harmonic frame periods, with the only restriction that the frame periods are multiples of the oscillator tick, e.g., multiples of 8 ns in a device running at 125 MHz. For each non-harmonic frame period, the *TTEthernet/ARINC664* devices support harmonic frame periods.

The integration of time-triggered traffic with unsynchronized traffic imposes additional constraints on the time-triggered schedule. When time-triggered frames are scheduled back-to-back, there is a potential threat that rate-constrained frames will not be delivered within their required temporal bounds. Hence, for the integration of synchronized and unsynchronized traffic the time-triggered schedule should be sparse. We call this property of a time-triggered schedule, the schedule "porosity." For schedule porosity, we have to insert blank intervals, intervals not assigned to time-triggered traffic, into the time-triggered schedule. We can distinguish three approaches to achieving a required level of schedule porosity: *a priori schedule variation*, *a posteriori schedule variation* and *schedule interpretation*. The a priori schedule variation takes the requirements on the blank intervals already as input to the scheduling process. Blank intervals will not be assigned time-triggered frames. The a posteriori approach generates a time-triggered schedule and inserts blank intervals by post-processing. Schedule interpretation means that the scheduled entities, the communication slots, are already sufficiently large to allow the communication of time-triggered as well as rate-constrained frames.

8.3.3.2 Rate-Constrained Parameters

A rate-constrained frame f_i on a dataflow link $[v_k, v_l]$ that is the unidirectional connection between two *TTEthernet* devices v_k and v_l), $f_i^{[v_k,v_l]}$, is fully temporally specified by the following tuple:

$$f_i^{[v_k,v_l]} = \{f_i.rate, f_i.length\} \tag{8.10}$$

$f_i.rate$ is the maximum rate with which frame f_i may be generated. In ARINC664-p7, for example, typical frame rates are between 1 ms and 128 ms. A non-faulty sender of rate-constrained frames will source two succeeding instances of frame f_i only with a minimum of $\dfrac{1}{f_i.rate}$ time-units in between. The upper bound of temporal distance of two succeeding instances of a frame f_i is not bounded. As discussed in Section 8.2.5.1, a switch that supports rate-constrained traffic will execute a rate-enforcing algorithm to prevent a faulty device from exceeding its defined frame rates.

The rate-constrained frame rates and frame lengths are *a priori* given. It is the task of a checker tool to off-line calculate bounds on the maximum latency and jitter, as well as on the worst-case memory usage in the system.

8.3.3.3 Best-Effort Parameters

Best-effort traffic implements the standard Ethernet communication paradigm. Hence, the temporal characteristics of best-effort traffic beyond standard Ethernet is also not part of the *TTEthernet* parameters set.

8.4 Communication Interface

A *TTEthernet* end system consists of a controller and a host. The *TTEthernet* controller is in charge of executing the *TTEthernet* services as discussed throughout this chapter and specified in the SAE AS6802 standard. The host executes the system applications. Host and controller communicate via the communication interface. The communication interface itself consists of an application-programming interface (API) and an implementation-specific interface.

The API defines the functions of the communication interface, such as functions for transmitting or receiving TT frames, and *TTEthernet* vendors have defined the API for *TTEthernet* outside the scope of the SAE AS6802 standard. It ensures that applications can easily be migrated to different *TTEthernet* realizations with small overhead. The implementation-specific interface depends on a particular realization of *TTEthernet*. It addresses the physical connection, data structures and buffering/queueing capabilities. We discuss the implementation-specific interface next.

TTEthernet end systems are available with PCI (Peripheral Component Interconnect) and PCIe (Peripheral Component Interconnect Express) as well as System-on-Chip (SoC) solutions. In addition to the standard PCI and PCIe form factors, these devices also come in PMC (PCI Mezzanine Card) and XMC (Switched Mezzanine Card) form factors.

From the buffering/queueing perspective, the communication interface of *TTEthernet/ARINC664* complies with the ARINC 664-p7 standard which specifies "communication ports" in sampling or queueing mode (in short sampling and queueing ports) and "service access points" (SAP) ports. Sampling and queueing ports are defined by the ARINC 653 standard.

The data structures communicated between host and controller differ with respect to the traffic class. For rate-constrained and time-triggered frame transmissions, the host sends the respective virtual link id, the EtherType and the Ethernet payload to the controller which then assembles the complete Ethernet frame. Also, the controller can be configured to manage a sequence number for each rate-constrained and time-triggered frame. This sequence number is then used as specified by ARINC 664-p7. Best-effort traffic is sent as Ethernet-compliant frame from the host to the controller starting with the MAC field and ending with the payload field. The controller will then calculate and add the frame-check sequence.

8.5 Validation and Verification Efforts

8.5.1 Formal Verification and Analysis

TTEthernet has been designed for use in ultra-highly dependable systems. Therefore, the assessment of its core algorithms by means of formal methods has been a primary goal right from the beginning of its industrial development. In particular, the following aspects of *TTEthernet* have been addressed within the CoMMiCS project[2]: The permanence function, the compression function and clock synchronization algorithms, as well as the overall interplay of the synchronization services. In addition to that, clock synchronization proofs from the ROBUS protocol have been adapted for *TTEthernet*.

The formal assessment of the permanence function is a model-checking study using the SAL model-checker. In particular, the formal model is based on continuous time represented as real values and uses the "calendar automata" approach [83]. The verification procedure uses the SAL infinite-bounded model checker `sal-inf-bmc` and *k*-induction. The procedure is highly automatized and the verification runs take a few seconds to complete. The formal studies show that the permanence point in time happens in a receiver a constant duration after the dispatch point in time in the sender. This is in a perfect world, free of digitalization effects, measurement errors when crossing clock boundaries, and similar. For the real world we have extended the model. Here we have shown that the permanence point in time in the receiver happens within a calculable interval defined by the accumulated errors. Again, the temporal position of this interval starts a constant delay after the dispatch point in time in the sender.

The formal verification of the compression function uses the same modelling paradigms as the permanence function. However, the actual proof is more complicated and requires abstraction techniques also presented in [83]. For abstraction we

[2]European Community's Seventh Framework Programme (FP7/2007-2013) under grant agreement *n°*236701

have to define abstract system states and transitions between them by hand. We then prove that the abstraction is valid. This means that for every transition in the original state machine there exists either a transition between abstract states or the system remains in the current abstract state. SAL helps in the definition of the abstraction by providing a counterexample if an inconsistency between the actual state machines and the abstraction is present. Once the abstraction is verified, we can use the abstraction in the proof of the actual properties of interest. For the compression function, we have proven a termination property, an agreement property and two timing properties [310]. The termination property shows that the compression function produces a compressed point in time when triggered. The agreement property proves that the collection phase in the compression function collects the PCFs from all correct SMs. Finally, the timing properties relate the permanence points in time to the compressed point in time and show that the fault-tolerant median is correctly executed.

The interplay of the synchronization services, in particular the coldstart, integration, clique detection and restart services, is significant for the correct operation of a time-triggered protocol. The state space it constructs, especially in the presence of faulty components, is huge. In order to keep the problem to a reasonable size, we have used a discrete notion of time in granularity of slots and use the finite bounded model checker sal-bmc in the formal assessment. Here we initialize the system in a broad range of initial states and calculate the time it takes to stabilize into a synchronous state. The results for the two fault-tolerant configurations of *TTEthernet* are presented in [306].

8.5.2 Certified Development Process

Most application areas of safety-critical and mixed-criticality systems require a development process according to a quality assurance standard. *TTEthernet*, having originally come from the aerospace domain, has been developed according to the DO 178b standard for software and the DO 254 standard for hardware. However, most quality assurance standards have a similar structure involving the following design steps: Requirements definition, conceptual design, detailed design, requirements-based testing and independent peer review. These steps have been executed for *TTEthernet*.

8.5.3 Model-Based Testing

In addition to the requirements-based testing within the certified development process, *TTEthernet* has also been subject to model-based testing. For model-based testing, we use the SAL models developed for the formal verification to also generate particular test cases. Here, we set the SAL model into a particular initial system state and simulate the algorithmic execution for a configurable number of steps, for which SAL is capable of producing the simulation trace. Also, an actual *TTEthernet* network is set in the same initial state and allowed to run for a configurable duration. The execution trace on the hardware is captured as well and cross-compared with the simulation trace. By showing equivalency between the simulation trace and the execution trace, the formal models are validated.

8.6   Example Configurations and Implementations

TTEthernet has been designed as cross-industry communication infrastructure supporting the implementation of low-cost field-bus-like networks up to ultra-highly dependable systems. In this section we discuss two particular configurations of *TTEthernet* representing the extremes of the possible configuration space: A master-based configuration and a dual-fault tolerant configuration. Furthermore, we present a system-of-systems configuration.

Current *TTEthernet* implementations are available in FPGA and in software. We will summarize their key characteristics in this chapter. An ASIC implementation of *TTEthernet* is being finalized.

8.6.1   Configurations

TTEthernet network configurations may differ with respect to the number of end systems and the number of switches, as well as their connecting topology. In particular for the synchronization algorithms, configurations are determined by the assignment of synchronization roles to *TTEthernet* devices. As a reminder, we have introduced three roles in the synchronization algorithms of *TTEthernet*: The Synchronization Master (SM), the Compression Master (CM) and the Synchronization Client (SC). We will discuss different configurations with regard to these aspects. More general characteristics of time-triggered networks include the quality of oscillators used in the *TTEthernet* devices, the wire speed and associated physical layer (Section 8.2.1.3), the communication interface (Section 8.4) and the activated fault-containment measures (Section 8.2.5).

8.6.1.1   Master-Based Configuration

The most basic *TTEthernet* configuration consists of an SM and a CM where both synchronization roles may be integrated into a single device. From this configuration, classic single-master multiple-slave networks, as depicted in Figure 8.16, can be built. This example network consists of one SM, one switch configured as CM and six SCs (two switches and four end systems). End systems and switches are connected in a tree topology and there is no redundant channel present.

SM1 sources PCFs which are relayed by CM1. As there is only one SM, a compressed PCF generated by CM1 will be the same as before compression. Hence, a simplified device (integrating SM and CM) that sources PCFs with a fixed period of the integration cycle is sufficient as steady synchronization source for the SCs in the network.

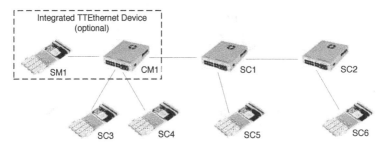

FIGURE 8.16
Master-Based Configuration with Three Switches Connected in Multi-Hop Topology

This single-master configuration is attractive for low-cost applications with a requirement of Quality-of-Service such as real-time performance. This configuration can be enhanced by safety mechanisms like the activation of the central guardian function in the switches or by the high-integrity design of key devices (Section 8.2.5). Hence, given that the appropriate fault-isolation measures are in place, even this basic configuration is appropriate for fail-safe applications (applications that can enter a safe state upon failure in which no protocol operation is required).

Availability, on the other hand, is limited: The failure of an arbitrary number of SCs will not affect the services as provided by *TTEthernet*. However, as in all single-master based systems, the loss of the SM or CM means a loss of the synchronization source and so of the *TTEthernet* protocol services. Availability requires the implementation of a sufficient degree of redundancy. We discuss a highly-dependable and highly-available configuration next.

8.6.1.2 Dual-Fault Tolerant Configuration

Highly-dependable systems such as civil airplanes or manned spacecrafts are fail-operational systems. These systems have to remain operational even in the presence of failures. Figure 8.17 shows a redundant *TTEthernet* network that tolerates two faulty devices without degradation of the *TTEthernet* services. The network consists of three redundant channels and five end systems. Each channel is formed by a single switch configured as CM and all five end systems operate as SMs. Furthermore, the SMs and CMs have to be high-integrity devices supporting an inconsistent-omission failure mode (Section 8.2.5.2). In this configuration, the failure of any two devices is masked by the *TTEthernet* services without quality degradation.

8.6.1.3 System-of-Systems Configuration

TTEthernet also provides a priority-based mechanism to realize system-of-system architectures.

Figure 8.18 shows a network architecture consisting of two *TTEthernet* subnetworks, a high-priority subnetwork and a low-priority subnetwork. The priority of the respective network is stored in the configuration data of each *TTEthernet* device

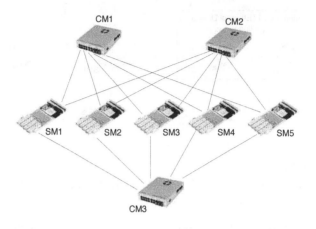

FIGURE 8.17
Dual-Fault Tolerant Configuration with Three Redundant Channels and Five Synchronization Masters

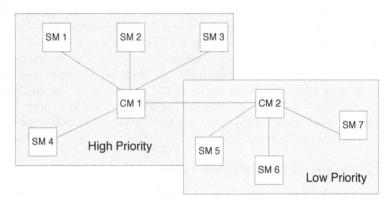

FIGURE 8.18
TTEthernet Systems-of-Systems Configuration Consisting of Two Subnetworks (© 2009 IEEE [359])

as well as in the Sync_Priority field in the PCFs. A *TTEthernet* end system can be configured to automatically synchronize to the highest priority PCF it receives. Alternatively, the change of a *TTEthernet* current priority to a higher priority can demand host interaction.

This priority mechanism supports the full operation of parts of the network, for example, to realize power-down modes. In the example in Figure 8.18, either one of the subnetworks can be shut down. Once it is powered on again, both sub-networks synchronize either automatically or upon host acknowledgment.

8.6.2 Implementations

TTEthernet specifies a set of services that can be implemented on top of standard Ethernet. End systems realize these services either in the form of a software stack on top of commercial-off-the-shelf controllers or in the form of dedicated FPGA solutions. Currently, *TTEthernet* switches are implemented in FPGA. Furthermore, ASICs for *TTEthernet* end systems and *TTEthernet* switches are being developed at the time of this writing.

The FPGA-based solutions of *TTEthernet* mainly differ with respect to the communication speed they support. The 100 Mbit/s FPGA-based version of the *TTEthernet* switch is specified as follows:

- Eight 100 Mbit/s and one 1 Gbit/s uplink port

- Guaranteed real-time delivery and microsecond synchronization

- Legacy Ethernet devices can synchronize to network time base without knowing about *TTEthernet*

- Support for legacy and best-effort traffic

- Standard TCP/IP protocols and applications can be used

- Flexibility for customer-specific extensions (ALTERA Cyclone III FPGA)

- Digital I/O for triggering measurements

- Dimensions: 170x121x55 mm; Weight: 800 g; Operating temperature: 0C to +70C; storage: -40C to +85C

- Robust housing

The 1 Gbit/s version of *TTEthernet* switch has the following characteristics:

- Four 1 Gbit/s copper/fiber ports

- Message schedules and routing information stored in internal ROM (loaded by TTE-Load download tool)

- Hardware-based on Altera COTS board and PHY daughter board

- 8 Gbit/s full-duplex bandwidth

- Multi-hop capable

- Single synchronization domain

A 12-ports *TTEthernet* switch is under development as well. This version provides 24 Gbit/s full-duplex cross-sectional bandwidth. The *TTEthernet* switch variants are depicted in Figure 8.19.

ᵀᵀᴱDevelopment Switch   ᵀᵀᴱDevelopment Switch   ᵀᵀᴱDevelopment Switch
1 Gbit/s 12 Ports   1 Gbit/s   100 Mbit/s

FIGURE 8.19
TTEthernet Switches

ᵀᵀᴱPCI Card   ᵀᵀᴱPCIe Card   ᵀᵀᴱPMC Card   ᵀᵀᴱXMC Card

FIGURE 8.20
TTEthernet FPGA-Based End Systems in Different Form Factors

TTEthernet end systems have been realized in FPGA and as a software stack. The FPGA solutions come in different form factors and are depicted in Figure 8.20.

The software stack realizing the *TTEthernet* services is depicted in Figure 8.21. This stack is designed for a 100 Mbit/s software-based *TTEthernet* end system running a single *TTEthernet* channel. COTS Ethernet Controllers can be used. In an example setup on an Intel ATOM running at 1.6 GHz with 1 GB of RAM and 0.5 MB cache, the stack was implemented on Standard Linux.

Using a cluster cycle of three milliseconds and communicating ten time-triggered messages (1500 bytes each), a total network load of 40 Mbit/s has been generated. In this configuration, the software stack used about three percent of the ATOM CPU.

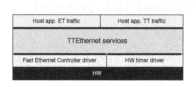
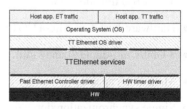

System without OS support   System with OS support

FIGURE 8.21
TTEthernet Software Stack with and without Operating Systems Support

9

TTCAN

R. Kammerer

Vienna University of Technology

CONTENTS

9.1   Protocol Overview

Controller Area Network (CAN) [149] provides an inexpensive and robust network technology which is widely deployed in many application domains such as automotive, avionic and industrial control systems. For example, present day cars contain multiple CAN buses deployed for different domains such as comfort or powertrain subsystems [93, 119]. Properties of CAN that have lead to its success include its simplicity, high flexibility, efficiency and low cost. Adversely, CAN exhibits limitations with respect to reliability, diagnosis and scalability.

In classic CAN networks, the communication has two main characteristics: The communication is event-triggered, and the shared medium is a bus. Both of these properties lead to limitations of the conventional CAN protocol in the context of dependable embedded systems. An example of a hazard to reliability is the missing fault isolation for babbling idiot failures [327]. A faulty CAN node can disrupt the communication abilities of all other nodes by continuously transmitting high-priority messages. Besides these limitations which have their origin in the bus-based topology, conventional CAN does not provide time-triggered concepts like dedicated communication slots for real-time communication, because conventional CAN does not make use of a priori knowledge about communication activities.

Due to the event-triggered nature of CAN, peak loads can occur if nodes in the network try to send messages at the same point in time. The arbitration mechanism of CAN performs sequential transmission of all messages according to their *identifier priority*. This arbitration makes a scheduling analysis of the whole system difficult because worst-case peak loads have to be considered for the analysis. Additionally, the temporal behavior of sending nodes is often unknown or not defined.

Time-Triggered CAN (TTCAN) [148] uses the concept of cyclic communication, divided into slots to implement time-triggered behavior. The standard requires that all activity assigned to one slot (including interrupt handling) is finished until the next slot starts, whereas one message can be assigned to several slots.

It is important to note that TTCAN adds an *additional layer* before the existing standard CAN layers. The physical layer and data link layer of CAN (specified in [150]) are kept unchanged. Mapped to the ISO-OSI model, TTCAN resides on layer 5, the session layer. Within this layer, TTCAN is divided into two different modes of operation. While level 1 and level 2 both support time-triggered behavior, the capabilities of these levels differ in key properties. The most outstanding difference is the notion of a *global time*, that only exists in level 2, and allows a much more fine grained synchronization of the nodes in a TTCAN network.

The following sections discuss the key properties of the TTCAN protocol, where both levels of TTCAN are taken into consideration.

9.2   Protocol Services

9.2.1   Communication Services

According to the standard [148], TTCAN specifies a serial communication protocol that supports distributed real-time control and multiplexing for use within road vehicles. The underlying CAN-protocol is unchanged, therefore TTCAN uses the same arbitration and medium access control mechanisms (MAC) as conventional CAN.

TTCAN level 1 provides cyclic message transfer, while level 2 adds the notion of a global time. The key element for the periodic and cyclic communication is the *reference message*, which is sent by a *time master*. TTCAN supports up to eight alternative time masters, where only one is allowed to be the current time master. The reference message is sent by the current time master at the beginning of each *basic cycle*. Figure 9.1 shows the execution of basic cycles over time.

Every basic cycle is divided into windows of communication activity. The sum of all basic cycles forms a *system matrix*. Basic cycles are the rows of the system matrix. Additionally, note that the number of basic cycles in the system matrix shall be a

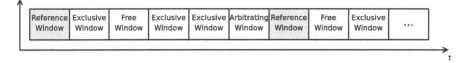

FIGURE 9.1
Basic Cycle in TTCAN [148]

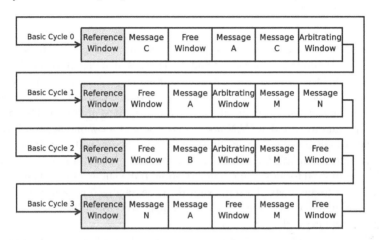

FIGURE 9.2
Communication Matrix in TTCAN [148]

power of two. The variable *Cycle_Count* refers to the current basic cycle in the system matrix. *Cycle_Count* starts from zero and has a maximum value of *Cycle_Count_Max*.

Figure 9.2 gives an overview about the communication matrix used in TTCAN. A message (e.g., message C in basic cycle 0) can be assigned to one or more time windows. The columns in the system matrix form so-called *communication columns* which are important for the periodicity of a message. The two important factors for the periodicity of a message are the number of basic cycles and the number of communication columns.

As shown in Figure 9.2, TTCAN supports different types of communication windows:

- Reference window

- Exclusive window

- Arbitrating window

- Free time window

Reference windows are used to flag the start of a basic cycle. The format of reference messages and their importance for TTCAN are discussed in Section 9.2.2.

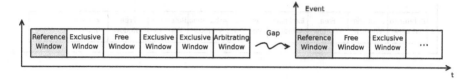

FIGURE 9.3
Event-Synchronized Basic Cycles in TTCAN

Exclusive windows are assigned to specific, periodic messages where no other communication is allowed in the same window. There is no competition on the CAN bus within an exclusive window. *Arbitrating windows* are slots where all CAN nodes are allowed to communicate. Conflicts are resolved by the conventional CAN identifier arbitration mechanism (e.g., the message with the highest priority will be transfered). It is important to note that CAN nodes may not start a communication if the bus is not idle. To guarantee that nodes finish their transmission within their assigned window, there is a *time window* (Tx_Enable) at the start of the arbitrating window where nodes are allowed to start their communication. If the attempt to send fails, nodes are not allowed to automatically retransmit their message and have to wait for another arbitrating window. *Free time windows* are reserved slots which will be eventually used in future network extensions.

TTCAN supports two different methods of basic cycle synchronization, namely time-triggered and event-synchronized. In time-triggered synchronization, the time master sends the reference messages in equidistant time slots. Additionally, TTCAN supports an event-synchronized transmission of reference messages. In this case, the start of a basic cycle is synchronized to a specific event in the current time master. The gap is only allowed between the end of one basic cycle and the start of the next basic cycle (i.e., its reference message). If there is a gap between basic cycle n and basic cycle $n + 1$, the time master has to announce the gap in the reference message of basic cycle n. For this purpose, the reference message, which will be described in detail in Section 9.2.2, contains a single bit flag *Next_is_Gap*. Event-synchronized transmission is shown in Figure 9.3.

9.2.2 Clock Synchronization

The key element for synchronization in TTCAN is the *reference message* that is sent at the beginning of every basic cycle. The following section discusses the structure and the relevance of the reference message and the achieved synchronization quality for both levels of TTCAN.

The TTCAN standard [148] states that a reference message shall be a data frame characterized by a specific CAN identifier which shall be received and accepted by every FSE[1] (frame synchronization entity) except the time master, which is the

---

[1]Every CAN controller in a time-triggered CAN network has its own FSE

sender of the reference message. For reasons of simplification, FSEs are simply called nodes in the following. Level 1 and level 2 differ in the length of their corresponding reference messages. For level 1, the data length shall be at least one byte; for level 2 it shall have a length of at least four bytes. All bits of the identifier field, except three, are used to characterize the reference message itself. The three least significant bits are used to define the priority of up to eight (2^3) time masters. The reference messages of both levels have in common, that they shall include the number of the current basic cycle (Cycle_Count) and one bit that signals that there will be a gap between the current and the next reference message (Next_is_Gap), which is used for event-synchronized communication. In general, the reference message is transmitted periodically, with the exception of event-synchronized communication, or if the transmission of a reference message is disturbed. In the latter case, the time master is allowed to retransmit the reference message. In both levels, the reference message may be extended to eight CAN data bytes. Reserved bits shall be transmitted as logical zero and shall be ignored by the receivers. Additionally, both levels have in common that the most significant bit (bit 7) is transmitted first.

The format of reference messages for level 1 and level 2 are shown in Figure 9.4 and Figure 9.5, respectively. In both figures grayed boxes represent optional bits in the reference message, which shall be transmitted as logical zero.

7	6	5	4	3	2	1	0
Next is_Gap	Reserved	Cycle Count (5)	Cycle Count (4)	Cycle Count (3)	Cycle Count (2)	Cycle Count (1)	Cycle Count (0)

FIGURE 9.4
Reference Message in TTCAN Level 1

All timing in TTCAN, for level 1 and level 2, is controlled by the local clock, where the resolution for time is the network time unit (NTU). For the simpler level 1, the NTU is the nominal CAN bit time. The local time is implemented as a simple incrementing counter that contains 16 bits. In level 1, this counter is increased by one every NTU.

For level 2 communication, the reference message contains two additional fields, namely Master_Ref_Mark (MRM), which is a timestamp measured in global time, and Disc_Bit, which flags that there is a discontinuity in the global time. If the transmission of a reference message is disturbed, the time master shall retransmit the message and the Master_Ref_Mark is updated. The structure of the reference message for TTCAN level 2 is shown in Figure 9.5.

The first byte of a reference message in level 2 contains the same bits as the reference message for a level 1 message. The second byte contains the Disc_Bit, which flags a discontinuity in the global time, and the NTU_Res, which is the resolution of the network time unit (NTU). TTCAN supports four additional bits for a more fine grained specification of the NTU_Res. If a node does not support these additional bits, they shall be transmitted as logical zero. The third and fourth bytes contain the low and high byte of the Master_Ref_Mark.

	7	6	5	4	3	2	1	0
Byte 1:	Next is_Gap	Reserved	Cycle Count (5)	Cycle Count (4)	Cycle Count (3)	Cycle Count (2)	Cycle Count (1)	Cycle Count (0)

	7	6	5	4	3	2	1	0
Byte 2:	NTU Res (6)	NTU Res (5)	NTU Res (4)	NTU Res (3)	NTU Res (2)	NTU Res (1)	NTU Res (0)	Disc Bit

	7	6	5	4	3	2	1	0
Byte 3:	MRM (7)	MRM(6)	MRM (5)	MRM (4)	MRM (3)	MRM (2)	MRM (1)	MRM (0)

	7	6	5	4	3	2	1	0
Byte 4:	MRM (15)	MRM(14)	MRM (13)	MRM (12)	MRM (11)	MRM (10)	MRM (9)	MRM (8)

FIGURE 9.5
Reference Message in TTCAN Level 2

In TTCAN, there are three different notions of time [126], namely local time, cycle time and global time. Each of these times and their importance for TTCAN will be discussed in the following.

Local Time

As in level 1, the time in level 2 is measured in network time units (NTUs). In contrast to level 1, the NTU in level 2 is a fraction of the physical second. This is required to correct the slight differences in the local clock oscillators of the nodes and to synchronize them to the global time. For level 2, the counter for the local time shall contain at least 19 bits, where the 16 most significant bits represent whole NTU ticks. The rest of the counter bits are reserved for the fractional parts of the NTU. Therefore, the counter counts in units of $\frac{NTU}{2^n}$ if NTU_Res contains n bits. If the counter is incremented 2^n times, this is equivalent to 1 NTU. For example if 3 bits are reserved for the fractional part of the NTU, the counter has to increment 2^3 times to increment one NTU. Every node has a basic time unit (e.g., the frequency of its local clock oscillator). To set this basic time unit in relation to the NTU, there exists the *time unit ratio* (TUR). This usually non-integer value specifies the ratio between the length of a NTU and the length of a basic time unit. The current value of TUR (TUR_Actual) is responsible for the velocity of the NTU counter. The TTCAN standard does not specify the implementation or the data representation of the TUR. The TUR has an essential role for clock synchronization in level 2. Whenever there is a difference between the local time and the global time transmitted by the time master, the TUR gets adjusted. The generation of the *Local_Time* is shown in Figure 9.6.

Cycle Time

The time within one basic cycle is measured in the *Cycle_Time* parameter. At each start of a data frame (SOF) or remote frame the node saves its current value of the local time to Sync_Mark. Whenever the node receives a reference message, it saves the current value of Sync_Mark to Ref_Mark. Therefore, the value of Ref_Mark contains the timestamp of that point in time of the start of the last reference message. The

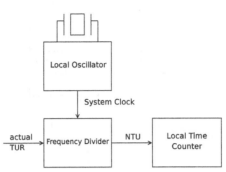

FIGURE 9.6
Local Time Generation in TTCAN [126]

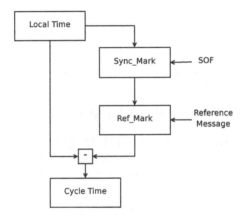

FIGURE 9.7
Cycle Time Generation in TTCAN [126]

Cycle_Time is the difference between the local time of the node and its Ref_Mark. Cycle_Time has no fractional part; therefore, only the 16 most significant bits contribute to the Cycle_Time. The generation of the Cycle_Time is shown in Figure 9.7.

Global Time

In TTCAN level 1, the common time base is the Cycle_Time which is restarted at each basic cycle. In this sense, level 1 has a global time with the horizon of one basic cycle. Level 2 adds a more fine grained concept of global time which is used to calibrate the local time base of each node in the TTCAN network. To compensate a clock drift in the nodes in level 2, the TUR value is adjusted to the current time master's view of the global time at every reception of a reference message. At a pulse of frame synchronization, a node stores its local view of the global time as its Global_Sync_Mark, which has at least 19 bits. The current time master transmits its Global_Sync_Mark (i.e., its view of the global time) as the bits reserved for

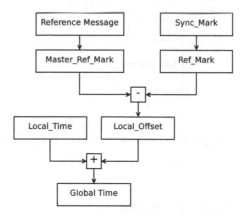

FIGURE 9.8
Global Time Generation in TTCAN [126]

Global_Ref_Mark in its reference message. Whenever a node receives a reference message, it calculates a local offset:

$$Local\_Offset = Master\_Ref\_Mark - Ref\_Mark$$

The generation of the global time in TTCAN is shown in Figure 9.8. The current time master's view of Local_Offset remains constant. Every node starts with a Local_Offset of zero. Changes in the value of the Local_Offset show that there is drift between the local view of time and the global view which has to be corrected. This clock rate correction is done by adjusting the NTU value according to the following formula [126]:

$$df = \frac{Ref\_Mark - Ref\_Mark_{previous}}{Master\_Ref\_Mark - Master\_Ref\_Mark_{previous}}$$
$$TUR = df * TUR_{previous}$$

TTCAN considers the 16 most significant bits of local time and local offset of the current time master as its *global time*. The local time and local offset of a node are an approximation of the global time.

The time master in TTCAN is allowed to use *external clock synchronization*. The TTCAN standard describes the following three means:

- *Frequency adjustment:* In this method, the external time period is used as the base for the NTU. To adapt the NTU, the length of the external time period is used as the TUR value. First, the time master writes the new value to TUR_Adjust. At the beginning of the next basic cycle, TUR_Actual gets overwritten by TUR_Adjust.

- *Phase adjustment:* Phase adjustment can be done by continuous frequency adjustment or by inserting a discontinuity in the global time. In the latter case, the time master has to set the Disc_Bit to inform the nodes that a discontinuity has to be expected. At the beginning of the next basic cycle, the time master adds the difference between the desired global time and the actual current time to its local offset to influence the global time.

- *Adjustment of Cycle_Time:* If this synchronization scheme of external clock synchronization is applied, the time master sets the Next_is_Gap flag in its reference message to inform the nodes that the next basic frame will be event synchronized. After that, the time master waits as long as required by the external clock before it starts sending the next reference message.

9.2.3 Sending and Receiving Messages in TTCAN

As described before, the communication in TTCAN is organized with the help of the system matrix. Within the system matrix, the rows form the basic cycles, where each of these basic cycles starts with a reference message. When it comes to sending and receiving of messages, two parameters are of utmost importance: Cycle_Time and Tx_Trigger. A basic cycle is divided into columns that form the time slots when a message shall be sent. Time within a basic cycle is measured in Cycle_Time. *Tx_Trigger* specifies that instant in time when a message corresponding to a time slot shall be sent. A Tx_Trigger contains the following information:

- A reference to a message that shall be sent.

- Activation time mark, which is the column of the system matrix.

- Row of the system matrix measured in Cycle_Count.

- Repeat factor: Position of the same column in which it will be sent next. The value is a power of two.

A vital parameter for TTCAN and the sending of messages is the *Tx_Enable* parameter. A message has to be sent within a specific time window after the start of its corresponding time slot. If the transmission would start too late, the massage would overlap with the start of the next slot. To avoid this erroneous behavior, there exists the Tx_Enable parameter. It specifies a window from the start of the corresponding time slot (Tx_Trigger) within the transmission of a message is allowed and safe. This window is specified as a number of nominal CAN bit times with a range from 1 to 16. If a node cannot start the transmission of a message within the Tx_Enable window, the transmission may not be started at all. In this case, the node has to wait for the next Tx_Enable to send the message.

The concept of the Rx_Trigger is related to the Tx_Trigger. It contains the same information as a Tx_Trigger, but specifies the point in time when a received message has to be completed and verified.

There are two additional triggers that are important in a TTCAN system: Tx_Ref_Trigger and Watch_Trigger. Tx_Ref_Trigger is a special Tx_Trigger that refers to reference messages and is only used in potential time masters. Whenever this trigger is reached, a potential time master tries to send a reference message. Within a strictly time-triggered TTCAN system (no Next_is_Gap bits), one Tx_Ref_Trigger is enough. If event-synchronized transmission of basic cycles is planned, a second Tx_Ref_Trigger is used. The first trigger is used for the periodic transmission of reference messages, where the second restarts the first one whenever the event that was flagged with the Next_is_Gap bit did not occur or was missed.

Watch_Trigger is also used by nodes which are not potential time masters. These nodes will be called slave nodes in the following. Watch_Trigger is reached if there is no reference message on the bus longer than a specified threshold. This could be the case on a disturbed bus. Like the Tx_Ref_Trigger, there shall be two Watch_Triggers to support event-synchronized communication. When a Watch_Trigger is reached, the application shall be informed about the erroneous behavior in the system. During startup, the Watch_Trigger is disabled until the first successful transmission or reception of a reference message.

9.2.4 Restart, Re-Integration, Integration

TTCAN differentiates between the startup behavior of a time master and its fault tolerance capabilities. This section describes the startup, whereas Section 9.2.6 gives an overview about fault tolerance in TTCAN.

At system startup (or a reset), all potential time masters try to become the current time master and all of them try to send a reference message. There are two parameters that influence the decision which potential time master becomes the current time master. The potential time master with the highest priority shall use the CAN ID with the highest priority (in relation to the other potential time master). As specified by the TTCAN protocol, CAN IDs of time masters only differ in the three least significant bits. The rest of the bits reserved for the CAN ID are equal for all potential time masters. Additionally, there is a parameter which specifies how long a potential time master is idle before it tries to send a reference message. The potential time master with the highest priority has the shortest delay before it tries to send its initial reference message.

Both of these parameters, the priority of a potential time master and the time a potential time master waits for its initial transmission, influence the decision which potential time master will become the current time master and how long it takes to establish a new time master.

Figure 9.9 shows a state machine of the startup protocol and the establishment of a new time master. State 1 is the initial state, which is reached for example after a hardware reset. Transition 0 is always taken. State 2 is reached by slaves as well as potential time masters. If the node is a slave and it receives a reference message on the bus, transition 1 is taken and the node is in state 4, the slave state. If the node is in state 2 and it is a potential time master, two transitions may be taken. First of all, a potential time master will try to send its own reference message. The best case for the potential time master is that it reads its own reference message on the bus. In this case, the potential time master takes transition 3 and becomes the new

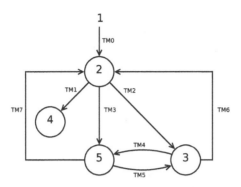

FIGURE 9.9
Master/Slave Relation in TTCAN [148]

time master (state 5). On the other hand, a potential time master in state 2 can read a reference message that contains an ID that is not equal to its own. In this case, the potential time master takes transition 2 and becomes a *backup time master* (state 3). From this state, a potential time master can become the current time master with transition 4 whenever it reads its own ID in the reference message. This can happen if the prior time master fails and a backup time master can now successfully transmit its own reference message. However, a current time master can also be degraded to a backup time master. This happens if it is the current time master (state 5) and reads a reference message with a higher priority on the bus. In case of an error, the time master as well as backup time masters can be reset to state 2.

After the initialization, but before the synchronization has finished, each node refers to its local time as the global time of the system. The parameter Local_Offset is initially set to zero. The actual time master tries to establish its view of the global time as the global time in the TTCAN system. Therefore, it sends its own Global_Sync_Mark as the Master_Ref_Mark in its reference message. An important fact is that a potential time master that becomes the actual time master (e.g., when the old time master fails) keeps its value for the Local_Offset. This is used to avoid discontinuities of the global time.

The synchronization of a node can also be described with the help of a state machine (refer to Figure 9.10). A node is in state 1 after a hardware reset. From this state, transition 0 always occurs. In state 2, the node is not yet synchronized (Sync_Off). After leaving an internal configuration mode and setting Cycle_Time to zero, the node is in state 3. From state 3, the node can either take a transition 2 to state 4, where it is In_Gap, or transition 3 to be in schedule (In_Schedule). For both transitions TS2 and TS3, the node has to receive two successive reference messages where the last of the two did not contain a Disc_Bit. For TS2, the last reference message has to contain a set Next_is_Gap bit, whereas for TS3 Next_is_Gap has to be unset. A node can leave the In_Schedule state with transition TS4 whenever a reference message contains a set Next_is_Gap bit. When a node is in the In_Gap state

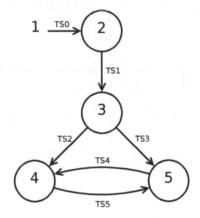

FIGURE 9.10
Synchronization in TTCAN [148]

(state 4), it can reach a synchronized state (state 5) whenever it receives a reference message where the Next_Is_Gap bit is unset.

9.2.5 Diagnostic Services

TTCAN provides simple means for diagnosing the communication on the CAN bus. A *Watch_Trigger* is used to monitor the reception of reference messages. TTCAN uses two different Watch_Triggers, one for the reception of periodic messages and one for event-synchronized messages. If there is no reference message for a specified time, the Watch_Trigger is activated and an error handling procedure is started. Additionally, the application shall be notified. During initialization, the Watch_Trigger is disabled and gets activated upon a successful reception of a reference message.

As stated before, TTCAN provides an additional layer above the conventional CAN protocol. All means of CAN bus error handling support the error handling of TTCAN.

Error Counters in CAN

CAN provides a receive error counter as well as a transmit error counter which can be used also in TTCAN. The values of these counters are vital for the CAN node because they define the error state of the node. CAN defines the following three error states for a node:

- Error active: The node is allowed to send active error frames

- Error passive: The node sends passive error frames

- Bus off: The node does not take part in bus communication

According to the state of the error counter of the CAN node, it can send active or

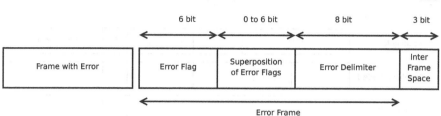

FIGURE 9.11
Error Frame in CAN [356]

passive error frames. When a node is highly disturbed, and therefore has a high error count, it is only allowed to send passive error frames. Within these frames, the node is only allowed to send recessive error flags. Therefore, a node in passive error state cannot disrupt the communication of the bus, because its recessive error flags cannot disturb the communication of working nodes.

When the corresponding error counter has a value between 0 and 96, the node is in *error active state*. If the error counter exceeds 96, it reaches a *warning* limit, where the node sets an error flag and generates an interrupt. Between 97 and 127 the bus is heavily disturbed and the node is still in error active state. Between 128 and 255 the node is in *error passive* state and above 255 the node reaches *bus off* state.

Error Frames in CAN

Whenever a node detects an error on the bus, it sends an error frame on the bus. These error frames contain a six bit error flag. Additional 0 to 6 bits are reserved for the superposition of further error flags. After the error flags, an error frame contains eight bits used for the error delimiter. The delimiter contains eight recessive bits to allow a restart of the communication after the superposition of error flags. After the delimiter, three bits are reserved for an inter frame space. The fields of an error frame are shown in Figure 9.11. Superposition of error flags occurs in the following situation: A receiver detects an error on the bus and starts to send an error frame. Within this error frame, the other receivers will detect a violation in the bit stuffing of CAN. Therefore, the other receivers will send error frames themselves and their first six bits will lead to a superposition of error flags in the first error frame. Figure 9.12 shows the superposition of error flags. In this case, receiver 1 detected the error on the bus first and started to send its error frame. After the first six bits, receiver 2 also detected an error and started to send its error frame with its error flags. This shows one important mechanism of CAN, the *globalization of local errors*. Whenever a node detects an error on the bus, it will try to establish a global error state on the bus.

Bit stuffing in CAN

Bit stuffing is used in CAN whenever five consecutive bits of the same polarity occur in the CAN bit stream. In this case, one bit of the opposite polarity will be inserted. In CAN there are fields with a fixed bit pattern which are not part of the bit stuffing.

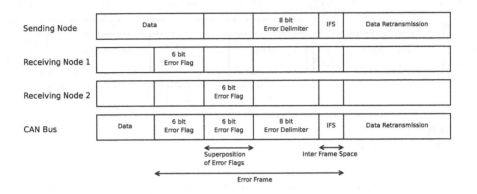

FIGURE 9.12
Error Handling in CAN [356]

In CAN the fields for start of frame (SOF), the arbitration field, the control field, the data field and the CRC field are part of the bit stuffing area. The bit stuffing property of CAN is used to flag errors on the bus, because whenever six consecutive bits of the same polarity are read on the bus, all nodes will interpret this as an error.

9.2.6   Error Detection and Fault Isolation

Fault isolation and tolerance is of utmost importance whenever a network is used in a safety relevant context. For traditional CAN, it is possible to retransmit a message if an error occurs during the transmission of the message. This simple mechanism cannot be used for TTCAN because it has to follow the schedule for its messages. Therefore, in TTCAN the automatic retransmission of messages is turned off.

The TTCAN standard itself provides error detection mechanisms and a classification of error severity. It also states that active fault confinement shall be left to a higher layer or to the application. The node shall provide error detection and failsilence [148].

TTCAN defines the following errors and their severity. Note that the standard does not specify where these bits should be saved (i.e., this could be in registers of a node).

- Scheduling_Error_1 (S1): An error of this kind is flagged whenever the difference between the highest message status count (MSC) and the lowest MSC of all messages is larger than 2. Additionally, a scheduling error is flagged if the MSC of an exclusive receive message object is 7. The MSC is used for periodic messages in exclusive time windows. The MSC has a range from 0 to 7. The MSC is updated at the Rx_Trigger event that corresponds to that specific message. If the message is successfully received, the MSC is decremented by one. Otherwise, it is incremented by one.

- Tx_Underflow (S1): This bit is set whenever the count for transmitted messages is lower than its expected value.

- Scheduling_Error_2 (S2): Whenever a MSC of a transmit message object reaches 7, this error bit will be set.

- Tx_Overflow (S2): This flag will be set whenever the expected maximum value (Expected_Tx_Trigger) is reached and there are still Tx_Trigger events.

- Application_Watchdog (S3): The application watchdog gets reset by the application itself via a Host_Alive_Sign. If the application fails to set this sign, an error will be flagged.

- CAN_Bus_Off (S3): This error is flagged whenever the node went off the bus due to CAN-specific errors.

- Config_Error (S3): Slots in TTCAN have a specific time window where the transmission of the data has to start. If the actual point in time exceeds this specified limit, a Config_Error will be flagged.

- Watch_Trigger_Reached (S3): This error flag will be set whenever a reference message is missing.

The TTCAN standard specifies the following severity classes:

- S0: No error.

- S1: Warning - the application will be informed, the reaction is application specific.

- S2: Error - the application will be informed, and all transmissions in exclusive and arbitrating windows will be disabled. Potential time masters are still allowed to transmit reference messages with the maximal allowed offset.

- S3: Severe - the application will be informed, and all bus activities will be stopped. In this case, an update of the configuration is required.

In addition to the fault handling specified by the TTCAN standard itself, there are several approaches for how to add fault tolerance to TTCAN. One simple approach is described in [44], where each frame is sent twice.

A different approach is to add a *mailbox system* to TTCAN [213]. This is done by reserving an exclusive window, the mailbox, at the end of each basic cycle to retransmit messages that could not be sent in their dedicated slot, as shown in Figure 9.13. The basic idea is to add a software supervisor layer above the TTCAN controller that manages additional send and receive queues. Whenever a message cannot be transmitted in its dedicated slot, this message is scheduled to be transmitted in the mailbox slot at the end of the basic cycle. This solution assumes that bit error rates

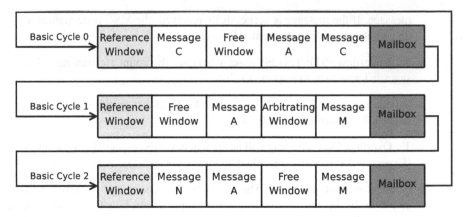

FIGURE 9.13
System Matrix with Mailbox

occur in the magnitude of 10^{-7}. Therefore, a basic cycle would have to be $2 * 10^7$ messages long to corrupt two messages within one basic cycle. This would require a basic cycle of about 2.2 hours. The authors conclude that one mailbox is enough to provide sufficient fault tolerance. Whenever an error occurs, the supervisor sends an error frame on the bus. After that, it enqueues the failed message to the send queue and transmits the message in the mailbox slot. At the start of the system matrix (and at every iteration), the send queue gets flushed to prevent queue overflows.

TTCAN provides fault-tolerance by combining two (or more) TTCAN buses [231]. A system of two TTCAN buses is considered a *coupled TTCAN pair* if there is at least one "gateway" node connected to both TTCAN buses. A system of TTCAN buses is considered as coupled if there exists a path from the start bus via coupled nodes to the end bus. More formally [231]: bus_a and bus_b are TTCAN coupled if there exists a sequence $(bus_1, \ldots, bus_n)$ of TTCAN buses with $bus_1 = bus_a$ and $bus_n = bus_b$ where (bus_i, bus_{i+1}) are coupled TTCAN pairs $\forall i = 1, \ldots, n-1$. Now a fault tolerant TTCAN network is a system of TTCAN buses where every two of them are TTCAN coupled. Figure 9.14 shows a coupled TTCAN system where bus_a and bus_c are coupled.

The main problem of providing a coupled, and therefore fault-tolerant, TTCAN system is the synchronization between the different buses. Reference [231] states three different problems of synchronization, which are related to the different times used in TTCAN. With the synchronization mechanisms described in the paper, it is possible to synchronize a system of two or more TTCAN buses and therefore provide fault-tolerance in the system: Phase synchronization of cycle time, phase synchronization of global time and rate synchronization.

Phase Synchronization of Cycle Time

To recapitulate, the cycle time is used to measure the time between two successive

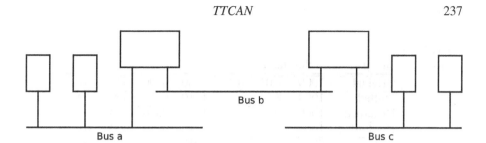

FIGURE 9.14
Coupled TTCAN Buses [231]

reference messages. At every start of a basic cycle, a reference mark is saved and *cycle time = local time − local reference mark*. All points in time in TTCAN are specified according to the cycle time. The simplest solution to phase synchronize two (or more) buses is to use one bus as a "master bus" and synchronize the second bus to the first one. The situation can further be simplified if the "gateway" node — the node connected to both buses — is a time master for the second bus. To initiate the desired phase relation, the time master has to set its Next_Is_Gap bit and stall the next reference message until the desired offset. This is the simplest method to solve this issue. If the gateway node is not a time master on the second bus, a simple meta protocol can be used. In this case, the gateway node transmits the desired phase shift in a TTCAN message to its time master and the time master will take care of the rest. This method of "master buses" can be extended to n TTCAN buses, where the first bus is the master bus and every bus i for $i = 2, \ldots, n$ synchronizes to bus $i − 1$.

Phase Synchronization of Global Time

To synchronize the phase of the global time of two time masters on distinct buses, the same methods as for the phase synchronization of the cycle time can be used. The gateway node calculates the desired phase for the slave global time and informs the master node. As well as for the synchronization of the cycle time, this can be simplified if the gateway node is the time master of the slave bus. The new global time can then be announced to the slave bus. This is done by setting the Disc_Bit to flag a discontinuity in the global time.

Rate Synchronization

To synchronize the rates of two TTCAN buses, first the rates of the two (or more) buses have to be measured. One option is to measure the same physical time interval on both buses. From this measurement, the ratio for the slave bus can be calculated. This value will then be sent to the local time master. This time master will then update its TUR value and within the next basic cycle the TTCAN bus will be rate synchronized.

For details about the different synchronization methods, please refer to [231].

In addition to the presented TTCAN specific methods of fault isolation, classi-

TABLE 9.1
CRC—Hamming distance shortfall [356].

Original	1000	0001	1011	1100	0100	0011	01
Stuffed	1000	0010	1101	1110	0010	0001	101
Disturbed	1000	0110	1101	1110	0010	0000	101
De-Stuffed	1000	0110	1101	1110	0010	0000	01
CRC	0000	0111	0110	0010	0110	0011	00

cal CAN error detection mechanisms of CRC-checksums can be used in TTCAN networks.

CRC in CAN

CAN uses a 15 bit cyclic redundancy check for its messages. The generator polynomial used for CAN is $x^{15} + x^{14} + x^{10} + x^8 + x^7 + x^4 + x^3 + 1$. The polynomial has a guaranteed minimal hamming distance of 6, which means that up to five random bit failures can be detected. Due to bit stuffing, there can be the situation that only two disturbed bits lead to to a situation that cannot be detected by the CRC check. For an example, refer to Table 9.1.

9.2.7 Configuration Services

The TTCAN standard specifies three main configuration interfaces, where each of these interfaces is divided into sub-interfaces that allow the configuration of protocol settings. The TTCAN standard demands that all interfaces shall be lockable against random changes, whereas reading settings is always possible. The following paragraphs provide an overview of the configuration interfaces and their most important variables. For a complete list, refer to the corresponding section of the TTCAN standard [148]. Note that the standard does not specify the location of these interfaces or how to access these interfaces. One practical solution is to store the values into registers and provide an API to access these interfaces.

General Configuration Interfaces

As the name states, this interface is used to set general protocol parameters for a TTCAN node. Settings include the configuration of the TUR, the operation mode of the node (configuration, CAN communication, time-triggered communication and event-synchronized time-triggered communication), its role as a slave or a potential master, if external clock synchronization shall be used, which interrupt sources shall be masked and an 8-bit value for the application watchdog limit.

The main role of the interface is the configuration of the system matrix. With this interface, the system engineer can specify message objects and the corresponding triggers. The triggers contain a reference to a valid message, a time mark when they should be activated, the position in the transmission column with its first activation and their repeat factor. The standard states that the configuration can be read and written during the initialization phase, and shall be locked during the time-triggered communication activity.

Application Interfaces

The application interfaces provide information about the current state of the node to the application.

Accessible information includes the priority (3 bit) of the current time master, the master state, the current value of the global time in NTUs, the Cycle_Time, the number of the current basic cycle and the actual value of the TUR (TUR_Actual), which is read-only. Additionally, the interface is used to configure operational and error detection interrupt sources. Operational interrupt sources are among others the start of a basic cycle, the start of the system matrix, an interrupt that occurs if the global time wraps or a change in the local master state. Error detection interrupt sources are for example over- and underflows of the Tx-Counter, the application watchdog or a CAN_Bus_Off interrupt.

Optional Interfaces

Optional interfaces are not required to comply to the TTCAN standard. Examples are interfaces that signal that the node is out of synchronization, or that the Disc_Bit should not be used and to prevent a discontinuity in the global time. Additionally, there could be an interface to specify the maximum synchronization deviation.

9.3 Protocol Parameterization

TTCAN is a highly flexible protocol, which allows the system engineer to configure a TTCAN system according to the requirements.

TTCAN Level

As described in previous sections, TTCAN offers two levels which differ mainly in the synchronization quality. Where in level 1, time is measured solely with the Cycle_Time and no drift compensation is provided, TTCAN level 2 provides the notion of a global time. The first and most basic decision is the level of TTCAN which should be used. All time masters in the network have to use the same level. Time slaves in a level 2 network are free to use level 1, whereas level 2 communication is not allowed in a level 1 network. TTCAN controllers provide a flag to specify the level of TTCAN (e.g., Operation Mode Register).

System Matrix/Tx_Trigger

When it comes to TTCAN parameterization, the system matrix and the Tx_Trigger

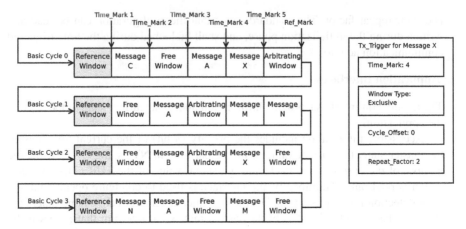

FIGURE 9.15
Tx_Trigger for Exclusive Message X [148]

parameter are of utmost importance. The system matrix defines the communication schedule of the TTCAN system. Every row in the matrix starts with a reference message and the rows in the matrix specify a time slot in which a message will be sent. Messages themselves are specified in a message object. As specified by the TTCAN standard, a message object shall provide storage for one LLC frame (logic link control) together with control and status information. These message objects are linked to the system matrix with the help of Tx_Triggers (see Figure 9.15).

Application Watchdog

Every application in TTCAN can specify an application watchdog. If the application does not set the Host_Alive_Sign parameter, an interrupt will be raised. The watchdog can be parameterized with the Appl_Watchdog_Limit parameter which is an eight bit value specifying the period in NTUs.

Potential Time Master

The system engineer can parameterize the role of a node. It can be in the role of a time slave or it is a potential time master. In the second case, up to eight potential time masters can be specified. Potential time masters are configured with two parameters. First the priority of the potential time master, and second its Initial_Ref_Offset. This parameter specifies the time after which a potential time master tries to send a reference message.

Interrupts

For error indication, TTCAN specifies an interrupt status vector (Interrupt_Status_Vector), which reserves one bit for every error detection mechanism. The engineer can enable every interrupt source with Interrupt_Enable. These interrupts include operational as well as error detection interrupt sources. To name a few, these operational interrupt sources include the start of a basic cycle, the start of the system matrix, the discontinuity flag or a wrap of the global time. Error detection interrupt sources include Tx_Overflow, Tx_Underflow or CAN_Bus_Off. This list is not intended to be exhaustive. For further details, refer to the standard [148].

Bit-Rate

The engineer has the ability to configure the bit-rate of the time-triggered CAN system. Maximum bit-rates for TTCAN are the same as for conventional CAN (up to 1 Mbit/s).

9.4 Communication Interface

The TTCAN standard itself does not stipulate a specific communication interface. It is up to the provider of the IP module to implement an interface for its users. Like for classic CAN, the interface can range from a simple hardware register based interface to a full programming API. As there is a limited range of TTCAN IP core providers (see Section 9.6), the following description of the TTCAN communication interface is based on the quasi reference implementation of Bosch [125].

Bosch's implementation of the TTCAN IP core has two application modes. As TTCAN requires classic CAN (i.e., for MAC control), the default mode is classical CAN operation. It is up to the user to switch into TTCAN mode. Bosch provides a simple register based interface for configuration, status retrieval, as well as message transmission and reception. To write to TTCAN's configuration registers, two requirements have to be fulfilled. First the *Init* bit has to be set in the CAN control register. In addition to *Init*, the *CCE* (Configuration Change Enabled) flag has to be set. If both of these requirements are met, *TTMode* can be set to *configuration mode*. Entering as well as leaving the configuration mode requires the flags *init* and *CCE* to be set. If a node is in configuration mode, its configuration registers are readable and writable. As an example, consider the *TM* flag in the TT operation mode register which is used to specify if a node is a potential time master or not, or the *MPr* flag (3 bits wide), which specifies the priority of a potential time master. If a node is a potential time master, a seven bit value specified in NTUs can be written which defines how long a potential master, which is not the current time master, waits until it tries to send a reference message. Additional settings in the operation mode register are *L2* to define the TTCAN level and *EECS* which enables external clock synchronization.

For time-triggered CAN operation, further protocol specific registers have to be configured. This includes the *trigger memory* which contains information of up to 32 message triggers, and the configuration of *message objects* (e.g., transmit objects). These registers are vital for TTCAN operation. While message objects contain the content of a message, the triggers specify the points in time of message transmission.

Reading and writing contents is relatively simple, as the content can be accessed at the address of the corresponding message object.

After all protocol parameters are set, the last step is to set the *TTMode*. These two bits define if the node operates in classic CAN mode, is in configuration mode, operates strictly time triggered or is in event synchronized time-triggered mode. In the two latter cases, a node will start its synchronized TTCAN operation. It is important to note that switching *TTMode* to one of the operational modes has to be the last step because the configuration can only be changed if *TTMode* is set to configuration mode.

As stated before, the communication interface is not covered by the TTCAN standard itself and is therefore highly implementation dependent. For further details concerning Bosh's TTCAN IP core, please refer to the user manual [125].

9.5   Validation and Verification Efforts

This section describes a selected set of academic papers that have been published on TTCAN validation and verification.

Saha and Roy [289] describe a model based approach to check the *startup* of TTCAN. In their paper, the SAL [74] model checker is used. The paper validates several liveness properties of fault-free and fault-injected startup scenarios. In the case of a fault-free startup, the number of potential time masters varies from 2 to 10. The liveness properties varies from the case where the highest priority potential time master will eventually become the current time master to a case where all potential time masters which are not the current time master will eventually move to a synchronized state. Additionally, fault-tolerant startup is examined. These properties vary from the simple case that once the current time master becomes faulty, one among all the other potential time masters will eventually be the new current time master to cases where once the current time master becomes faulty, all the potential time masters which will not be the current time master will eventually go to a synchronized state.

In [32] an experimental setup with two boards manufactured by Phytec and endowed with Bosch TTCAN controllers were set up. At first the duration of basic cycles was measured. The TTCAN controller was configured to send messages at a bandwidth of 1 Mbit/s. Within the 128 samples that were taken, the basic cycle duration spans between 0.999275ms to 0.999687ms with a mean value of 0.999451ms and a mean error with respect to the nominal value of about $0.55\mu s$. These results show that the controllers worked within their specification of the TTCAN standard which requires that the difference in length of basic cycles has to be less than one NTU. Additionally, the paper presents experimental data on synchronization of TTCAN networks. In the experimental setup, two TTCAN networks were investigated. In this experimental setup, the synchronization of two buses requires more than one basic cycle, whereas the synchronization was in a good bound. There was no shifting between the basic cycles of the two buses.

Saha and Roy [288] present a model of TTCAN, where properties of the protocol were checked with the model checker Spin [134]. To represent time explicitly a special version of Spin, DT-Spin [41, 42] was used. This version of Spin uses a discrete time model. A basic cycle divided into five time windows was used: The first slot for the reference message, the second slot as an exclusive slot for node 1, the third slot free, the fourth slot as an arbitration slot and the fifth as an exclusive slot for node 2. Note that in this setup node 0 has no dedicated slot. With this setup, vital properties of TTCAN have been proved. It was shown that the master eventually writes to the bus in the first slot and that node 1 and node 2 eventually write in the second and fifth slot. For node 0, this property cannot be proved because node 0 did not have a dedicated slot. Additionally, data consistency was shown. All the nodes eventually read whatever the time master writes to the bus. An important property is that no node writes to the bus in the third slot. For additional verification results, refer to the paper [288].

Leen and Heffernan [195] present a formal model of TTCAN, which is used to verify properties of TTCAN with the use of the UPPAAL model checker [28, 3]. Under the assumptions that the medium does not introduce errors, that all messages exchanged are fixed in length and that all clocks are assumed to proceed at the same rate (constant NTU), a formal model of TTCAN was abstracted from the text based specification. The final model defines a system of 10 timed automata, representing two potential time master nodes, a time receiving node and a CAN physical layer in the context of level 1 TTCAN implementation [195]. Leen et al. have shown that their model will never inadvertently enter an undesired error state. Additionally, deadlock free operation has been proved as well as the expected behavior of the error state automata. For the later tests, nodes have been removed from the network.

9.6   Example Configurations and Implementations

Standalone TTCAN controllers are not that widespread in the market. One reason might be the limitation in speed (about 1 Mbit/s), which makes TTCAN not the first choice for safety-critical real-time applications in the automotive industry. Additionally, competing protocols such as FlexRay recently got more attention from manufacturers.

Bosch, the main driving force behind TTCAN, provides a TTCAN IP core written in VHDL on RTL level prepared for synthesis. According to the product description, the TTCAN IP module provides all features of time-triggered communication that are specified in the TTCAN standard, including event-synchronized time-triggered communication, global time and clock drift compensation.

In addition to the "reference implementation" from Bosch, stand-alone TTCAN controllers are rare. Infineon provides a *MultiCAN* chip which has at least support for TTCAN level 2. Atmel produces 8-bit micro controllers which have full standard CAN support and provide initial TTCAN support like independent message objects (in the case of the TC89C51CC01 15 message objects).

In academics the GAST (General Application Development Boards for Safety Critical Time-Triggered Systems) project[2] provides a controller board which features a TTCAN controller card. The setup was used in a master's thesis to evaluate the performance of TTCAN [99].

Note that TTCAN — especially level 1 — has low hardware requirements. Basically every CAN controller that supports a single-shot mode or is able to support the cancellation of a transmission request is able to support TTCAN level 1 only in software.[3]

---

[2]http://www.chl.chalmers.se/gast/
[3]http://www.can-cia.org

10

LIN

W. Elmenreich

Lakeside Labs/University of Klagenfurt

CONTENTS

10.1   Protocol Overview

The LIN protocol was developed by a consortium of seven automotive partners (Audi, BMW, DaimlerChrysler, Volvo, Volkswagen, Motorola and VCT) as a complementary system to the widely used CAN bus [15]. In 2003, many updates to reflect the latest off-the-shelf microcontrollers as well as inputs from the SAE Task Force resulted in the definition of LIN 2.0. In fact, LIN 2.0 was a complete rework of the existing LIN 1.3 standard, but backward compatible. The most recent version of the specification is LIN 2.1 [200].

10.2   Protocol Services

10.2.1   Communication Services

The LIN protocol is built upon UART frames as basic communication units. The used UART format consists of a start bit (always low), eight message bits in non-return-to-zero (NRZ) encoding, an optional parity bit and a stop bit (always high). LIN does not use a parity bit, thus encodes one data byte with 10 bits.

In order to support nodes with unstable clocks, a periodic synchronization pattern is provided. This allows the use of microcontrollers with internal RC oscillators for the implementation of the nodes. Such RC oscillators change their clock frequency with varying temperature and supply voltage so that they require frequent resynchronization.

The communication is initiated by a dedicated master node; the smart transducers are considered the slave nodes. The LIN master follows a time-triggered message scheme. This means that the messages are scheduled to be transmitted at a predefined point in time. This feature guarantees a collision-free media access scheme and a predictable message ordering.

A main difference of LIN to other time-triggered protocols is that the slave nodes are implemented in an event-triggered way. This eases protocol implementation; however, since LIN nodes are not aware of the time triggered schedule, measurement and calculation tasks must be done on demand or in an unsynchronized way.

Each message in LIN is encapsulated in a single message cycle. The message cycle is initiated by the master and contains two parts, the frame header sent by the master and the frame response, which encompasses the actual message and a checksum field. The frame header contains a sync brake (allowing the slave to recognize the beginning of a new message), a sync field with a regular bit pattern for clock synchronization and an identifier field defining the content type and length of the frame response message. The identifier is encoded by six bits (allowing 64 different message types) followed by two additional bits for error detection. Figure 10.1 depicts the frame layout of a LIN message cycle.

The frame response contains up to eight data bytes and a checksum byte. Since an addressed slave does not know that it has to send a message before the reception of the respective frame header, the response time of a slave is specified within a

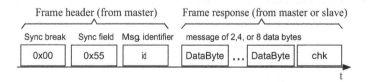

FIGURE 10.1
LIN Frame Format

time window of 140% of the nominal length of the response frame. This gives the node some time to answer, for example to perform a measurement on demand, but introduces a noticeable message jitter for the frame response.

From the slave's view, the LIN protocol is a plain polling protocol, since the slaves only react to the frame header from the master. It is the master's task to issue the respective frame headers for each message according to a scheduling table. The configuration of the network must ensure that each message has exactly one producer. Several slaves can subscribe to a particular message.

10.3   LIN 2.x

In 2003, many updates to reflect the latest off-the-shelf microcontrollers as well as inputs from the SAE Task Force resulted in the definition of LIN 2.0. In fact, LIN 2.0 was a complete rework of the existing LIN 1.3 standard, but backward compatible, allowing for the integration of new LIN 2.0 master/slave nodes in existing LIN 1.2/1.3 clusters (however, a cluster with LIN 2.0 slaves requires a LIN 2.0 master, due to the slave nodes' improved configuration capabilities). New features introduced in LIN 2.0 are an enhanced checksum, sporadic and event-triggered communication frames, improved network management (status, diagnostics) according to ISO 14230-3 / ISO 14229-1 standards, automatic baudrate detection, standardized LIN product ID for each node and an updated configuration language to reflect the changes.

In addition to the unconditional frames (frames sent whenever scheduled according to the schedule table) provided by LIN 1.3, LIN 2.0 introduces *event-triggered frames* and *sporadic frames*.

Similar to unconditional frames, event-triggered frames begin with the master task transmitting a frame header. However, corresponding slave tasks only provide their frame response when the signals transmitted in the data fields have changed. Unlike unconditional frames, multiple slave tasks can provide the frame response to a single event-triggered frame, assuming that not all signals have actually changed. In the case of two or more slave tasks writing the same frame response, the master node has to detect the collision and resolve it by sequentially polling (i.e., send unconditional frames) the involved slave nodes. Event-triggered frames were introduced to improve the handling of rare-event data changes by reducing the bus traffic overhead involved with sequential polling.

Sporadic frames follow a similar approach. They use a reserved slot in the scheduling table, however, the master task only generates a frame header when necessary, i.e., involved signals have changed their values. As this single slot is usually shared by multiple sporadic frames (assuming that not all of them are sent simultaneously), conflicts can occur. These conflicts are resolved using a priority-based approach: Frames with higher priority overrule those with lower priority.

In November 2006, a revision 2.1 of the LIN specification was published by the LIN consortium [200]. LIN 2.1 revised the configuration interface (e. g., it features a function for assigning multiple identifiers at once, which allows a more efficient configuration than with LIN 2.0). Furthermore, LIN 2.1 defines time-out parameters based on ISO 15765-3 [152] for transmissions spanning multiple frames.

10.3.1 Clock Synchronization

LIN builds on a master-slave concept for coordinating the cluster. However, a LIN slave is not explicitly aware of a global time, thus the protocol does not support synchronization of actions which are not explicitly triggered by the master, for example time-triggered synchronized measurements [90]. Instead, coordinated actions on slaves are initiated by the master based on schedule tables stored at the master.

In order to support nodes with unstable clocks, the master sends a synchronization pattern which allows the slaves to adjust their internal clock speed to the master's clock. This allows the use of microcontrollers with internal RC oscillators for the implementation of the nodes. Such RC oscillators change their clock frequency with varying temperature and supply voltage so that they require frequent resynchronization.

10.3.2 Restart, Re-Integration, Integration

The simple model of LIN slaves supports an easy restart and re-integration of nodes into a cluster in most cases provided the master node has not crashed.

If a node crashes and restarts, it will synchronize to the next frame from the master and respond if the frame was addressing itself. Thus, the node will be temporarily unavailable, but integrate smoothly when it comes back provided that the node's configuration had been preserved in some persistent memory.

However, if a node is new or a node's state is reset, there is the need for several configuration steps (assigning node diagnostic address, set slave configuration, etc.[1]) until the node can participate in the communication.

10.3.3 Diagnostic Services

The monitoring of the timing parameters depends on the diagnostic class of the slave nodes [200]: Diagnostic class I slaves are typically simple smart transducers requiring none or a very low amount of diagnostic functionality. Actuator control, sensor reading and fault memory handling is done by the master. Diagnostic class II are nodes with identification support, i. e., node identification, reading and writing data parameters. This functionality is typically required by vehicle manufacturers. Apart from the identification support, actuator control, sensor reading and fault memory handling is done by the master like for diagnostic class I nodes. Diagnostic class III slave nodes have internal fault memory, along with associated reading and clearing services.

---

[1]For details, see [200], chapter 7.3 Node Configuration and Identification.

Each LIN 2.0 node possesses its *LIN Product Identification*. This unique number is stored in the microcontroller's ROM and encodes information about this node.

- Supplier ID: Assigned to each supplier by the LIN Consortium

- Function ID: Assigned to each node by supplier

- Variant field: Modified whenever the product is changed but its function is unaltered

In addition to signal-bearing messages, LIN 2.0 provides *diagnostic messages*. These messages use two reserved identifiers (0x3c, 0x3d). Diagnostic messages use a new format in their frame response called *PDU* (Packet Data Unit). There are two different PDU types: *Requests* (issued by the client node) and *responses* (issued by the server node).

The LIN 2.0 configuration mode is used to set up LIN 2.0 slave nodes in a cluster. Configuration requests use SID values between 0xb0 and 0xb4. There is a set of mandatory requests that all LIN 2.0 nodes have to implement as well as a set of optional requests. Mandatory requests are:

- Assign Frame Identifier: This request can be used to set a valid (protected) identifier for the specified frame.

- Read By Identifier: This request can be used to obtain supplier identity and other properties from the addressed slave node.

Optional requests are:

- Assign NAD: Assigns a new address to the specified node. Can be used to resolve address conflicts.

- Conditional Change NAD: Allows master node to detect unknown slave nodes.

- Data Dump: Supplier specific (should be used with care).

10.3.4   Error Detection and Fault Isolation

LIN was invented for non-critical body electronic services, and therefore is not designed as a fault-tolerant protocol. Thus, a single fault of the master, a babbling node or a failure of the communication line can lead to the failure of the network. When used in a dependable system, a LIN cluster must be part of a fault-containment region [176].

Error detection was a weakness in LIN 1.3 because of the protection with only two extra bits for the message identifier of the header message, giving an error detection probability of only $1 - 1/2^2 = 0.75$. The header message is critical for communication, since an undetected erroneous header message may cause a wrong slave to answer with a syntactically correct message.

The response frame of LIN is protected by a check byte, giving an error detection probability of $1 - 1/2^8 = 0.996$.

LIN 2.0 solves the problem of the weak error detection of the header message by specifying an enhanced checksum for the frame response that includes the protected message identifier in the checksum calculation. An erroneous header message may still cause a wrong slave to answer with a frame response, but the checksum of the frame response will be different from the expected one. Thus, there are 10 bits used for error detection of the header message, giving an error detection probability of $1 - 1/2^{10} = 0.999$. However, note that there is still a chance of 75 % that an erroneous header message might cause the wrong slave to trigger an action at the wrong time.

The error protection of unconditional frames in LIN 2.0 is identical to LIN 1.3, thus rendering also an error detection probability of 0.996.

10.3.5 Configuration Services and Protocol Parameterization

LIN clusters are configured during the design stage using the *LIN Configuration Language*. This language can be used to create a *LIN description file* (LDF). The LDF describes the complete LIN network. Its syntax is deliberately not specified to allow for vendor specific implementations.

In addition to the LIN configuration language and LDF, which are the most important tools to design a LIN cluster, the LIN specification defines a (mandatory) interface to software device drivers written in C. Also, many tools exist that can parse a LDF and generate driver modules by themselves. The LIN C API provides a signal based interaction between the application and the LIN core (also called *core API*).

The *node capability file* (NCF) (since LIN 2.0 [16]) provides a standardized description of off-the-shelf slave nodes. This supports an automatic plug-and-play with slave nodes in a cluster. The NCF is structured as follows:

- The node's name.

- General compatibility properties, e. g., the supported protocol version, bit rates, and the LIN product identification. This unique number is also stored in the microcontroller's ROM and links the actual device with its NCF. It consists of three parts: Supplier ID (assigned to each supplier by the LIN Consortium), function ID (assigned to each node by supplier) and variant field (modified whenever the product is changed but its function is unaltered).

- The diagnostic definition specifying the properties for transport layer and configuration and the node's diagnostic class.

- Frame definitions. All frames that are published or subscribed by the node are declared. The declaration includes the name of the frame, its direction, the message ID to be used and the length of the frame in bytes. Optionally, the minimum period and the maximum period can be specified. Each frame may carry a number of signals. Therefore, the frame's declaration also includes the associated signals' definitions. Each signal has a name, and the following

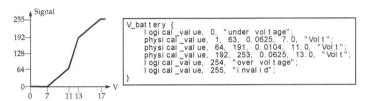

FIGURE 10.2
Example for a LIN Signal Definition

properties associated with it: **Init value** specifies the value used from power on until the first message from the publisher arrives. **Size** specifies the signal's size in bits. **Offset** specifies the position within the frame. **Encoding** specifies the signal's representation. The presentation may be given as a combination of the four choices *logical value physical value*, *BCD value* or *ASCII value*. Declarations of physical values include a valid value range (minimum and maximum), a scaling factor and an offset. Optionally, this can be accompanied by a textual description, mostly to document the value's physical unit. An example is given in Figure 10.2.

- Status management: This section specifies which published signals are to be monitored by the master in order to check if the slave is operating as expected.

- The free text section allows the inclusion of any help text, or more detailed, user–readable description.

The node capability file is a text file. The syntax is simple and similar to C. Properties are assigned using name = value; pairs. Subelements are grouped together using curly braces, equivalent to blocks in C.

Figure 10.3 depicts the role of NCFs and LDF in a LIN cluster. The LDF describes the complete LIN network (corresponding NCFs are parsed by the LIN cluster design tool into the LDF). The LDF is used by the LIN cluster generator to generate LIN related functions within particular nodes. The LDF is also used by the bus analyzer/emulator to support debugging of the LIN cluster [200].

The slave nodes are connected to the master node forming a LIN cluster. The corresponding node capability files are parsed by the LIN cluster design tool to generate a LIN description file (LDF) in the LIN cluster design process. The LDF is parsed by the LIN cluster generator to automatically generate LIN related functions in the desired nodes (the Master node and Slave3 node in the example shown in Figure 10.3). The LDF is also used by a LIN bus analyzer/emulator tool to allow for cluster debugging.

The development of a LIN cluster is partitioned into three phases (see Figure 10.3). During the *design phase*, individual NCFs are combined to create the LDF. This process is called *System Definition*. For nodes to be newly created, NCFs can be created either manually or via the help of a development tool. From the LDF, communication schedules and low–level drivers for all nodes in the cluster can be

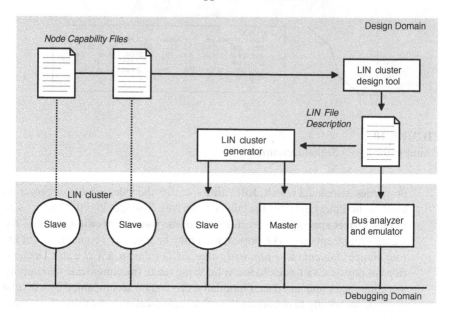

FIGURE 10.3
Development Phases in LIN (according to [200])

generated (*System Generation*). Based on the LDF, the LIN cluster can be emulated and debugged during the *Debugging and Node Emulation* phase. In the *System Assembly* phase, the final system is assembled physically, and put to service.

10.4 Communication Interface

The LIN 2.0 API consists of two sections. In addition to the LIN Core API that came with LIN 1.x, there is a new *LIN Diagnostic API* that is used for configuration and the diagnostic transport layer. The LIN node Configuration API is available only to the master node and mainly implements functions to perform diagnostics and configuration [16].

The LIN diagnostic transport layer is introduced to provide gatewaying functions between CAN and LIN slaves. It transports ISO diagnostic requests/responses and provides a simple *raw API*, as CAN ISO PDUs are very similar to LIN diagnostic frames. This allows us to manage LIN networks via a number of third party configuration tools, e. g., Vectors CANbedded LIN tool[346].

10.5 Validation and Verification Efforts

The LIN protocol is not fault-tolerant, thus a single fault of the master, a babbling node or failure of the communication line can lead to the failure of the network.

For particular implementations, there are efforts to provide verified LIN implementations (e. g., LIN core implementation for ASIC, Actel, Altera and Xilinx architectures [53]) or fail-safe LIN controllers [233].

10.6 Example Configurations and Implementations

The LIN bus system has been applied to various types of non-safety-critical car body applications, such as sun roof control, cruise control switch, windshield wipers, turning lights, climate control, radio controls, seat position motors, mirror, window lift, door locks, etc [317].

Since LIN does not depend on dedicated hardware, there exist many implementations for COTS hardware, e. g., for Microchip PIC [47] (LIN 1.0 slave), ATMEL microcontrollers [14] (LIN 1.0 slave), for Freescale's M68HC08 [58] (LIN 2.0 slave) or for the SPMC75F2313A board [319] (LIN 1.0 bus master). These solutions provide the source code for the embedded software to implement the basic LIN communication features together with a simple description of the necessary hardware setup.

11

TTP/A

W. Elmenreich

Lakeside Labs/University of Klagenfurt

CONTENTS

11.1   Protocol Overview

The TTP/A protocol is the low-cost field-bus protocol that is harmonized with the fault-tolerant system bus TTP/C of the time-triggered architecture (TTA). It is intended for the connection of smart sensors and actuators in embedded real-time systems in different application domains, e.g., industrial, automotive, etc. It is the objective of TTP/A to provide all services needed by a smart sensor, including timely communication, remote online diagnostics, and plug-and-play capability. While LIN is powered by industrial sponsors, TTP/A is an academic development started by the Technical University of Vienna, Austria, and then broadened to include the Technical University of Munich, Germany, and the University of Stuttgart, Germany. The first version of TTP/A was published at the SAE World Congress in 1995 [167]. Since then, TTP/A has been extended by a plug-and-participate function [89] and a unique interface scheme in the form of an Interface File System (IFS) [181].

11.2 OMG Smart Transducer Standard

TTP/A implements the Object Management Group (OMG) Smart Transducer Interface (STI) standard [241].

The STI standard defines a smart transducer system as a system comprising several clusters with transducer nodes connected to a bus. Via a master node, each cluster is connected to a CORBA gateway. The master nodes of each cluster share a synchronized time that supports coordinated actions (e. g., synchronized measurements) over transducer nodes in several clusters. Each cluster can address up to 250 smart transducers that communicate via a cluster-wide broadcast communication channel. There may be redundant shadow masters to support fault tolerance. One active master controls the communication within a cluster (in the following sections the term master refers to the active master unless stated otherwise). Since smart transducers are controlled by the master, they are called slave nodes. Figure 11.1 depicts an example for such a smart transducer system consisting of three clusters.

It is possible to monitor the smart transducer system via the CORBA interface without disturbing the real-time traffic.

The STI standard is very flexible concerning the hardware requirements for smart transducer nodes, since it only requires a minimum agreed set of services for a smart transducer implementation, thus supporting low-cost implementations of smart transducers, while allowing optional implementation of additional standard features.

The information transfer between a smart transducer and its client is achieved by sharing information that is contained in an internal IFS, which is encapsulated in each smart transducer.

11.3 Interface File System (IFS)

The IFS [181] provides a unique addressing scheme to all relevant data in the smart transducer network, i. e., transducer data, configuration data, self-describing information and internal state reports of a smart transducer. The values that are mapped into the IFS are organized in a static file structure that is organized hierarchically representing the network structure (Table 11.1).

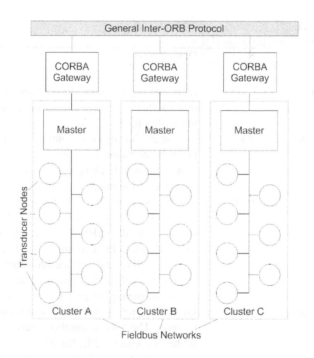

FIGURE 11.1
Multi-Cluster Architecture with CORBA Gateway

TABLE 11.1

Hierarchical structure of an IFS address

Element	Size	Description
Cluster name	8 bit	Identifies a particular cluster. Native communication (without routing) among nodes is only possible within the same cluster.
Node alias	8 bit	The node alias or logical name selects a particular node. Some values have an associated special function, e. g., alias 0 addresses all nodes of a cluster in a broadcast manner.
File name	6 bit	The file name addresses a certain file within a node. A subset of files, the system files, have a special meaning in all nodes. Each service of a node is mapped onto a file containing sections for the service providing and service requesting linking interface as well as for configuration/planning and diagnosis/management data.
Record number	8 bit	Each file has a statically assigned number of records. The record number addresses the record within the selected file. Each record contains 4 data bytes. Note that each file contains only the necessary number of records, thus, the number of addressable records is statically defined for each file.

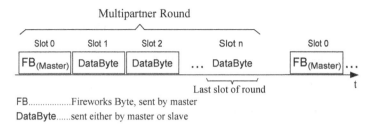

FIGURE 11.2
Example for a TTP/A Multipartner Round

11.4 Protocol Services

11.4.1 Communication Services

Communication is organized into rounds consisting of several messages. Each communication round is started by the master with a so-called fireworks byte. The fireworks byte defines the type of the round and is a reference signal for clock synchronization. The protocol supports eight different firework bytes encoded in a message of one byte using a redundant bit code [120] with a hamming distance of 4 supporting error detection. One particular fireworks byte is a regular bit pattern, which is also used by slave nodes with an imprecise on-chip oscillator for startup synchronization. This bit pattern is identical to the sync pattern used in LIN.

Generally, there are two types of rounds:

Multipartner round: This round consists of a configuration dependent number of slots and an assigned sender node for each slot. The configuration of a round is defined in a datastructure called ROund Descriptor List (RODL). The RODL defines which node transmits in a certain slot, the operation in each individual slot and the receiving nodes of a slot. RODLs must be configured in the slave nodes prior to the execution of the corresponding multipartner round. An example for a multipartner round is depicted in Figure 11.2.

Master/slave round: A master/slave round is a special round with a fixed layout that establishes a connection between the master and a particular slave for accessing data of the node's IFS, e.g., the RODL information. In a master/slave round the master addresses a data record using a hierarchical IFS address and specifies an action like reading of, writing on or executing that record.

The master/slave rounds are used for diagnostics and configuration of the smart transducer nodes. The periodical multipartner rounds provide a predictable real-time communication among the nodes. The master/slave rounds allow a point-to-point connection to a particular node for configuration, maintenance, and diagnosis purposes. Each node is assigned a logical node ID that is used in the master/slave rounds

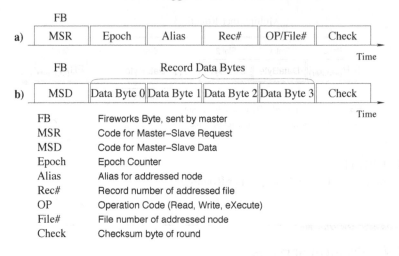

FIGURE 11.3
TTP/A Master/Slave Round

for addressing a node within a cluster. The logical IDs of TTP/A nodes can be assigned either at compile time or online when the node is integrated into the cluster. The online-assignment of logical node IDs is called baptizing. The baptizing algorithm is performed by the master and is based on binary search. It makes use of the unique identification number of every TTP/A node, the conditional setting of node identifiers by executing a special record in the file system of the node and the ability to detect simultaneous bus access of multiple nodes. The baptizing enables true plug and play but is not mandatory in the TTP/A standard.

Figure 11.3 depicts the fixed layout of a master/slave round. A master/slave round consists of two parts. In the addressing part (Figure 11.3 a), the action and the memory addressing is encoded in three parameter bytes.

In a further part (Figure 11.3 b), the addressed data bytes are transmitted between master and slave. The fireworks byte (MSD) is always sent by the master, while the data bytes are either sent by master or slave depending on the action defined in the addressing part.

The last byte of each round contains a checksum byte that protects the communication from bus failures. Master/slave rounds have idempotent semantics, thus it is possible to repeat the action in case of communication failures.

A master/slave round has a fixed layout. The address scheme is derived from the IFS. At startup the master uses master/slave (MS) rounds for determining the types of the connected nodes and configuring them. The multipartner (MP) round is intended to establish a periodical, predictable and efficient real-time communication. To support a diagnosis and maintenance access concurrent to the real-time traffic, master/slave rounds are scheduled periodically between multipartner rounds as depicted in Figure 11.4 in order to enable maintenance and monitoring activities during system operation without a probe effect.

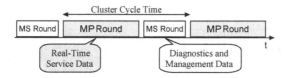

FIGURE 11.4
Recommended TTP/A Schedule

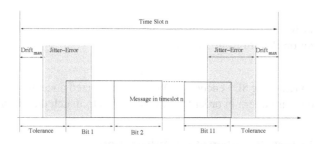

FIGURE 11.5
Tolerance Time in a Slot within a TTP/A Round

11.4.2 Clock Synchronization

TTP/A uses a periodic central master clock synchronization. The master starts a so-called *epoch* by issuing a fireworks byte that also defines the communication mode for the following round. The slave nodes restart their timer at the beginning of an epoch and organize their operations, i. e., sending and receiving frames and executing tasks in a lattice of slots. The size of a lot is predefined based on the requirement for transmitting one byte of data including an 8o1 framing information (except for fireworks bytes which are sent using 8E1 encoding, but have the same timing) as well as some tolerance buffer to account for timing differences between the nodes (see Figure 11.5). The fireworks byte from the master serves as a *state correction* of all the nodes' clocks. In order to cope with clock rate differences, a node can adjust the rate of its clock using frames sent from a trustable time source, e. g., from the master or a node with exact time. Slots containing such reference timing information may contain ordinary user data, but are especially marked in the RODL of the slave nodes.

The combination of rate and state correction supports even nodes with very imprecise oscillators, which may have a drift rate of up to $10^{-3}\frac{s}{s}$ and a high drift rate change of $10^{-1}\frac{s}{s^2}$. At a baud rate of $19200\frac{bit}{s}$, there must be a reference time frame at least every 17 slots in order to keep the slot-wise synchronization for nodes with such imprecise oscillators [88].

During operation, the above-described rate and state correction is taking place periodically within each TTP/A round. In order to initially synchronize a slave node's clock, the TTP/A master is using a specific bit pattern as one of its fireworks bytes (see Figure 11.6). An already synchronized node will interpret this pattern as a value

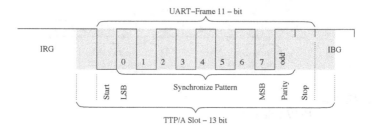

FIGURE 11.6
Synchronization Pattern from TTP/A Master

of 0x55 encoded using 8E1 encoding. An unsynchronized node, however, can search for such a regular pattern in order to adjust its clock to match the given baud rate.

11.4.3   Restart, Re-Integration, Integration

A major concept of time-triggered networks is that they maintain a minimum system state in order to allow for a fast restart of the networks or parts of it. The communication system of TTP/A does not require to store state information between successive rounds, except for the master/slave round, which keeps a state between the Master/Slave Address MSA and the Master/Slave Data MSD rounds. Thus, a re-integrating node, given it still has the RODL information, will just wait until it receives the next fireworks byte in order to participate in the communication.

Integrating new nodes needs special support, since a major design goal of TTP/A was that the communication for node setup may not disturb the real-time operation of the already running ensemble. Therefore, the trivial solution where new nodes simply report their presence by broadcasting a message containing their ID, is out of question since the broadcasts of unsynchronized new nodes might disturb the deterministic real-time service. Therefore, the TTP/A master polls for new nodes using a *binary search operation* followed by a *baptize operation* that assigns a local identifier to the detected nodes [89]. The local identifier consists of an 8-bit value and allows a unique addressing of a node within the same cluster.

11.4.4   Diagnostic Services

TTP/A specifies a *diagnostic and management* interface which establishes a connection to a particular smart transducer node and allows reading or modifying of specific IFS records. Most sensors need parameterization and calibration at startup and continuously collect diagnostic information to support maintenance activities. For example, a remote maintenance console can request diagnostic information from a certain sensor. The diagnostic and management interface is supported by the master/slave rounds of the protocol and is usually not time-critical.

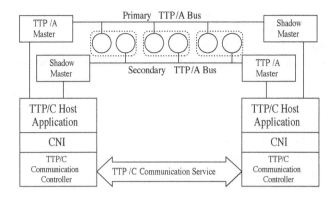

FIGURE 11.7
Integrated Architecture with Two TTP/C Nodes and TTP/A Networks

11.4.5 Fault Isolation

TTP/A per se is not fault-tolerant, a single fault of the master, a babbling node or failure of the communication line can lead to the failure of the network. However, the deterministic behavior of TTP/A allows it to be integrated as subsystems to hierarchical time-triggered systems, where each TTP/A network will form a single fault containment region.

For example by integrating TTP/A with a TTP/C network, both systems complement each other in order to support low-cost smart transducers within a highly dependable system. Since both protocols follow a strict time-triggered schedule, the TTP/A and TTP/C system can be synchronized in order to support synchronous interaction of all nodes in the network.

Figure 11.7 depicts a fault-tolerant architecture with two TTP/C nodes and two TTP/A networks containing a set of transducers. Each TTP/C node controls vice versa one master node of one TTP/A network and a shadow node to the other TTP/A network. The dotted ovals around the transducers indicate that these transducers are redundantly measuring the same real-time entity. The given example tolerates an arbitrary node failure of any node in the network. Since measurements from different sensors of the same real-time entity are not replica deterministic, it is necessary to run an agreement protocol between the TTP/C application in order to get a consistent system state.

11.4.6 Configuration Services and Protocol Parameterization

TTP/A specifies a *configuration and planning* (CP) interface allowing the integration and setup of newly connected nodes. It is used to generate the "glue" in the network that enables the components of the network to interact in the intended way. For a time-triggered system, this is mainly a consistent communication schedule that is consistent among all participating nodes. Usually, the CP interface is not time-critical, but it can coexist with an operational real-time service. In this case, the new communication schedule is written to an inactive RODL until the configuration process is complete. Then, the master switches to the new schedule.

Together with the diagnostics and management interface, the configuration and planning interface provides access to the interface file system and, therefore, a possibility to parameterize and configure nodes. Download of application software is not provided within the TTP/A protocol or the smart transducer interface. A respective boot loader for reprogramming a node would have to use a proprietary protocol. Typically, TTP/A nodes integrate protocol software and application software in a single microcontroller, which makes a separation for the application parts to be reprogrammed during operation difficult.

11.5 Communication Interface

For unique addressing of the slave's internals, all relevant data of a TTP/A node, like round definitions, application specific parameters and I/O properties, are organized into a structure called IFS [181]. The IFS is structured in a record-oriented format. Each record is addressable separately by master/slave rounds.

The IFS was introduced for two reasons:

- Provide a consistent view of the transducer properties.

- Decouple subsystems from the point of view of temporal control.

All nodes contain several files that can be accessed over the TTP/A protocol in a unified manner. The minimal setup for a smart transducer is:

Round Descriptor List (RODL): Each node contains at least one and up to six RODLs that contain TDMA schedules for the TTP/A multipartner rounds.

Configuration File: This file contains an eight-bit alias which is the slave's name within its cluster and may have extra configuration data within.

Documentation File: Each node is assigned a unique identifier (physical name) stored in this file. This physical name is a 64-bit integer assigned invariably to each node. The number is the concatenation of a series number, which identifies the node's type, and the serial number, distinguishing nodes of equal type. Optionally this documentation file contains the ASCII text of a uniform resource locator (URL) pointing to a file containing the node's data sheet. Documentation files are read-only from the master's viewpoint.

To support ultra-low-cost implementations of TTP/A slave nodes, it is also possible to omit the implementation of the file system and hard-code the TDMA schedule

FIGURE 11.8
Smart Transducer Based on Atmel 4433 Microcontroller with Distance Sensor Attached (scale in centimeters)

for the TTP/A multipartner rounds. Such a node would not respond to any master/slave round and does not support configurability. It is possible to build heterogeneous networks with ultra-low-cost nodes and configurable nodes together but this might have a negative effect for the system overview because the maintenance program is "blind" on the ultra-low-cost nodes.

11.6 Validation and Verification Efforts

TTP/A follows a correct-by-design approach, and thus allows us to verify its relevant properties due to its open and simple design. In [311] a formal model of the TTP/A master, slaves and channel was created and formally checked in order to prove the recovery of the real-time service part of a TTP/A system from a transient fault back to correct operation.

11.7 Example Configurations and Implementations

11.7.1 TTP/A Slave Nodes

TTP/A slave nodes are typically built as smart transducers [87], thus featuring a microcontroller that instruments a sensor or an actuator.

Figure 11.8 depicts the hardware of a smart transducer implementation based on an Atmel AVR AT90S4433 microcontroller and an attached distance sensor. This type of controller offers 4K Byte of Flash memory and 128 Byte of SRAM. The physical network interface has been implemented by an ISO 9141 k-line bus, which is a single wire bus supporting a communication speed up to 50 kBps. The wires to the left of the photo contain the bus line and the power supply.

Table 11.2 gives an overview on the resource requirements for smart transducer implementations in Atmel AVR, Microchip PIC and ARM RISC microcontrollers.

All three implementations provide a Baud Rate of at least 19.2 kbps. As physical layer, an ISO 9141 k-line is supported. For the Atmel AT90S4433, a maximum performance of 58.8 kbps had been tested on an RS485 physical layer.

TABLE 11.2
Resource requirements and performance of time-triggered smart transducer interface implementations (from [335]).

Microcontroller	Used FLASH	Used RAM	Clock Speed	Max.Speed
Atmel AT90S4433	2672B	63B	7.37 MHz	58.8 kbps
Microchip PIC	2275B	50B	8.0 MHz	19.2 kbps
ARM RISC	8kB	n.k.	32.0 MHz	19.2 kbps

The implementations of these microcontrollers show that due to the low hardware requirements of the time-triggered smart transducer interface it should be possible to implement the protocol on nearly all available microcontrollers with similar features, like the Atmel or Microchip microcontroller types, that is 4KB of Flash ROM and 128 Byte of RAM memory.

A software implementation for Atmel AVR8 microcontrollers is available at http://www.vmars.tuwien.ac.at/ttpa/ as open source under the Berkeley Public License.

11.7.2 TTP/A Master

Every node that is able to run the slave protocol can also run the master protocol. So for deploying a stand-alone TTP/A network, the same type of nodes can be used for master and slaves. In order to utilize the maintenance and configuration features, the master node is implemented as a gateway including one or more interfaces to other networks.

Within several research projects, implementations of TTP/A master nodes with gateway function have been created [254, 236, 158]. There exist also some prototype implementations of the TTP/A master protocol for TTP/C nodes. The most modular solution comes in the form of a PCMCIA card [38] which can be used as an I/O card in a TTP/C node (or any other computer system with a PCMCIA interface). Figure 11.9 depicts the size of the PCMCIA gateway card (without cover). The card hosts a microcontroller and provides interfaces to a TTP/A and CAN [39] bus. The microcontroller can be configured via the PCMCIA interface and runs the TTP/A master protocol.

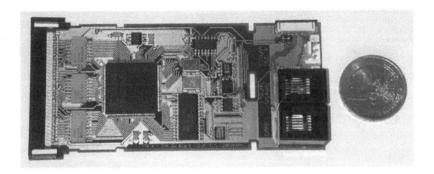

FIGURE 11.9
PCMCIA Gateway Card in Comparison to Size of 2 Euro Coin

12

BRAIN

M. Paulitsch

EADS

B. Hall

Honeywell

K.R. Driscoll

Honeywell

CONTENTS

12.1   Protocol Overview

The BRAIN (Braided Ring Availability Integrity Network) is a novel communication architecture supporting fault-tolerant time-triggered communication. As the name suggests, the BRAIN is built upon a braided-ring topology. This topology augments the standard ring topology with increased connectivity. In addition to the "direct link" connections between a node and its immediate neighboring nodes (as is used in simple rings), a braided-ring node also is connected to its neighbor's neighbor via a link called the braid or skip link (see Figure 12.1). The BRAIN utilizes the additional connectivity to achieve both high-coverage integrity and availability concurrently. This is in contrast to previous braided rings, which use these additional links only for availability. The BRAIN can use almost any existing Local Area Network (LAN) technology to implement its communication links, including any of the IEEE 802.3 Ethernet variants.

The BRAIN uses the least amount of hardware to achieve single fault tolerance (including Byzantine failure) of any known data network. The BRAIN also can tolerate most cases of two benign faults with no additional redundancy. The BRAIN topology enables adjacent nodes to collaboratively form self-checking pairs (SCPs). This allows standard simplex computational hardware to be run-time configured into high-integrity fail-silent computational platforms, which provides the high fault coverage for processing that one would find in architectures supported by SAFEbus (see Chapter 7) but without requiring any special SCP hardware for the processors.

The BRAIN's benefits derive from its time-triggered data flow and its use of high-coverage fault tolerance.

12.1.1   Development History and Design Goals

The BRAIN was originally conceived as a field-bus type protocol, targeting low-bandwidth applications. Acknowledging the simplicity of high-coverage self-checking architectures, such as SAFEbus (see Chapter 7), the BRAIN was conceived to apply these techniques to a field-bus protocol. The intent was to deliver SAFEbus levels of integrity and availability within a field-bus environment. Exploiting lessons learned from the application of other field-bus protocols in safety-relevant applications, the BRAIN has been designed with two principles in mind: (1) limited reliance on inline error detection mechanisms such as Cyclic Redundancy Codes (CRCs) for error detection needs and (2) the avoidance of dedicated guardian hardware commonly used in other field-bus protocols targeting safety-relevant deployment [122]. The BRAIN has evolved to become a more general purpose architecture, and it continues to evolve.

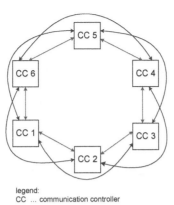

legend:
CC ... communication controller

FIGURE 12.1
BRAIN's Braided-Ring Basic Architecture

Limited reliance on coding-based techniques for error detection. It can be argued that the error detection properties of code-based inline error detection like CRCs are not strong enough to cover the failure modes of active interstages. Active interstages are defined as relaying stations with active circuits (e.g., integrated circuits). These typically are a network switch or hub in a communication architecture [250]. Codes like CRCs have been designed for good error detection coverage of typical failure modes of wires, such as occasional bit flips and burst errors. The failure modes of interstages and the potential for correlated failures are unlikely to fit these assumptions. Hence, the goal of the BRAIN is to not rely on coverage of CRCs or similar error detection schemes for interstages. This is described in more detail in [250].

Avoidance of dedicated guardian hardware. The requirement for local or central guardian hardware constitutes an intolerable overhead in cost-constrained applications. They also introduce complexities when the guardian action needs to be verified within a deployed system. Similarly, each fault-tolerance feature needs scrubbing logic – logic that tests whether the fault-tolerance mechanism is still operational. Such scrubbing logic needs to detect latent faults of the logic implementing the fault tolerance, in order to ensure the logic is available when needed. More allegorically explained, scrubbing ensures that the guard is awake and well when really needed. Distributed, built-in guardian logic as deployed in the BRAIN can be more easily scrubbed for latent faults during operation than central or local guardians. Central guardians may additionally introduce architectural constraints, for example by restricting the order of component power-up, etc. Therefore, a core goal of the BRAIN was to remove the need for any dedicated guardian components.

The BRAIN was conceived as a broadcast flooding network targeting network bandwidth on the order of 5 to 20 Mbit/s. At these speeds, the intra-node propagation

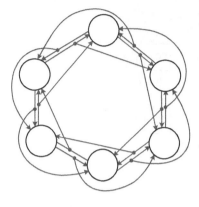

(a) full-duplex BRAIN configuration   (b) half-duplex BRAIN configuration

FIGURE 12.2
The Two BRAIN Connection Duplex Configurations

delay is minimal, comprising only a few bits of elasticity delay encountered at each link as the message floods around the segmented medium. A global, time-triggered, a priori agreed schedule coordinates the transmission sequencing, in a manner conceptually similar to the media access of a typical time-triggered bus, i.e., Time-Division Multiple Access (TDMA). For targeting very low-end applications, the BRAIN can use half-duplex instead of full-duplex BRAIN links. The half-duplex BRAIN sends data in both directions on each link medium, while the full-duplex BRAIN sends data through unidirectional link media. Half-duplex operation is described in [123]. The half-duplex configuration uses less physical layer hardware, at the expense of cutting effective bandwidth by about half.

Targeting medium to high performance applications, a recent development in the evolution of the BRAIN uses store-and-forward propagation to be able to more efficiently use emerging high-speed serial protocols, such as the gigabit per second (Gbs) Ethernet. However, the algorithms for this higher-speed variant of the BRAIN were still undergoing formal analysis at the time of this publication. Therefore, the details of this development are not discussed here.

Fault Tolerance and Fault Hypothesis

With point-to-point links, the BRAIN topology implicitly addresses the spatial and other physical layer damage issues that impact bus topologies. The BRAIN topology is able to tolerate complete loss of communications at any single geographic location on the ring. Similarly, any single node may fail and drop out from the BRAIN without adversely impacting the system level communication (availability and integrity) guarantees. In addition to these passive failures, the BRAIN protocol also has been designed to tolerate an active malicious fault. The BRAIN mitigates such faults by leveraging bit-for-bit comparisons between the independent data paths that exist between the skip and direct links, and/or the two opposing directions of traversing the ring (clockwise and counterclockwise), see Section 12.2.1. In addition, the BRAIN employs a "brother's keeper" guardian action, where each node may act as a guardian for its adjacent neighbors. Note that this is consistent with the original design goals, since it supports the deployment of guardian actions without incurring additional component overheads.

The BRAIN originally was designed to tolerate at least one faulty node.[1] A node's failure mode may be either i) passive (i.e., fail-stop) on any combination of its ingress or egress links, or ii) actively malicious, where the node acts actively to corrupt data and/or disrupt protocol operation, e.g., by masquerading as another node or babbling. In practice, the connectivity of the BRAIN provides tolerance to many more benign faults, providing service under at least two passive node faults or link faults. However, to tolerate the active fault, the initial field-bus BRAIN requires the full connectivity of the non-faulty nodes. A discussion of the BRAIN's sensitivity to node and link failures together with a comparison of the BRAIN's policies with other protocols is given in [248].

Note that, similar to SAFEbus, the BRAIN has not been designed to tolerate dual active malicious faults, i.e., where two nodes act in a coordinated fashion to corrupt data flow and/or disrupt the system. This fault hypothesis is backed by two observations. Firstly, statistical analysis shows that it is highly unlikely that two faults manifest in identical correlated active failure modes in two independent[2] devices within the very tight time window of self-checking action. Secondly, the combination of the self-checking forwarding action, the periodically scrubbed guardian function and potential reconfiguration actions of Section 12.3 are sufficient to tolerate an active fault for the time of exposure between scrubs.

12.1.2 Minimal Overhead Replication and Input Agreement

In Byzantine-tolerant real-time systems, the bandwidth and scheduling overheads associated with the Byzantine data exchange and agreement often constitute significant network, system and software burdens. This observation created another design goal of the BRAIN, which was to provide a low-overhead framework to support task replication and data agreement. To this end, the BRAIN introduces additional mechanisms to contain node-sourced Byzantine failures; for example, preventing a node from supplying different data on each of its outgoing directions. With such fault containment in place, an inconsistent omission fault model may be assumed for those messages needing agreement. This reduces the expenses of higher-level agreement exchanges required for such messages typically used in other Byzantine tolerant protocols. In place of the fault message exchange and voting typically used in Byzantine agreement, a simpler agreement on reception status may be implemented. This is a hierarchical Byzantine agreement mechanism, similar to the one developed for

---

[1]Later variants are targeting tolerance to two faults.
[2]This independence can be made arbitrarily large — separate boxes, separate power supplies, etc.

SAFEbus. The BRAIN also encompasses hardware assistance mechanisms for this exchange that does not need software interaction (see Section 12.4.1).

12.2 Protocol Mechanisms and Services

12.2.1 High-Integrity Data Propagation

The core service of the BRAIN is the fault-tolerant, high-integrity, high-availability broadcast distribution of data. This mechanism provides the foundation for all of the higher-level protocol services. Abstractly, the BRAIN data propagation scheme is very simple and can be viewed independently of the higher-level protocol services. These techniques may also be applicable to other time-triggered protocols such as FlexRay (see Chapter 6), TTP (see Chapter 5), or Time-Triggered Ethernet (see Chapter 8).

A conceptual representation of the BRAIN's high-integrity data propagation is illustrated in Figure 12.3. At a high level, the BRAIN is best viewed as offering two concurrent modes of data propagation, with both modes collaborating to maximize data integrity and data availability. The two modes are Self Checking Data Relay and Independent Path Data Integrity Reconstitution.

12.2.1.1 Self-Checking Data Relay

The BRAIN's first propagation mode focuses on inline integrity failure detection, i.e., the possible corruption of data during a node's data relaying action and availability in the event of a single fault. As illustrated by the long outer arrows in Figure 12.3, a sending node transmits its message in both directions around the ring ("clockwise" and "counterclockwise"). Broadcasting a message in both directions provides availability. A message will be delivered successfully if either one of the directions is intact (the "Availability OR" in the figure). The independence of these two paths ensures that there will always be one success path available from any arbitrary sending node to any arbitrary receiving node, given a single fault assumption.

If a receiving node gets two copies of a message (one from each direction) that are different, it must decide which one is good. This decision is supported by the use of message integrity status provided by an integrity scheme employed within each direction. This integrity scheme prevents a message from being corrupted in transit without the message being flagged as suspect. For every potentially faulty component within the data propagation path, there exists completely independent hardware, which checks that this component has not caused any corruption. If corruption is detected, the message is flagged as suspect (i.e., has lost its integrity guarantee). One instance of the mechanism supporting this scheme is shown in Figure 12.3 as the "Integrity AND" function for the counterclockwise direction. This is replicated for every node in both directions.

As a message traverses the ring, it tries to pass through every node using the direct links. Each node is also bypassed by a skip link. For the general case, in a particular direction, a node receives two inputs, one from the direct link and one from the skip link. Each node compares the data it receives from its direct link with the data it receives from its neighbor's neighbor via the skip link. This bit-for-bit comparison of all data transmitted ensures that any data corruption injected by a faulty node is immediately detected. For example, consider the node marked "faulty" of Figure 12.3 corrupting the message transmission from the sending node as it is being relayed. The erroneous data from the faulty node is immediately detected by the Integrity AND comparison function performed at the next downstream node, which compares the direct link output from the faulty node with the original data (sending node) available on the skip link. This skip-direct comparison is performed by all nodes as the data is propagated around the ring. Hence, data corruption at any point can be immediately detected. The status of the propagation integrity comparison is signaled by setting the value of a status flag at the end of the message. This flag must follow all of the bits within the message that are checked for integrity.

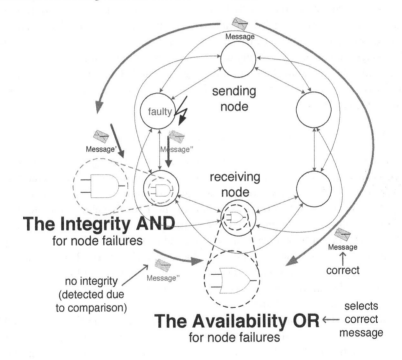

FIGURE 12.3
Conceptual Brain Operation for Tolerating One Arbitrary Fault

The implementation details for the integrity status flag depend on the particular physical layer technology used to implement the links (e.g., Ethernet) and whether the flag is equivalent to a single binary value (i.e., message integrity is intact vs. message integrity is not trusted) or is a vector of such binary values, one for each node through which the message passes. For vector implementations, the vector can either be a fixed length (with one element of the vector for each node in the ring) or the vector can be variable length (growing in size by one element for every node through which the message passes). Vector implementations enable precise diagnostics to isolate where in the propagation path the errors are induced. However, it increases the message overhead and may not be compatible with some physical layer technologies. For the single binary implementations, an aggregate status of all up-stream comparisons can be carried by a shared status flag.

It is essential that the setting of the integrity status flag (single binary value or vector) by all repeating nodes is strictly one-way. That is, repeating nodes are limited to either propagating the current integrity status value or setting it to "not trusted," i.e., they are not allowed to change a non-trusted status back to one that says the message has integrity. To enforce this one-way behavior, the value of the integrity status is also checked in the direct-skip link comparison described above. If either link says that the message integrity is no longer trusted, the relayed message will have its integrity status flag say that the integrity is no longer trusted. In framed formats, such as Ethernet, the loss-of-integrity status can be signaled by truncating the frame (e.g., deleting the end-of-frame marker), forcing a detectable error at the end of message (e.g., using the wrong disparity in 8B/10B encoding) or forcing a frame check sequence to be invalid (e.g., inverting Ethernet's CRC field) which causes the frame to be detectability erroneous by the physical technology used to implement the links.

12.2.1.2   Independent Path Data Integrity Reconstitution

The BRAIN's second propagation mode focuses on tolerating a second benign (fail-stop or omission) failure. The BRAIN is able to tolerate a benign second fault without any increase in redundancy. This provides the BRAIN an additional degree of fault tolerance. However, the current evolution of the BRAIN cannot tolerate an active fault and an arbitrary benign fault at the same time. The first propagation mode provides for fail-op/fail-stop operation. The second propagation mode adds fail-op/fail-op/fail-stop operation for benign faults. This is worst-case. The BRAIN can tolerate many more faults for most cases. For example, the BRAIN can tolerate any number of benign node failures, as long as no two or more failed nodes are adjacent when three or more nodes have failed.

This second propagation mode reverses the locations of the "Integrity AND" and "Availability OR" mechanisms. See Figure 12.4 and compare it to Figure 12.3. In this mode, the skip and direct links provide increased channel availability, while the Integrity AND is performed by comparing the data a node receives from each direction. If the data received from both directions matches bit for bit and the data from neither direction violates the link data encoding nor framing checks, the integrity of the data received may be "reconstituted." This allows the data to be used, despite having the integrity flags from both directions showing that integrity is suspect. The situation of getting a correct high-integrity message via integrity reconstitution may happen only if there are multiple benign node failures (fail-omission or broken links). By

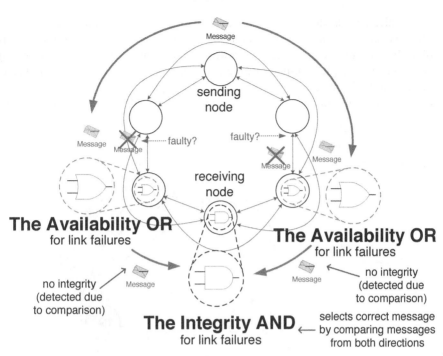

FIGURE 12.4
Conceptual BRAIN Operation for a Second Benign Failure

leveraging the reconstitution of integrity, a system may lose multiple skip and direct link connections, yet still be able to deliver high-integrity data to the application.

In the second propagation mode, the Availability OR is implemented by each node selecting the data from either link to be forwarded, with the loss of integrity status signaled if the two incoming links do not match bit for bit. This selection can be arbitrary in most cases. The exception rules for selection are discussed with the rules for the time sequenced guardian roles in Section 12.3.1.2.

It is emphasized that both propagation modes of the BRAIN co-exist simulta-neously and there are not any specific mode selections nor additional bandwidth required to support them both. Therefore, there are three paths that may constitute a high-integrity data reception, either clockwise or counterclockwise with the inline integrity status confirmed, or the reconstituted path from the comparison of both di-rections.

12.2.1.3 Self-Checking Processor Pair Broadcast

The connectivity of the BRAIN that enables the data relaying policies of the BRAIN, as described above, also can be used to compare the output of two adjacent nodes. This allows for adjacent nodes to be configured into high-integrity *message-based self-checking pairs*. Implementing the paired actions is as simple as configuring the communication schedule to make the two halves of a pair transmit in a shared slot (the time allocated on the media to transmit one message). The synchronous nature of the BRAIN and the high-integrity forwarding mechanism ensure that the receiving nodes receive a single high-integrity message when the data sent from the two halves of the pair are identical. Such a configuration is depicted in Figure 12.5, where the copies of message "msg a" appears as the copies of a single message at all receivers.

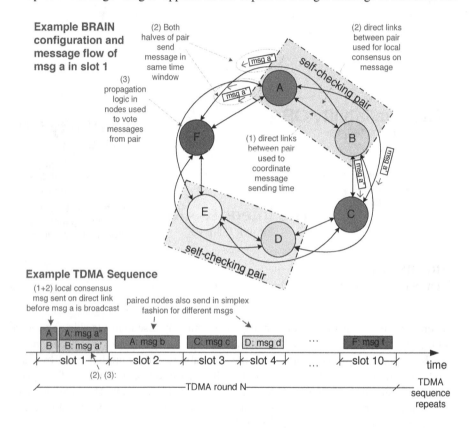

FIGURE 12.5
Self-Checking Pairs

Using the time-triggered network communication primitives to implement the high-integrity computational comparison presents significant advantages when contrasted with other self-checking approaches. Traditionally, high-integrity computational hardware platforms have required specialized design to "clock step" or "lock-step" the processing platform. In modern processors, such lock-step comparisons can often introduce a significant run-time performance degradation. In addition, with the emergence of multi-core processing engines, the complexity associated with such comparisons may impact the very feasibility of the lock-step approach. The loosely-coupled time-triggered message-stepping comparison[3] of the BRAIN does not suffer from performance degradation, since the comparison is made at the communication line rate as a by product of the BRAIN's normal message propagation mechanisms. Additionally, the communications-based comparison approach may arguably offer superior fault coverage, since it detects failures that occur within the communication hardware itself. It should also be noted that the BRAIN message-stepping scheme enables nodes to selectively perform self-checking or non-self-checking message transmissions. As illustrated in Figure 12.5, nodes can operate as message-stepping pairs, during one TDMA slot (see nodes A and B in slot 1), and the same nodes can make independent simplex transmissions in other slots (node A in slot 2). This scheme may provide improved processing hardware utilization, since it provides the freedom to replicate only safety-relevant software tasks and does not require the full replication of all computation on the self-checking host.

There are advantages in reducing the tight redundancy coupling required for processor integrity mechanisms. For example, in missions and applications where spare resources are scarce or the logistics pipeline is expensive, systems designers often strive for "generic sparing" strategies that enable common reconfigurable hardware to be used in more than one application or mission role. Hence, the ability to configure standard COTS processing hardware into high-integrity fail-silent computational pairs via a simple change of the TDMA schedule may be very attractive. Additionally, each half of the pair may use dissimilar hardware, providing a path for generic design failure mitigation. A triple modular redundant variant of the self-checking scheme is also possible, as presented in Figure 12.6. Using such a configuration, it may be possible to guarantee fail-operational behavior with a generic processing fault, while using minimal hardware.

To implement the message-stepping pairs, input agreement is required between each half of the pair to ensure that a fault entering the two halves of the pair cannot force each half of the pair to diverge or disagree. To implement such agreement, the direct connections between the paired nodes may be used. Section 12.4.1 outlines the use of agreement hardware.

12.2.2   Clock Synchronization, Startup and Clique Resolution

The self-checking data propagation mechanism described above is extensively utilized by the BRAIN's clock synchronization and startup protocols. Within a BRAIN network, adjacent nodes act as self-checking pairs to send protocol synchronization messages. Unlike other protocols that utilize distributed algorithms to vote and converge output from multiple "peer" nodes, the BRAIN adopts a hierarchical protocol strategy. In place of a fault-tolerant convergence function, the BRAIN employs a fault-tolerant priority-based selection function to select a master clock pair. The selection of a pair is performed once each synchronization interval. It is, therefore, tolerant to dynamic system membership changes and will reselect an alternative driv-

---

[3]Integrity comparisons are done on a message-by-message basis rather than on every clock tick of the processor.

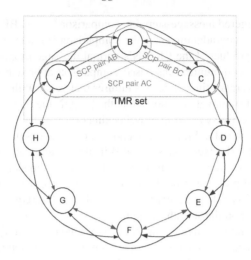

FIGURE 12.6
Triple-Modular Replication (TMR) Deployed on BRAIN

ing pair when the highest-priority selected master fails. Similarly, when the higher-priority pair recovers, the BRAIN protocol incorporates mechanisms to ensure that the recovering pair will assume the running timeline before it resumes its mastership. As is generally true for the creation of BRAIN self-checking pairs, clock synchronization self-checking pairs are created simply by including this information in the schedule tables. Nodes that are not configured to be in the set of potential self-checking pair clock masters are called "slaves." The protocol messages sourced from the self-checking pairs are simple, comprising two fields as noted in Table 12.1. In the BRAIN, the same message format is used for synchronization and integration, hence the messages are termed Synchronization-INTegration (SINT) messages.

TABLE 12.1 SINT message fields.

Field	Description
TDMA POSITION	The TDMA slot counter
PRIORITY	The priority of the sending self-checking pair

Within the SINT message, the TDMA POSITION field enables nodes to agree and integrate on the TDMA schedule phase. The PRIORITY field is used to arbitrate among clock-pair masters, in the event that the pairs disagree on timing and/or schedule phase. The PRIORITY is statically allocated to pairs at design time and the pair with the highest priority wins. Following any detected contention, the highest priority pair will continue running; the lower priority pairs will yield and resynchronize according to the content of the higher priority pair's SINT message. The timing of the highest priority SINT messages also is used for global synchronization, with lower priority pairs and slave nodes correcting their local timing with respect to the highest priority master's SINT message arrival. As discussed in Chapter 4, Section 4.2, there is a relationship between precision and drift of synchronized time. The more frequently an algorithm synchronizes, the tighter the precision but, frequent re-synchronization may lead to a larger drift of the global time (affecting accuracy). The BRAIN master/slave-based approach performs better than classical distributed algorithms because it maximizes accuracy (eliminates the integration of "read error" suffered by peer synchronization algorithms) and minimizes jitter impact. The time between scheduled SINT messages (the synchronization interval) is determined at design time as an engineering trade-off between better clock precision and minimizing bandwidth overhead, which can be different for every application.

12.2.2.1 Self-Checking Master Coordination

The accuracy and validity of the BRAIN's leader-elect master/slave synchronization approach is dependent on the fault coverage of the master clocks. For this reason, the BRAIN introduces special qualification logic to cross-check a pair's health prior to them initiating any protocol traffic. In the BRAIN, all self-checking pairs are formed from adjacent nodes. Hence, each half of a master-clock pair is connected by the direct link (private link) that connects the nodes. The pairs use this link to initiate a pairing rendezvous. This rendezvous is performed by each pair once per synchronization interval and is scheduled to occur just prior to the sending of the associated pair's SINT frames. This ensures that the clocks are checked for accuracy before each half of the pair initiates SINT transmission. To achieve this, as part of a pair's rendezvousing action, each half of the pair monitors the timing of its other half. Subsequently, one half of a pair will not support a SINT transmission if the timing of its partner's clock fails beyond a configured tolerance.

To ensure clock correctness at start-up, the rendezvous procedure is also performed as part of the protocol startup. On power up, the nodes of each self-checking pair perform a rendezvous prior to entering a listen-timeout period, i.e., the period during which the nodes listen and wait to detect running TDMA traffic. If the self-checking pair nodes complete the listen-timeout without detecting traffic, the nodes rendezvous again on completion of the listen period. Thus, the duration of the listen period constitutes the period over which the startup clock frequency monitoring occurs. To ensure that the clocks are operating within the configured tolerance, both halves of the pair monitor and cross-check the duration of the listen period as indicated by its partner.

Should listen activity result in a change of schedule phase, i.e., following the reception of a SINT frame from another master, each half of the pair inhibits SINT transmission for the first synchronization interval following integration. This ensures that a full synchronization interval duration is used for clock cross-checking prior to the pair sourcing of any coordinated SINT messages. Similarly, if one half of a pair has successfully integrated into running TDMA traffic, it will not support its partner in the establishment of a disjoint-phase clique. Figure 12.7 illustrates the rendezvous procedure, showing the sequence of SCP rendezvous (REND) messages used to align their respective timelines.

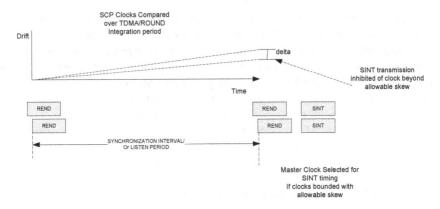

FIGURE 12.7
A Self-Checking Pair Clock Monitoring and Rendezvous

To maximize clock accuracy, the two halves of a pair do not perform mutual clock averaging. Instead, one of the pair's clocks is used as the master clock, and all protocol timing is derived from this single clock. The clock of the other half of the pair is used only for monitoring, to ensure that the master clock remains within the expected accuracy. As discussed earlier in Chapter 4, Section 4.2, this improves the clock accuracy.

12.2.2.2 Connectivity Building and Clique Aggregation

The self-checking leader-elect synchronization strategy described above is conceptually very simple. However, it is not sufficient, by itself, to address the multi-hop and segmented-media topology of the BRAIN architecture. Without mitigation, the segmentation of the BRAIN may cause startup clique formation. This vulnerability is illustrated in Figure 12.8, which depicts a hypothetical scenario of how two cliques could be formed. If the skip links around nodes A and E have failed and these two nodes are powered-on later than the other nodes, disjoint cliques can form on each side (because no messages can get past the locations of nodes A and E until they power-up). Once either of these nodes power-up, a communication path is created between the two cliques and clique resolution is required. As described in [314], existing clique resolution protocols (that are based on the properties of a globally perceived broadcast medium) cannot resolve such clique scenarios in all cases. This is because the disjoint phases of the nonaligned TDMA schedules can cause "collisions" which prevent nodes from one TDMA timeline from hearing messages sent by nodes on another TDMA timeline. For example, in Figure 12.8 the power-up timing of nodes A and E are such that node A joins the right-hand clique (H, G, and F) and node E joins the left-hand clique (B, C, and D). The nodes A, B, E and F are called frontier nodes because they are the "outermost" nodes on their respective cliques. Frontier nodes belong to one clique, but they may hear messages from another clique. These messages are rejected because they don't meet the timeline constraints of the clique to which the frontier node belongs.

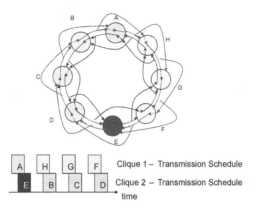

FIGURE 12.8
A BRAIN Clique Scenario

To address these issues, the BRAIN approaches the startup problem differently from previous clique resolution protocols. In place of a monolithic protocol that targets both synchronization and clique resolution, the BRAIN splits the protocol action into two layers. The BRAIN introduces a novel foundation layer in the form of a constructive connectivity-building and clique-aggregation (CBCA) algorithm that is used to build network connectivity and clique allegiance between adjacent frontier nodes. As such, the CBCA algorithm is conceived to run underneath the normal TMDA protocol activity. This means that it is unable to use, and does not need to use, the TDMA coordinated activity to implement its signaling and/or link arbitration. For this reason, the foundation of the CBCA algorithm is a simple asynchronous medium arbitration mechanism. For half-duplex rings, nodes that share an adjacent link are configured with relatively prime periods of transmission, as illustrated in Figure 12.9. By using relatively prime periods, for example 200 us and 500 us as shown, it can be assured that should a collision occur during one transmission, the next transmission will be collision free (in the fault-free case). Therefore, periodically successful transmissions will occur from each direction. For full-duplex rings, the "collisions" occur within frontier nodes, which receive messages that are out of sync with its TDMA timeline. This form of collision prevents these out-of-sync messages from being forwarded into the clique.

Similar to SINT messages, the CBCA messages leverage self-checking pair protection, where adjacent nodes collaborate to signal the CBCA messages. Each frontier node and its nearest neighbor within the same clique form a pair for sending identical CBCA messages across the frontier.

The transmission of CBCA messages is timed relative to the SINT transmission schedule of the initiating clique. The decision to transmit CBCA messages is made

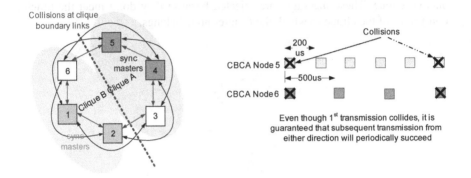

FIGURE 12.9
CBCA Link Arbitration Resolves Asynchronous TMDA Clique Boundaries

by each node determining if it is on a frontier. The full procedure is illustrated in Figure 12.10.

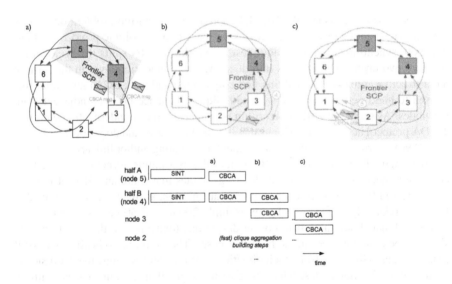

FIGURE 12.10
CBCA Clique Aggregation

In a simple startup scenario (to illustrate the CBCA process), one master-clock self-checking pair of nodes happens to start up before the other nodes, e.g., nodes 4 and 5 in Figure 12.10. At the beginning of this scenario, these nodes are the master pair and the frontier nodes. The CBCA process is initiated from this pair, following the transmission of its SINT message. Because the nodes adjacent to the pair have not sent or relayed any messages indicating that they are in the same clique as the master pair, the master pair assumes that they are frontier nodes and initiate the coordinated CBCA transmissions, sending repetitive CBCA messages on both skip and direct links. The CBCA message comprises two fields as illustrated in Table 12.2. The TIME_TO_NEXT_SINT field is updated for each CBCA transmission, relative to the sourcing-clique's clock-master's SINT timing interval, with the values decrementing as the time of the next SINT transition approaches.

TABLE 12.2 CBCA fields.

Field	Description
MASTER PRIORITY	The priority of clock-master to which the frontier node is synchronized
TIME TO NEXT SINT	A count-down field that denotes time, or number of remaining CBCA transmissions that may occur before the next scheduled SINT message

When a node adjacent to the master pair powers up and sees identical CBCA messages on both its direct and skip links, it joins the master's clique. If the newly connected node had already been part of a different clique (instead of this power-up scenario), the node would have compared the priority of the sending clique (as sent in its CBCA message's MASTER PRIORITY field) with the priority of the clique to which it was already synchronized. If the CBCA message had a higher priority, the node would join the higher priority clique and synchronize to its schedule, i.e., defecting from their previous clique/schedule. Following a node's defection or integration, it extends the self-checking frontier action with its immediate neighbor not already in the higher priority clique. This propagates the clique status and associated priority to its neighbors, which expands the frontier out by one node. This CBCA action continues for a duration proportional to the value of the TIME TO NEXT SINT field received in the CBCA messages. All CBCA action is timed to cease before the next scheduled SINT message. During the CBCA propagation in frontier expansions, no other traffic is allowed on the newly converted links. Thus, prior to the next SINT message transmission, the media among all newly joined nodes is reserved to enable clean SINT message propagation through all of them. Note that, as the CBCA mechanism executes, each node records the status of which neighboring nodes have joined its clique. Then, the nodes cease CBCA activity on those links that have indicated membership in its clique, following the next SINT transmission. Therefore, normal TDMA activity can commence between all nodes confirmed to be in the same clique, immediately following the next SINT message transmission.

12.2.2.3 Synchronous Mode Clique Aggregation Breakthrough

The CBCA mechanism above is sufficient to mitigate all cliques given a BRAIN with a single node fault. However, to improve robustness, additional mechanisms have been added to resolve dual-link and dual-benign-node failure scenarios. In such scenarios, it is necessary to utilize the integrity reconstitution mechanisms of the BRAIN, to qualify integrity of the CBCA status information by using information received from the two different ring directions. To facilitate CBCA integrity reconstitution, special CBCA "breakthough" slots are added to the normal TDMA schedule. In a breakthrough slot, each node at a frontier of the clique sends the received CBCA priority of the neighboring clique, via this TDMA slot. Unlike normal TDMA data or synchronization slots, these slots are not assigned to individual nodes or pairs and are not protected by Brother's Keeper Guardianship (see next section). Instead, all nodes are permitted to send in these slots. Hence, a suitable data authentication scheme is required to prevent a single maliciously faulty node from signaling erroneous clique activity. Such a scheme is described in Section 12.3.2.2.

Given suitable authentication of messages received during the CBCA breakthrough slot, all nodes can perform the CBCA priority comparison. If a higher priority clique is detected via CBCA reconstitution, nodes yield by ceasing to execute their current schedule and prepare to receive the SINT message from the confirmed higher priority SINT source. Therefore, by using the CBCA breakthrough mechanism, it is possible to resolve cliques even in the presence of two link faults or two benign node faults.

12.3   Fault Isolation

12.3.1   Time-Triggered Sequenced Guardian Roles

The basic high-integrity data-propagation mechanisms described in previous sections are sufficient to protect data during its transmission on the BRAIN fabric. However, additional mechanisms are required to qualify data as it enters the BRAIN, to ensure that the BRAIN's data integrity is consistent for all member nodes. These additional mechanisms can be viewed as guardian roles that cross-check and police data as it enters the BRAIN. The specific roles are selected in accordance with the TDMA schedule and are performed by the active transmitting nodes' immediate neighbors (direct links) and neighbors' neighbors (skip links). Hence, it is called Brother's Keeper Guardianship. In synchronous operation, the nodes adjacent to the currently scheduled transmitter implement guardian enforcement actions. Thus, guardianship can be pictured as moving around the ring as the TDMA communication sequence progresses, with Brother's Keeper Guardians only standing on either side of the currently transmitting node. Note that the guardian, being an independent neighboring node, ensures that guardian action is fully independent of the transmitter it is guarding, even if it is embodied in the communications controller hardware. This gives all the benefits of fully independent redundant guardian hardware, without requiring the addition of any redundant hardware components (such as central guardians or dedicated hardware components at each transmitter). Note that using the Brother's

Keeper Guardian strategy, the early limitations restricting slot order and slot size for bus topology protocols such as TTP and FlexRay can be removed.

The generic guardian role implements the TDMA selection of the schedule transmitted, granting its neighbor access to initiate a transmission, during its assigned transmission slot. This mechanism prevents a node, disturbing traffic outside its schedule slot. Additional guardian roles, described below, may also be selected according to a node's relative placement to the active sender and/or self-checking senders, and the data consistency requirements for the schedule slot.

12.3.1.1 Directional Integrity Exchange

The directional integrity mechanism is illustrated in Figure 12.11. It is performed by nodes adjacent to an active simplex sender to ensure that the sender sends consistent data in each direction, i.e., the copies of the data sent in each direction are identical. For a TDMA slot in which a simplex node is scheduled to transmit, each node on either side of the sender receives data from the sender via their direct links and immediately relays it to the partner guardian on the other side of the sending node via the skip links that connect them. The guardian nodes also perform an integrity comparison of the data transmitted directly from the sending node with the data reflected via the other guardian. Data integrity is signaled using the BRAIN's main data integrity mechanism, i.e., by setting the integrity status trailing the transmission. This guardian action ensures that the transmitting node sends consistently in both directions. If the data sent in the two different directions is not identical, both copies of the transmitted data are marked with integrity loss.

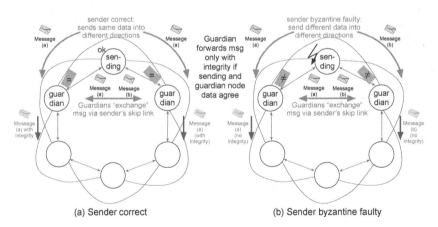

(a) Sender correct   (b) Sender byzantine faulty

FIGURE 12.11
Consistent Data Guardian Exchange

12.3.1.2   Skip Guardian Link Forwarding

The directional integrity mechanism described is sufficient to guarantee that data broadcast in both directions around the ring is the same. However, it has the disadvantage that it requires a guardian influence on the data transmissions in both directions of the BRAIN. Therefore, under guardian failure, the directional integrity mechanism may impact communications availability (but, integrity *is* assured). For example, consider the faulty guardian (FG) in Figure 12.12. This guardian corrupts the data in both the clockwise and counterclockwise directions, i.e., the data it is forwarding directly from the sender and the data it is reflecting to the adjacent guardian located on the other side of the sender. In this case, the integrity is flagged as lost in both directions. However, given that data integrity can be reconstituted, this situation can be mitigated by the nodes adjacent to the guardians. These nodes bias their selection of the data they forward such that in the event of data conflict/mis-compare the data from the skip link is forwarded, i.e., the data from the link directly connected to the sender. This action overrides the normal Availability OR selection of the standard high-integrity data propagation logic.

As illustrated in Figure 12.12, this mitigation is performed by nodes on each side of the two guardians. Using this mechanism, the data from the sender propagates, even with a faulty guardian. If a consistent transmission is made in both directions, the data reconstitution mechanism performed by all nodes as part of data reception will accept the data as good, despite the loss of integrity status induced by the faulty guardian component.

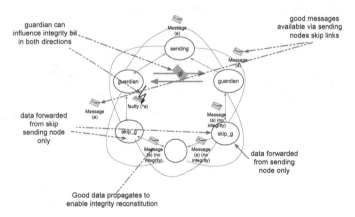

FIGURE 12.12
Skip Guardian Action Mitigate

12.3.1.3   Self-Checking Pair Neighbor Guardian

For the BRAIN's self-checking pair assumptions to be valid, the output from each half of the pair must be cross-compared as part of any data validity evaluation. Therefore, the BRAIN needs to implement enforcement mechanisms to ensure that such

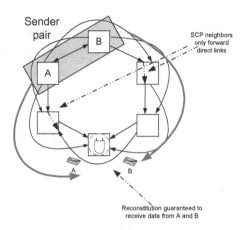

FIGURE 12.13
Self-Checking Pair Guardian Action

an evaluation is always done. The required roles of self-checking pair neighbors are illustrated in Figure 12.13. To ensure that data from each half of the pair can be utilized by the integrity reconstitution logic, the data flow from the pair is restricted such that the nodes adjacent to the self-checking pair only propagate data from the direct links, with the data received from the skip links used solely for data integrity comparison. This data flow restriction exists even in the event of the direct link being "dead," i.e., indicating no data. As illustrated in Figure 12.13, this restriction prevents integrity reconstitution from evaluating the data from a single node (which would be a violation of the integrity separation guarantees for self-checking pairs).

12.3.2 Asynchronous Guardian Roles

12.3.2.1 Startup Enforcement

Prior to entering synchronous operation, the time-triggered guardian roles above are not applicable. Hence, the BRAIN deploys alternative enforcement actions to provide the protection during protocol startup. Following power-up, the startup guardian action is performed by all nodes. Before protocol startup, the SINT message is the only message expected to flow around the BRAIN. As outlined above, the SINT message is a very small message, comprising the frame ID and clock master priority (as well as information to avoid a single node imitating a SCP (which is discussed later). Therefore, the guardian is only required to limit the size and rate of SINT message activity and ensure that no erroneous nodes send messages larger than the SINT message. The behavior is illustrated in Figure 12.14.

Initially, a node performs a race-based arbitration on skip and direct links in each

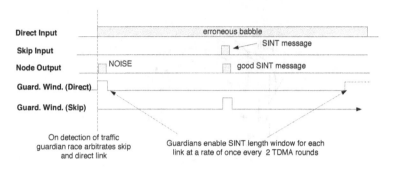

FIGURE 12.14
Startup Guardian Action

direction. Once a link (i.e., direct or skip) is identified as the winner, the guardian enables data propagation for a window slightly larger than the SINT message. Once the SINT message window has expired, the guardian blocks the link, truncating any activity that is still in process. The winning link is then blocked for a period that is larger than two rounds of the minimum TDMA synchronization cycle. This blocking action can be viewed as a toggle-based enforcement where, during the blocking period, the guardian allows only the non-winning link to take part in arbitration. Once the blocking timeout expires, the winning link is re-enabled to take part in arbitration action.

12.3.2.2 Source Authentication

In the asynchronous startup mode and/or in time-triggered slots where guardian action is disabled (for example, the CBCA breakthrough slots of Section 12.2.2.3), an additional source authentication mechanism is required to prevent a node from masquerading as two independent sources. Without authentication, a faulty node would be able to unduly influence the system and potentially disrupt protocol operation. Many authentication schemes are possible. The key requirement for all schemes is to ensure that data from a single node is not used for both inputs in cross-checking. The initial authentication scheme of the BRAIN utilized a simple "port stamping" scheme. This mechanism required frontier nodes to write their ID into the message, together with the port of entry (i.e., skip or direct), upon reception. The topology of the BRAIN ensures that the port stamps are independently applied from each direction. Since all data used for integrity reconstitution is a function of two directions, this simple port stamping scheme is single-fault tolerant, requiring two nodes to agree before the data is validated. Alternative authentication algorithms in the BRAIN can be based on a hop count [122]. It should be noted that in all cases only the port stamp or hop count data is used to implement the authentication, and no application data entering the BRAIN has influence on the authentication decision.

12.3.2.3 Additional Guardian Fault Containment Behavior

Short-Circuit Detection

One vulnerability flow through guardian mechanisms, such as those used by the BRAIN, is undetected short-circuits between links or within nodes, that may render the node's respective guardian ineffective (e.g., by the short-circuit bypassing the blocking or flag-setting actions of a guardian). Although such failures may be mitigated by the active fault scrubbing of the guardian hardware, the overhead and system complications of in situ guardian testing is often non-trivial, potentially introducing significant software and system complications. The period of guardian scrubbing can also become a limit of system reliability. Therefore, the BRAIN implements continuous scrubbing. This scrubbing uses dissimilar link encoding, which is similar to the bus encoding of SAFEbus [81]. Each link that might short to another link in a way that could bypass a guardian function is given a trivial "cryptographic key" that is different from the keys used by any of the other links to which it might short. These keys are used to "encrypt" each link's data streams using some trivial algorithms such as XOR. If the physical layer technology uses block-coded symbols, such as 8B/10B, the "encryption" is done before the block encoding, in order to maintain the desired spectral characteristics of the block encoding. Each encrypted signal that meets in some geographic locality must have a different key. Six keys are sufficient for most physical layouts of the BRAIN. This means a three-bit key.

12.4 Diagnostic and Agreement Services

12.4.1 Host Task Set Agreement

The message-stepping self-checking pair configuration of the BRAIN is an efficient scheme for software replication and output comparison. However, for such a scheme to be effective, the replicated task sets on both halves of the self-checking pair processor need to be replica determinate. It is also important to prevent a fault from entering the pair and upsetting the pair's consistency (agreement of state between the two halves of the pair). Therefore, the BRAIN utilizes the direct link that connects configured self-checking nodes to implement input state agreement. Using this private link, both halves of the pair agree on the data they received. Given the inconsistent omission fault model of the BRAIN data exchange, this agreement mechanism is similar to the SAFEbus syndrome exchange, where the nodes simply agree on the received status. A full value exchange is not required. This mechanism ensures that any data not received consistently by both nodes is dropped by both nodes. Hence, data entering inconsistently is prevented from causing state divergence within the pair.

12.5   Validation and Verification Efforts

The braided-ring topology of the BRAIN is arguably optimal with respect to hardware and bandwidth usage for fault-tolerant systems. The reliability of braided-ring topologies has been investigated in the literature and research on this related to the BRAIN is described in [122]. Summarizing, the additional links provide significant additional reliability, especially for long mission times. In [248], the reliability of the braided-ring is compared to dual-star topologies, evaluated in the context of extended dispatch scenarios very similar to the time-limited dispatch in engine control systems as described in ARP5107 [287]. The results are that braided rings significantly outperform dual stars in typical architectures and deployment scenarios with respect to reliability, while using less hardware.

Certain aspects of the BRAIN, such as the BRAIN high-integrity forwarding logic, have been model-checked using SRI's SAL (Symbolic Analysis Laboratory).

12.6   Example Configurations, Implementations and Deployment Considerations

The BRAIN is still an evolving concept. It has a wide variety of possible implementations. It can use almost any physical layer technology to create its links. It can be half-duplex or full-duplex. The links within a single BRAIN can use multiple types of physical layer technologies, as long as the net speed and the type of duplexing are the same.

As discussed earlier, a high-speed variant of the BRAIN is also under development. Targeting higher speed protocols, such as 100Mbs or Gbs Ethernet, this variant of the BRAIN incorporates both store-and-forward and cut-though messaging. It is also targeted to tolerate multi-fault scenarios, i.e., to be tolerant to simultaneous active and benign node failures. Although many of the core protocol mechanisms of this BRAIN variant are the same as those discussed above, this new variant of the BRAIN replaces the sequenced guardian roles of Section 12.3.1 with a more aggressive source authentication scheme. Such a strategy enables greater availability to be leveraged from the same BRAIN connectivity and amount of hardware components while simultaneously reducing protocol complexity. Targeting full-duplex connectivity, this high-speed variant also removes the CBCA algorithm, since the possibility of store-and-forward messaging enables messages to propagate without regard for TDMA phase alignment.

A potential physical layer BRAIN deployment option is an optimization that is possible with respect to BRAIN media routing. Often, the conceptual diagrams of the BRAIN lead readers to consider that it introduces a large cabling overhead. However, the cabling of the BRAIN does *not* have to be routed separately, and the BRAIN does not suffer any loss of dependability when cable layout is optimized. For example, consider the physical wiring of Figure 12.15. Here the physical routing of skip links passes through the same shield as the direct links and the skip connections are routed through the neighboring nodes. This results in simple physical ring topology with respect to wiring and connections. It should also be noted that the number of connections for a BRAIN is also equivalent to a dual-star despite the BRAIN's greater dependability.

The BRAIN has yet to be deployed in series production or completely prototyped within a full-up system. However, the efficiencies offered by the BRAIN approach may be attractive to high-integrity control architectures. In addition, as Ethernet and Time-Triggered Ethernets become more pervasive, hybrid approaches of mixed BRAIN and COTS technology may present advantages; for example, by allowing a high dependability by-wire infrastructure to coexist with non-critical Ethernet. This may allow generic Ethernet to be used for system maintenance and loading, allowing for the use of COTS and standard laptop or desktop derived test equipment, while still providing very strong assurance and fault tolerance guarantees for the by-wire backbones. In addition, the BRAIN's ability to support message-stepping self-checking pairs, together with its guaranteed inconsistent omission fault handling for data exchange, may also allow for significant reduction of software complexity and associated overheads. With its various flavors and forms, a version of the BRAIN can be chosen to fit a wide variety of cost and performance targets, while providing a guaranteed level of fault tolerance with the least amount of hardware possible.

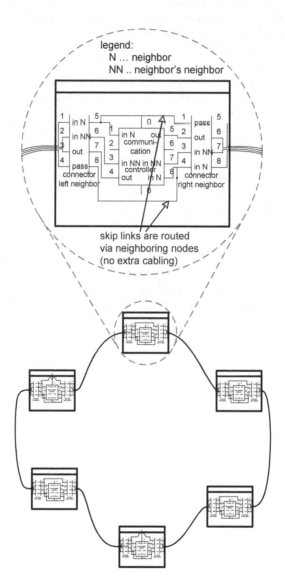

FIGURE 12.15
Cabling of BRAIN Routed within Neighboring Nodes

13

ASCB – Avionics Standard Communications Bus

M. Paulitsch

EADS

CONTENTS

13.1   Protocol Overview

ASCB stands for *Avionics Standard Communication Bus* and is a proprietary communication protocol for the avionics domain of general, business and regional aviation aircraft. The general aviation market comprises small multiple seater aircraft, the business aviation market turbo engine powered comfortable small aircraft for business clients, and regional aviation aircraft are single-aisle, medium aircraft flying regionally with up to about 100 passengers. Honeywell has developed ASCB for fault-tolerant periodic, real-time communication between avionic modules of its Primus Epic® avionics suit deployed in such aircraft. ASCB is the primary bus in Primus Epic. There are multiple ASCB versions available, some even standardized [108], the latest being version D (abbreviated ASCB-D). Detailed information on Primus Epic is available in [135, 222, 351, 61, 318, 348].

13.2   Protocol Services

ASCB-D is based on an architecture consisting of four Ethernet buses, two for each side of an airplane. Each of the Network Interface Controllers (NICs) is assigned to one side of the airplane. The NICs on the left side are connected to the left side primary, the right side primary and the left side secondary (backup) bus. The NICs on the right side are connected to the right side primary, the left side primary and the left side secondary (backup) bus. Each NIC can listen to and transmit messages on its two buses on its side. Each NIC can also listen to but not transmit on the primary bus of the other side. For example, a left side NIC can transmit and receive on the left side primary and backup bus, and listen-only on the right side primary. It is not connected to the right side backup bus. From the viewpoint of one NIC, the NICs on the same side are called *onside* NICs; NICs of the other side are called *xside* NICs, where *xside* stands for *cross-side*. Figure 13.1 depicts the structure of the main buses and connections of NICs.

13.2.1   Communication Services

ASCB-D is a TDMA-based protocol where a minor frame (the fastest period) is running at 80 Hz. Figure 13.2 depicts the bus traffic of one bus. First, master NICs send synchronization pulses (sync pulse) followed by periodic traffic (Ethernet frames). The beginning of each frame is called a *frame tick*. The physical layer of the buses is based on 10Mbit/s Ethernet using twinax cables and connectors. Figure 13.2 shows the minimum period (called *minor frame*) on the bus. Two sync pulses (special Ethernet frames with synchronization information) are the first frames on the bus. The two time servers that are connected to this bus send these sync pulses. After this, individual NICs (i.e., both the time servers as well as the time slaves) send based relative to frame tick according to a predefined dispatch schedule, where the frame tick is the beginning of a minor frame. The synchronized time of NICs is used for avoidance of collisions on the network. ASCB is part of the Primus Epic architecture, which does not leverage sub-frame timing (i.e., on actions based on time offsets within the minor period). Chapter 14.2 will explain details.

13.2.2   Clock Synchronization, Restart, Re-Integration and Integration

Clock synchronization, startup or re-integration in Primus Epic are basically the same, as ASCB deploys a master-selection protocol between four cooperating masters called *time servers*. Each of the NICs (including time servers) needs to select one time server. The arbitration on which of the four time servers is selected by other NICs is based on a protocol of cooperating time servers leveraging strike counters and predetermined unique constants in case of conflicts between multiple time servers (multi-master conflict) or no master is active (mastership transition). The time server that first or last counts to its predetermined constant depending on the strike

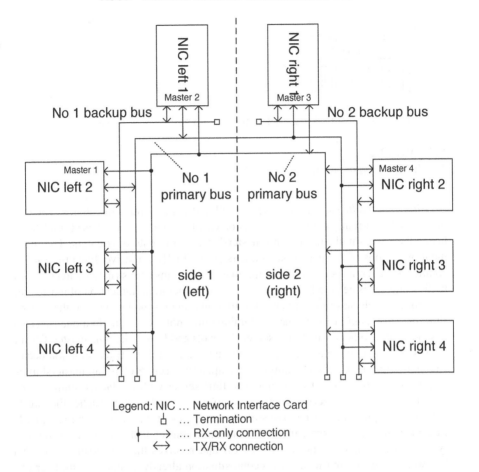

FIGURE 13.1
ASCB Architecture

counter communicates its decision of being the timing master, which is then followed by other time servers. Startup has to complete in 200 milliseconds [351].

In more detail, each time server sends on both of its onside buses and receives the messages on the xside primary bus. A time master only receives messages on one of the onside buses (backup or primary). A time master sends a sync pulse at an offset from its frame tick (i.e., the start of its frame period) based on its ID. Each time server can derive the time difference between another time server's frame tick and its own by measuring the time difference between that server's sync pulse reception time and the expected reception time. Each time server can choose to synchronize to another time server's sync pulse by adjusting its local clock to account for the measured time difference with another time server's sync pulse so that their frame ticks are synchronized.

The protocol provides two strike counters at each time server, called *mastership*

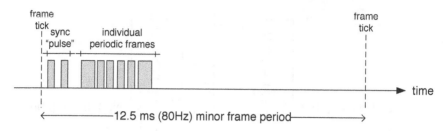

FIGURE 13.2
ASCB Minor Frame Period

transition and *multi-master* strike count, with individual unique thresholds. The *time master* is the currently active time server driving the common time-base for all NICs, hence the names multi-master and mastership transition strike counters. The increment operations of the two strike counters are mutually exclusive. Each time server counts for different periods (using the mastership transition strike counter) in case no time master is detected and sends a sync pulse including indication of mastership (claiming the elected time server is the time master) on the bus in case the mastership transition count reaches the local unique threshold. Once the first one succeeds in sending this, other correct time servers back off and follow the time master. During the period of no time master, time servers run autonomously. If the situation occurs where multiple time servers claim mastership, a time server claiming mastership increases its multi-master strike count. The time server with the largest multi-master threshold wins. A time server where the multi-master strike count reaches the threshold and multiple time masters are still present, gives up mastership and restarts. In case of connectivity issues, the synchronization master that does receive multiple synchronization masters does not back off and receives the mastership over other less well connected masters. This second situation already assumes multiple errors. In case this also fails, the remaining units choose one master and follow its local master. The protocol is not guaranteed to resolve multiple failure scenarios.

The *external timing master test* is performed at certain points in time and is a test that is based upon the receiving sync pulses from other time masters and the relationship of sync pulses to each other. Examples for relationships are tests for periodicity, temporal gap between successive synchronization signals, for errors in the order of arrival of synchronization signals and for absence of particular synchronization signals. The external time master test should prevent any time master from selecting a timing master with unacceptable timing integrity and detects the absence of the timing master. The *self test* implements a self-monitor and de-activation component in the algorithm. Each time server validates its own timing integrity and de-activates itself when it does not have the required integrity. The time server performs a sync pulse wraparound monitor of the frame counter, a frame tick periodicity monitor and a local clock integrity monitor. The *break sync action* in Figure 13.3 implies that a master that is sure that it is the correct master—based on these locally performed tests (external and self test)—does indicate this to other time masters via sending a synchronization signal claiming mastership or indicating that it wants to take over mastership. Figure 13.3 provides a graphical overview of the ASCB startup and clock synchronization algorithm.

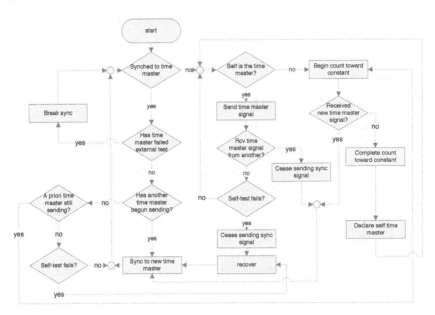

FIGURE 13.3
Flow Diagram of ASCB/D Synchronization, Integration, and Startup Algorithm [348]

Please note that the algorithm may have changed slightly due to reported incidents tracing back to synchronization [96].

13.2.3 Diagnostic Services

Examples of diagnostic services are already explained above in the synchronization section, where time servers check their correctness based on externally stimulated events. Similarly, NICs perform built-in extensive self-tests at power-up and some continuous self-tests during operation. Heartbeats for power and microcontrollers are deployed and are based, e.g., upon the frequency of data transmissions.

13.2.4 Fault Isolation

Fault isolation in ASCB is implemented by the separation of data buses and listen-only connections to primary buses as a first defense. As a second level of defense serve secondary data buses that are limited to one side only. Transceiver functions for different buses are separated.

13.2.5 Configuration Services

Configuration of ASCB is performed via a special network, a separate local-area connection. This is used to configure local configuration tables of networks and processors. This network can also be used for monitoring of operation.

13.3 Protocol Parameterization

The communication schedule is composed of multiple minor frames (12.5 milliseconds) in powers of two up to 1 second. Exact sending patterns are designed and automatically scheduled by a tool called ESCAPE (Essential System Configuration and Architecture for Primus Epic). Data groups, so called parameter groups, are handled as a unit, meaning that they are only used if the correctness and completeness checks of all elements of these groups are correct. A correctness check is, for example, whether the error detection codes associated with this data group are correct. A data group is a logical unit, which also implies that a group does not necessarily need to be transmitted in the same Ethernet frame.

Data elements sent on the network are scheduled at frame level (frame is used here in the sense of a frame period). That is, the location of a message (Ethernet frame) within the frame period does not matter for end-to-end latency calculations for applications. The data transmitted during a frame period is not immediately used within this frame period, but only in the next frame period. This allows deploying of very simple scheduling algorithms at the expense of delay. Effectively, the frame-period-level scheduling approach decouples any dependencies between the network timing table and operating system timing tables. Consequently, the delay from one data element in an application is always larger than a minor frame period. Actually, as this principle is deployed based on the frequency of the application, the delay is always larger than the frame period of the application at which the application is run.

13.4 Communication Interface

The communication interfaces in Primus Epic are generally using ping/pong buffers. These are double buffers, where one buffer is used by the application in one period, while the other buffer is used for updates by the network in the same period. This way network traffic does not overwrite the data that is currently used by the application and vice versa. At a predefined point in time, namely at the period boundaries, the role of the buffers changes and the other buffer that has been reserved by the application is now used by the network for sending and receiving.

13.5   Validation and Verification Efforts

ASCB core algorithms, such as clock synchronization, have been verified under the covered fault assumptions and is described in [351, 61]. ASCB has been developed using the common applicable aerospace standards for design including DO-178b.

13.6   Example Configurations and Implementations

ASCB has been used in the Primus Epic avionic platform in multiple aircrafts. Details are mentioned in Chapter 14.

14

Industrial Applications

M. Paulitsch

EADS Innovation Works

E. Schmidt

TTTech Automotive GmbH

C. Scherrer

Thales Rail Signalling Solutions

H. Kantz

Thales Rail Signalling Solutions

CONTENTS

14.1   Introduction

This chapter describes industrial applications of time-triggered communication. We present products and real-world systems, which are examples for the use of the time-triggered networks that were introduced in the previous chapters. In particular, we discuss the advantages and disadvantages of the properties of time-triggered communication protocols in these examples. The properties of time-triggered protocols such as composability, determinism and predictability are most useful in safety-critical applications, where these properties facilitate the realization of fault-tolerance, timeliness in all specified fault and load scenarios, rigid validation and certification. Therefore, we present examples of safety-relevant systems from the aerospace, automotive and railway domains. For each domain we outline a typical overall system architecture and explain the role of time-triggered communication networks in this architecture. We discuss the requirements of the domain and explain how these requirements are met by the time-triggered communication protocols.

14.2   Time-Triggered Communication in Aerospace

Historically, digital data communication and time sharing of bandwidth emerged as a way to reduce weight after digital computers were introduced in the 1970s replacing analog data communication and point-to-point networks between then often stand-alone "boxes" associated with aircraft functions. The three major aircraft avionics functions are Communication, Navigation, and Surveillance, where communication refers to communication of the cockpit and ground and not databuses. Today, these airplane functions are often integrated on computers in order to share not only databuses but also computing power. A concept often referred to as Integrated Modular Avionics (IMA) and expanding to include more than these traditional functions like utility control systems.

Digital databuses are (mostly) common standards confined to avionics addressing special physical, electrical and logical interfaces and interactions. They are used by avionics functions or subfunctions and lately also cabin and utility functions to send internal data between each other. Data may comprise sensor and actuator information (control traffic), the results of internal calculations, system commands, information from internal storage, audio and video data, graphics, in-flight-entertainment and passenger service data, relayed data or any information that may be generated by a computational device.

Compared to other ground or commercial communication, the difference is often related to dependability requirements (high integrity, availability and safety) due to safety-critical applications. Yet, not all dependability requirements are for safety-critical applications, but the cost of diverting a flight due to failure may create economic availability requirements. Another key characteristic of aerospace databuses is the amount of certification required to ensure that the very high level of integrity and safety required for aviation is reached and maintained.

Early databuses have been characterized by single transmitter and multiple transceivers. ARINC429 [10] is the most prominent example of this architecture. Such single-transmitter-and-multiple-transceiver architectures led to two basic classes of equipment. One with single transmitters and single receivers like a sensor or a switch panel. The second equipment class required information from many receive channels (like flight management) and led to many wire bundles (often 50 to 100 receive channels). Large wire bundles led to the additional desire to multisource databuses. These, however, potentially suffer from the fault effects problem that needed management by the network and system architect. Functions previously independent were then with a certain (low) probability related to each other and could lead to loss or corruption of data. Or even worse, with some finite probability a function could impersonate data of another (higher criticality) function.

One other aspect of fault effects is temporal behavior, the special focus of this book. In aerospace, databuses need to guarantee availability of the databus within specific time intervals for a function to meet its response time independent of faults being present.

14.2.1 Requirements

This section elaborates on the general and special network-related requirements in aerospace: First, major high-level requirements that characterize the aerospace operational and development environment; second, more network-specific ones. The general requirements are partially dependent on each other. Hence, in the classification below, categories shall not be seen in sole isolation when applied later on.

Long life-time. The aerospace industry is characterized by long life-time. Aircraft stay in operation for 30 years or more (sometimes even for over 50 years). While certain parts like avionics are upgraded and redesigned more often, the required operational life of electronics is still significantly higher than current consumer electronic life-time.

Safety culture. The safety culture in aerospace is driven by minimizing accident numbers and high publicity surrounding accidents due to large number of fatalities. This safety culture continuously refines cockpit features to increase situational awareness of many possible dangerous incidents (like avoidance of mid air collisions, potential runway overrun information, warning of controlled flight into terrain, flight envelop protection, gust detection).

Design process and certification. The small production numbers and serious consequences of any failure make the aerospace industry a highly regulated one with the certification process as a way to ensure quality assurance and to ensure that high dependability requirements are met.

High availability and reliability. High availability and reliability are driven by the safety requirement and economic considerations. Aircrafts' high cost require operational hours of as high as 50% of a day. Dispatch numbers of 99.5% or more for long operational sequences of a fortnight are common requirements for new aircraft. In other words, an aircraft shall complete a schedule of 14 days with 12 operational hours a day nearly all the time (only 0.5% or less are allowed maintenance actions). And this required in an environment with extremely high safety requirements.

Weight. Low weight is a critical design target, because fuel and ultimately money highly depend on weight. Often the term "SWaP" for "size, weight and power" is used. SWaP shall be minimized, which ultimately stems from the fact of requiring low weight designs.

Cost. Cost is a significant driver in general aviation and the helicopter market. Cost has also become a driver in recent years in the air transport area with the increased competition and reduced ticket prices. This cost aspect not only translates to decreased development cost and unit prices, but also decreased maintenance and service cost.

Automation. An aircraft becomes more and more automated to offload duties from the pilot and to reduce stress on the aircraft, but the pilot is ultimately responsible for flying the aircraft and in charge at all times and monitors safe operation. Increased automation requires functional integration and increased cooperation and, hence, increases communication needs between functions.

Specific databus functional and non-functional application requirements result from the above high-level operational and development requirements. Additionally, databuses now transport a multitude of different traffic classes on the same network, which are important to understand in order to achieve the best system viewpoint. The following presents an augmented list of databus requirements derived from above, experience in time-triggered network development and safety standards and Advisory Circular 20-156 [95]:

Long life-time and cabling. The initial installation and maintenance of wire harness is recognized as a significant life-cycle cost. Twenty to 30 years or more of service life contributes to the issues such as corrosion, chaffing and vibration that eventually degrade the wiring infrastructure. Electronic components also need long life-time (say, at least 10 years) in order for aerospace to be interesting and in order to assure maintenance support.

Any multi-source (more than one transmitting node) buses present an opportunity to further drastically reduce the number of aircraft wires, and thus save significant cost and weight. In addition to the reduction of cabling, the goal is to minimize the number of databus types in an airplane for development and maintenance reasons in order to minimize development cost and ease maintenance.

Safety. Databuses must comply with legal rules like FAR/CS 25.1309 and guidelines like ARP4754 and ARP4761 in order to determine how the databus will affect safe operation of the aircraft. This determination shall consider the following aspects:

Bus partitioning: The most significant challenge for multi-source topologies is in the fault-effects domain, as the network has become a shared resource. For databuses that highly integrate multiple functions one must accept that data of varying criticality will be transported by the new network. The importance of separation of traditional guidance and control parameters (those which keep the aircraft flying) is obvious. But, multi-source topologies must also assure that non-critical – and thus for cost reasons given less stringent scrutiny by certification processes – never disturb critical communication.

Databus availability and reliability: Databus availability and reliability requirements meet the safety requirements, like low-probabilistic failure targets, even when considering reconfiguration modes or mechanisms. Typically safety-related requirements for a databus (without replication considerations) are in the order of 10^{-5} to 10^{-7} failures per hour and less likely than 10^{-9} for replicated system buses.

Failure reporting: Failure detection, reporting and management features such as the use of redundancy features, detecting the loss of nodes, the support of transparent shadow nodes and the support of parallel nodes. A node is a databus component capable of processing, sending and receiving over the databus.

Common cause: Common cause (including common mode) and cascading failures: For each aircraft, a common cause analysis must be performed to address faults that go beyond databus fault analysis. This common cause analysis includes a zonal safety analysis, particular risk analysis and a common mode analysis. The location of line replaceable units (LRUs) and the routing of wiring between those components are critical in the achievement of acceptable zonal safety and determining the particular risk analyses. Redundant channels of the databus must not be routed through zones where a probable risk may cause loss of all channels. To protect databus applications, electrical and mechanical means to isolate databus channels shall be used. One common example of common cause considerations is power supply and lightning effects analysis. Given the latest developments of increased use of composite hulls, scrutiny in the area of electromagnetic effects and common cause influence is important.

Verification and validation: Databus requirements must be verified and validated, which includes DO-178B/ED12B [277], DO-254/ED-80 [278], DO-160E [279]. Verification should ensure that the requirements meet the implementation and validation that the requirements and implementation meet the intended design and match system requirements. DO-254/ED-80 concerns design assurance guidelines for complex electronic devices. DO-178B/ED12B ensures that the design and development assurance of software related to the databus and databus architecture is met with the respective safety levels. DO-160 ensures that emissions and susceptibility (like electromagnetic compatibility, lightning) are met and describes environmental conditions and test procedures. Properties of the electrical signals, cabling, shielding, interconnects, etc., play a critical role in susceptibility and emission of databuses. For validation purposes, functional tests of the integrated databus shall ensure that it meets its intended function. Tests should also include failure and recovery procedures and ensure that built-in tests function correctly. A simple example of a supporting functional test is when protocol messages on a databus have a special format and include basic protocol information. This alleviates the functional test by allowing information about the state of the network at all times.

Fault containment: Fault containment is one form of preventing common cause effects (especially cascading failures). Containment of faults concerns multiple aspects. Integrity of claimed addresses is one concern in order to prevent masquerading of equipment. This is an important— sometimes overlooked—requirement, as otherwise one function has an inadvertent effect on another function (more on this as part of the data integrity paragraph below). But also physical effects like over-voltage need to be considered to contain faults.

Data integrity. Many functions not directly in the path of traditional control impose criticality requirements. The situational awareness of the flight crew (e.g., displays) is a good example. Presenting misleading data to the flight crew is a prime concern. More subtle requirements emerge regarding pilot workload (latency, response time) or assuring the pilot's trust in the system (irregularities, unconfirmed faults). Integrity requirements deal with a system probability to introduce errors into the system, which for databuses can be end systems (also called nodes) and the physical media connecting the buses. In order to handle such a data integrity requirement, the databus needs to evaluate and adequately design the maximum error rate per byte or message expected for transmission, the effectiveness of error detection techniques, ensure that the time integrity is met and no stale or old data is able to remain or be on the network (data too long in the system), consider buffer overflow and underflow aspects, ordering integrity is assured (no sequencing errors), and authenticity of the sender is maintained. It is understood that no perfect system exists, but the residual probability of faulty operation is sufficiently low to be of no concern to the system.

Protection of integrity can, e.g., be achieved by a robust physical layer, but also with more intelligent techniques like repetitive transmission (temporal redundancy), successive sample filtering or periodic built-in test (BITE) for single channels. For periodic BITE, the integrity is checked periodically during operation. Also, multiple channels can be used for integrity or availability. While availability is more often the goal of replication, integrity can be the goal (a good example of integrity via replication is the braided link of the BRAIN described in this book in Chapter 12 or the replication for self-checking of SAFEbus described in Chapter 7).

A very important requirement to ensure integrity is ensuring that existing or latent fault mechanisms are regularly "scrubbed" for their correct operation. Otherwise, the long life-time of equipment in aerospace needs to consider that the monitor can become faulty, which—given the long life-time—is very likely. Again, testing the monitor equipment or function at regular intervals (e.g., when powering down the avionics) is a necessary step to ensure high integrity.

Availability. Safety-related availability requirements are described above. Additional availability requirements surface due to the cost of flight diversions or flight cancelation, which can approach a million dollars for loss of revenue and the expense of re-routing and providing food and lodging. Such availability requirements due to economic reasons are quantified as *aircraft dispatch targets*. The dispatch target is the probability that an aircraft will successfully dispatch and is likely 99.5% or more for modern aircraft. Acceptable dispatch is defined during design time by engineering analysis through a minimum equipment list (MEL). The MEL is additionally tightly related to safety. In addition, a quite complex supply chain and associated cost have led to requirements of extended dispatch to minimize having to store significant amounts of spare equipment at remote airports. As a consequence of allowing dispatch with faults, modern databases need to assure that dispatch despite a fault does not cause interferences, which is discussed by the fault containment requirement above.

Performance considerations. Performance considerations in databases are endless and include end-to-end delay, effective bandwidth use (effectiveness of data usage pattern needs mapped to actual physical bits and bytes), ability to schedule the data effectively on the network, ability of determining maximum traffic utilization for timely data delivery. Also physical layer properties need close attention, especially as databases often stretch through the whole aircraft with multiple interconnects due to different aircraft production sections (up to ten are possible) and lead to loss of signal strengths.

A property especially related to time-triggered protocols is the need of determinism and bounded latency. These are driven by audio and video requirements due to human factors, but also due to tight control loops that run over the network. Any variation may introduce additional test and/or functional mitigation (more robust control approaches).

Performance considerations also need to consider testing and loading. Flight test equipment often needs detailed insights into the data exchanged and hence high bandwidth data rates. Similarly, software or data (like maps) loading needs to be considered. Airliners want load times in the order of 10 minutes for volumes of a few hundred megabytes. Such requirements likely need parallel load sessions and sufficient bandwidth.

Tables 14.1 and 14.2 present some databus traffic classes of modern air transport airplanes. The figures are rough estimates only and depend on the actual system architecture. The table excludes in-flight and passenger entertainment services as these are likely separate from avionics type databases and follow different economical cycles. Yet, it should be mentioned that requirements for in-flight services – if including passenger announcements – have tight timing constraints and, hence, could be well suited to time-triggered networks.

Configuration management. Each individual aircraft databus configuration could potentially be and likely is unique. Due to this uniqueness, any changes to the databus configuration may result in an adverse effect on the databus performance and reliability. In order to prevent adverse effects, manufacturers must establish a configuration management process and guidelines for the databus. Such installations will also require maintenance personnel and installers to use manufacturers' approved configuration databases and tools when maintaining or re-establishing the airworthiness of the databus configuration. It is important to address the system configuration management items when developing your databus such as ensuring there are mechanisms of configuration control for continued operational safety and configuration control of the modifications by in-service maintenance or design changes.

Security assurance requirements. Many modern databuses introduce potential security risks that are not common in traditional networked systems. First, they are better known publicly and, hence, the vulnerability potential is greater than in the past. Second, many newer system architectures integrate many more functions and, hence, may be open to access by non-trusted personnel. Hence, access security and data protection of databuses should be addressed. When airborne systems interact with the outside world through a databus or network, they may become vulnerable to potential malicious attacks such as software viruses. Each databus should be evaluated for this potential risk and, depending on the risk, adopt appropriate security techniques and controls to protect access to the airborne software with techniques like encryption techniques, authentication and access control policies and intrusion detection. Similarly, critical information used and stored in airborne systems should be addressed by information verification methods to ensure that the data have not been corrupted during loading and storage (for example, using cyclic redundancy check and checksum) and use of audit trails to account for data accessed.

Internet protocol. An aircraft Flight Management System (FMS) is the flight crew's equivalent of an industrial control plant's control center or an engineer's workstation. The design and progressive improvement of such systems has demanded more and more computer resources. A strong argument exists that the architecture (and capabilities!) surrounding such systems be made similar to that of networked workstations, file servers and the like. More and more movement of data with Internet protocols is finding support in avionics due to widespread knowledge of engineers (important during development) and abstractions similar to conventional network architectures. Usage of known principles and concepts also ensures familiarity of the databus technology by the engineers during development.

Environment. High-speed avionics networks will likely be contained in the pressurized portion of an aircraft and need to meet the modest requirements of industry documents (DO-160E) and temperature -40 to +85, etc. Field-buses on aircraft have more severe requirements, however.

14.2.2 A General Discussion of Time-Triggered Communication to Meet Requirements

Many of the above-mentioned requirements are not specific to any bus approach – be it time-triggered or not. In the following, we discuss some potential advantages of time-triggered communication, but also potential drawbacks. First to the advantages:

- Time-triggered communication requires that system resource allocation needs to be made explicit early in the design phase. This is done by creating a schedule that is used to trigger events. This naturally fits the requirements in aerospace of early planning of resources. The very positively seen planning requirement and assignment of resource is sometimes taken against time-triggered communication and argued as imposing unnecessary planning too

TABLE 14.1
Exemplary traffic classes categorization with properties for regional or transport category airplane (Part 1).

Class	Typical Applications	Bandwidth	Smallest Latencies	Periodicity	Criticality
Traditional Control	actuation and sensing, flight control, flight guidance, air data, inertial navigation, radio altimeter	low volume traffic for each individual subclass; overall 1 to 2 Mbit/s	few milliseconds	100 Hz to 1 Hz	traffic is highly critical as applications include moving aircraft flight surfaces; fully testable; full separation; 10^{-9} failures/hour or less
Utility Control	environmental, cabin pressure, waste management, power distribution	low volume traffic individually; overall around 1 Mbit/s	ten to 100 milliseconds	100 Hz to 1 Hz	power distribution and cabin are highly critical, others - like waste management - have economical availability requirements that are due to minimization of cost due to flight diversions
Video	exterior cameras for ground maneuvering and airborne and ground situational awareness; interior cameras for monitoring passenger situations or supporting safety investigations	tens of Mbit/s (vary depending on compression and video technology); recording requires much lower bandwidth	hundred milliseconds or more (but compression of video may reduce network latency even further)	potentially monitor update rates (25-50Hz)	likely low; inherent fault tolerance in video due to slowness of human visual processing

TABLE 14.2
Exemplary traffic classes categorization with properties for regional or transport category airplane (Part 2).

Enhanced Vision Systems	various sensors to augment what the flight crew sees due to inclement weather, airborne pollutants, etc.	likely tens of Mbit/s	hundred milliseconds or lower; latencies may need to be lower than video traffic	potentially monitor update rates (25-50 Hz)	critical for landing and departure and ground maneuvering; likely 10^{-9} for short critical phases
Graphics	graphics deal with information displayed to the flight crew with sub-classifications for graphics, like traditional displays (parametric driven), remote clients, client-server (implementing paperless cockpit or moving images like weather radar and terrain views), and image-based communication support	few kilobits/sec for parametric data through multiple Mbit/s for some remote clients, to 20 Mbit/s for moving terrain systems (if graphics module and display are separated, then even much higher display rates)	latency of few milliseconds	displays in the order of 10 to 30 Hz	criticality extremely high for air data (10^{-9} or lower); high criticality for engine indications, navigation data, etc.
Audio	cockpit intercoms, crew intercoms, passenger announcements, digital radios, aural alerts, maintenance attachments	around 10Mbit/s (higher bandwidth due to higher quality compared to telephone, multi-channel, and voice inflections driven by application-level tolerance checks with side tone approaches	10 milliseconds for some channels	100 Hz	cockpit, aural alerts, and air-ground critical to 10^{-9}

early in the process. Such perception may be exacerbated by missing or less than optimal tooling support. Similarly, the requirement of creating schedules can be perceived as arduous when "playing around with" the system in early phases of system designs.

- It can only be argued that the requirement to create early resource allocation attempts strengthens the safety case and system development approach. For fairness reasons it needs to be mentioned that this design approach is by no means confined to time-triggered systems; early planning can easily be achieved by non-time-triggered systems.

- Time-triggered systems in general reach a good bandwidth utilization for systems with regular traffic as can be often seen in aerospace and control operations. Rate-constraint traffic, i.e., traffic that is periodic but not triggered by time (see Section 14.8), cannot even in the best case achieve similar maximum utilization levels due to missing alignment of sending times and related queueing needs and long sending queues. This increased bandwidth utilization obviously is only an advantage when communication bandwidth is in need and expensive.

- An integrated system approach where applications take advantage of the availability of coordinated behavior based on time allows restriction of the application-level testing cases to a few manageable parameters especially with respect to timing variations. Long possible variations for non-time-triggered systems require more test cases to cover all possible timing variations. Depending on the variations, this may reduce testing requirements significantly.

There are also the following perceived or actual disadvantages:

- Time-triggered communication requires a minimum set of operational nodes for fault-tolerant system-level algorithms such as synchronization and startup. This minimum set of nodes requires considerations that may become significant during certain system-level states such as shortly after power up or when considering maintenance modes where only sparse resources are available. All possible operational modes may complicate definition of available resources and may lead to design trade-offs between synchronization and power-up considerations that are only present for time-triggered operation.

- For time-triggered systems, the system fault tree likely includes faults related to synchronization. This leads to the impression that synchronization is an additional burden on the system. This can but does not have to be the case. It may be that the hazards are only more obvious with synchronization, but are naturally present for time-triggered and non-time-triggered systems.

- The rigidity of synchronization and the TDMA-based approach can be seen as a disadvantage. This perception is re-enforced especially by the unnecessary requirement of being only able to send once per TDMA round due to early bus guardian approaches (like in TTP/C; see Chapter 5). In such cases, often present asymmetric bandwidth needs – where sensors only need to send a few bytes and central computers need a lot of bandwidth – cannot be easily met or require a work-around or less efficient network schedules. This early restriction has been recognized and been removed by later generations of time-triggered systems such as FlexRay or TTEthernet.

- Similarly, future extensions of time-triggered schedules are argued to be much harder due to either having to change existing schedules and re-test existing system behavior or not having the freedom to change otherwise. While such arguments are true, time-triggered systems are not the only systems to "suffer" from such additional needs. Other system-level approaches have to retest all applications on the bus in all cases as their timing changes all the time. As such the perception of a restriction is more a planning requirement that if enough freedom for future extensions can be reserved in early versions of created schedules, time-triggered systems have an advantage.

Summarizing, the use of time-triggered communication in aerospace is not a requirement per se, but some requirements are more easily and more efficiently achieved with time-triggered systems. This is especially true if early in the design cycle, the system-level approach is clear and all involved parties are aware of the design needs of time-triggered systems. At the same time, time-triggered communication introduces system-level constraints into the architecture that may not always be easily dealt with. Time-triggered systems are perceived to introduce a relationship between components that may not be of advantage. Also, algorithms for synchronization of time-triggered systems generally require a minimum operational set of network resources (nodes). These introduced requirements need to be known for design of aircraft systems. If so, time-triggered communication can be a very efficient way to minimize resource requirements and allow reduced testing efforts during system-level integration.

14.2.3 Use of Time-Triggered Communication Networks in Aerospace and Space

This section describes a couple of cases where time-triggered communication is used to the advantage of the system. In nearly all of these approaches it becomes clear that the combined effort of all designers at different system levels leads to an advantage for the airplane design.

The following platforms are described in more detail below.

- Boeing 777 and the use of SAFEbus

- Primus Epic and its use of ASCB

- TTP's use in the Modular Aerospace Controller

- Orion and the use of TTEthernet

14.2.3.1 SAFEbus® in Boeing 777

The avionics architecture in the Boeing 777 is characterized by functional integration compared to predecessor aircraft. Classical avionics at that time had a federated avionics architecture, where a set of functions is implemented in one or more line replaceable units. In the Boeing 777, this has been replaced by a more integrated architecture where multiple sets of functions are combined in fewer line replaceable units (LRUs) all in common cabinets instead of many separate boxes. This Integrated Modular Avionics (IMA) is formed around the concept of powerful processing modules with a deterministic operating system that hosts many applications. This integration offers many benefits, such as lower weight, lower power consumption, increased reliability, less frequent maintenance and greater flexibility. Yet, because functions share hardware resources, greater care must be taken to ensure they will operate correctly, even if co-resident functions fail. Integration increases the risk that unwanted interactions among the functions residing on the shared hardware will lead to unforeseen failures. This is the engineering challenge of integrated architectures [136].

An integrated cabinet typically is larger than a federated LRU, but smaller than the sum of all LRUs that the cabinet replaces. The separate functions in a cabinet share a power supply, general and special input/output (I/O) units and processing resources. To increase availability or integrity, functions can be replicated in multiple line replaceable modules (LRMs) or in multiple cabinets. The attraction of the integrated architecture is the economies that can be achieved by sharing resources, like power, mechanical housing, processing and I/O.

The core avionics of Boeing 777, also called AIMS (Aircraft Information Management System), attempts to make the execution environment of each function in the cabinet as much like the environment in the discrete LRU as possible. Essentially, all shared resources in the cabinet are rigidly "partitioned" to ensure that one function cannot adversely affect another under any possible operating condition, including the occurrence of faults or design errors in the functions. The best way to ensure adequate partitioning is via strict deterministic control.

Functions must be partitioned both in space (memory and I/O) and in time. Deterministic control over the partitioning of space means that it can be guaranteed that no function can prevent another from obtaining adequate resources like memory space and that the resources are assigned to one function only. In the case where some memory space is assigned to one function, this means that it cannot be corrupted by another function. Pre-allocated memory areas with hardware-based memory protection such as memory management units (MMUs) prevent contention for memory space. Deterministic control over the partitioning of time means that it can be guaranteed that one function's changing demand for hardware resources will never prevent another function from obtaining a specified minimum level of service and that the timing of a function's access to these resources will not be affected by variable demand or by the failure of another function [79].

Figure 14.1 provides an overview of one of the Boeing 777 AIMS cabinets. The cabinet comprises three—at that time powerful—Core Processor Modules (CPMs) for computing tasks, running display, graphic and data conversion gateway applications. One module is dedicated especially to communication (including flight deck communication) and four identical modules to I/O. Other modules like local power modules support all modules located in a cabinet. Furthermore, three slots (for one CPM and two Input/Output Modules (IOMs)) are reserved for future growth. The replicated applications hosted on AIMS are as follows, along with the number of redundant copies of each application per shipset in parentheses [355]: Displays (4), Flight Management/Thrust Management (2), Central Maintenance (2), Data Communication Management (2), Flight Deck Communication (2), Airplane Condition Monitoring (1), Digital Flight Data Acquisition (2), and Data Conversion Gateway (4).

Applications in AIMS use the following shared platform:

1. Common processor, power supply, and mechanical housing

2. Common input/output ports

3. Common backplane bus (SAFEbus) to move data between CPMs and between CPMs and IOMs

4. Common operating system, built-in test (BIT) and utility software

The CPMs are based on AMD 29k RISC microprocessors using an AMD 29050 for AIMS-1 and HI-29KII for AIMS-2. Each CPM operates two processors in a lock-step configuration, which compares all output of compute applications for consistency similar to SAFEbus. If an error is detected, an error is flagged and all processing is halted ensuring that erroneous data is not propagated to other modules. Once errors are processed, errors are logged, and recovery is attempted. Figure 14.2 depicts the configuration of self-checking processors with SAFEbus.

The predictable time-driven operation of SAFEbus is extended to the processor and operating system. The scheduling of the operating systems is static and table-driven in order to ensure timely partitioning and minimize unwanted interactions between applications running on the same processor. A processor cyclically executes a schedule where different partitions (applications or parts of applications) are assigned different time slots and can use assigned hardware during this period. The time slots of applications (also called partitions) are fixed similar to the time slot of messages on the bus. This means that the dispatcher at partition level is purely time and table-driven without any other influence leading to rigid temporal separation. Within the partition time slices, one or more tasks (called processes in [143]) may run, which can be scheduled more flexibly and do share resources. Tasks of one partition may not require separation as they belong to the same partition and, hence, application. The principles of operation and standardized interfaces applied to AIMS have been standardized in ARINC 653 [143, 145, 144].

Scheduling of AIMS is described and proposed in [51]. Real-time repetition rates for applications range from 1 Hz to 80 Hz. Synchronization between the AIMS processors is used to control latency and allows the system to "hand over" data at points during the repetition cycle (also called *frame*). When data is shared during cycles and the processors and the network are synchronized, the latency can be minimized.

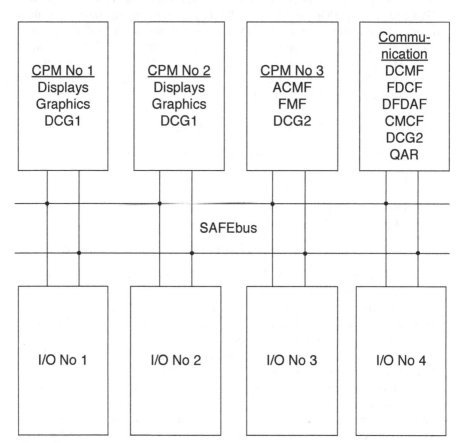

Legend:
DCG ... Data Conversion Gateway
ACMF ... Aircraft Condition Monitoring Function
FDCF ... Flight Deck Communications Management Function
DFDAF ... Digital Flight Data Acquisition Function
DCMF ... Data Communication Management Function
CMCF ... Central Maintenance Computer Function
QAR ... Quick Access Recorder
I/O ... Input/Output Module
CPM ... Core Processor Module

FIGURE 14.1
Overview of One AIMS Cabinet Modules and Function Allocation [227]

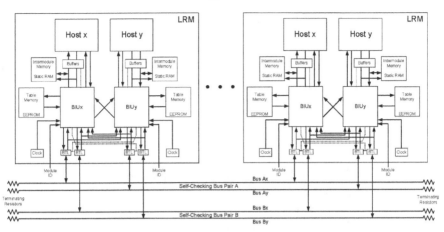

FIGURE 14.2
SAFEbus Architecture Overview (Courtesy of Honeywell)

Yet, computation and communication are dependent on each other when creating a schedule and can pose a significant challenge for getting a feasible computation and communication schedule that meets all deadlines and ensures data consistency (i.e., neither a process nor the network writes data to a common buffer at the same time). In order to get an idea of the scheduling problem in AIMS, about 3000 unique data items had to be moved with their required frequency in each cabinet. This resulted in 17,000 total messages per period in all cabinets including messages due to oversampling requirements for cabinet-external interactions.

The AIMS cabinets are just one part of the overall avionics. The integrated cabinets are connected to the airplane interfaces via a combination of ARINC 429, ARINC 629, display buses and discrete I/O channels (see Figure 14.3; note that for clarity ARINC 429 and discrete channels are not shown). In addition to the cabinet, other flight deck hardware elements that make up the AIMS system are six Flat Panel Display Units, three Control and Display Units, two Electronic Flight Instrument System (EFIS) Display Control Panels, one Display Select Panel, two Cursor Control Devices and two Display Remote Light Sensors. Not all are depicted in Figure 14.3. The system bus in Boeing 777 is mainly ARINC 629 [9], a non-time-triggered, but periodically driven system bus. Boeing 777 avionics is synchronous only within each cabinet (in the domain of ARINC 659), but not at an airplane level.

The synchronization of the bus and software also assists debugging and validation. First, due to the nature of SAFEbus' independent determinism, a function experiences the same system timing whether the cabinet is fully populated or nearly empty. Second, since each processor in a cabinet is synchronized to the bus, the functions running in different processors are implicitly synchronized and timing errors at application level will be more quickly exposed, making the system simpler to debug. In asynchronously scheduled multi-processor systems, such timing problems show up as intermittent failures, which can be very costly to track down and make it impos-

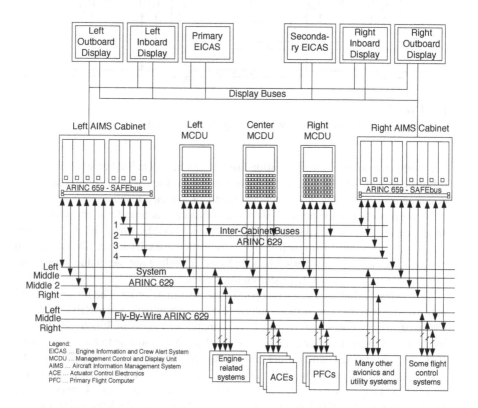

FIGURE 14.3
Boeing B 777 Avionics Overview (Courtesy of Honeywell)

sible to validate the system. Third, whenever the system is stopped or single-stepped, it passes through a succession of clearly defined states. The clear relationships between the processes, which are defined by the SAFEbus table, make it easier to trace behavior [79].

Many of the elements of Boeing 777 have led to ARINC standards such as the cabinet backplane, SAFEbus has been standardized as ARINC 659 [8]; the system bus as ARINC629 [9]; ARINC Report 651 "Design Guidance for Integrated Modular Avionics" is the top-level design guide for IMA.

Boeing 777 is said to be the first but not last plane to leverage IMA extensively. Related to SAFEbus, VIA (Versatile Integrated Avionics) is a derivative of AIMS and also uses SAFEbus on the Boeing 737NG, Boeing 717, McDonnell Douglas MD10, McDonnell Douglas MD90.

14.2.3.2   ASCB in Primus Epic®

Primus Epic is one of Honeywell's avionics platforms for general, business and regional aviation. The general aviation market comprises small, multiple-seater aircraft; the business aviation market mainly consists of turbo-engine-powered, comfortable, small aircraft for business clients; and regional aviation aircraft are single-aisle, medium aircraft flying regionally with up to about 100 passengers. Honeywell deploys the Primus Epic® avionics suite in such aircrafts. In one instance, Primus Epic is also deployed on a helicopter. The biggest market, however, is business aviation. Specifically, Primus Epic is used on the following certified airframes: Agusta AW139 (helicopter), Cessna Citation Sovereign, Dassault Falcon 2000DX, Dassault Falcon 2000LX, Dassault Falcon 900DX, Dassault Falcon 900EX, Dassault Falcon 900LX, Dassault Falcon 7X, Gulfstream 350, Gulfstream 450, Gulfstream 500, Gulfstream 550, Gulfstream 650, Raytheon Hawker 4000, Embraer 170/175, Embraer 190/195. Further aircrafts using Primus Epic are under development.

Primus Epic is an avionics suite consisting of single or multiple racks and cabinets with integrated circuit cards/modules, which are installed in slots in the cabinets. Each module can contain one or more functions. Each cabinet's configuration can vary by the number of modules installed in each cabinet and the functions loaded into the modules. There can be multiple racks and cabinets installed on the aircraft. The software loaded into the modules determines the functionality of the systems modules. The Modular Avionics Unit (MAU) is a hardware cabinet containing field-replaceable modules that represent the "building blocks" of the Primus Epic system. The MAU incorporates input/output (I/O), processing and database storage modules linked to ASCB and LAN aircraft-wide networks via the NIC modules. Primus Epic deploys a concept called *Virtual Backplane*, meaning that the physical location (e.g., cabinet) of a module is not important. All data generated by any one function within the system is "globally" available to any other function. This makes the MAU flexible and adaptive, allowing for more options in locating and mounting equipment on the aircraft. The integration of processing into a single processing unit (in the meaning of a card) means that it can be shared to perform multiple tasks previously requiring individual processors. This increases the integration results in improved power, weight, reliability, maintainability and volume. The Primus Epic system redundancy can support multiple redundancy concepts, among others a dual-dual replication arrangement for system redundancy [318].

Summarizing, Honeywell claims that deployment of Primus Epic with its Modular Avionics Unit (MAU), modular radio cabinets, and flat panel displays provides the following advantages over existing avionics suites mainly due to integration and improved technology deployment:

- Lower total acquisition costs

- Higher reliability, dispatchability

- Improved maintainability

- Lower weight, power

- Lower wire count, installation costs

- Increased functionality

- Large growth capability

- Lower pilot workload, increased safety

- Open architecture by supplying development kits to third party module suppliers

Figure 14.4 provides a high-level diagram of Primus Epic's compute and onboard communication architecture. Primus Epic in this example consists of multiple MAUs (Modular Avionics Units). ASCB-D is depicted as one network in Figure 14.4. The detailed four-bus diagram can be found in Figure 14.6. Furthermore, a LAN is shown, which is used for system maintenance, diagnostics, software and table loading tasks, and a printer. The LAN is an Ethernet 10Base2 network. In each MAU one or two NICs connect the system-level data sent on ASCB with the backplane databus, a version of Compact PCI (industrial version of PCI derated to 25 MHz being 32-bit-wide). Power modules provide local power to the modules in the cabinet. I/O cards provide interfaces to sensor and actuator data as well as integrate other more federated aircraft systems into the Primus Epic architecture.

Figure 14.5 shows a picture of one Primus Epic cabinet (MAU).

Figure 14.6 depicts an example Primus Epic deployment with two MAUs in the center. Many peripheral, display, utility, sensor and actuator systems/devices are also shown.

Primus Epic distinguishes itself from other avionics platforms in that it deploys a special operating system (OS) called DEOS, which stands for *Digital Engine Operating System*. DEOS is a DO-178B level A operating system deployed in FAA-certified environments. It allows multiple software applications/levels to execute on the same processor and provides time and space (memory, I/O) partitioning. Associated with DEOS is a consistent software development method throughout engineering using

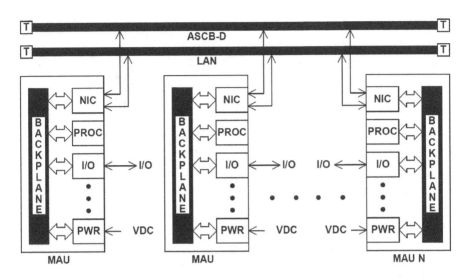

FIGURE 14.4
Primus Epic Buses (Courtesy of Honeywell)

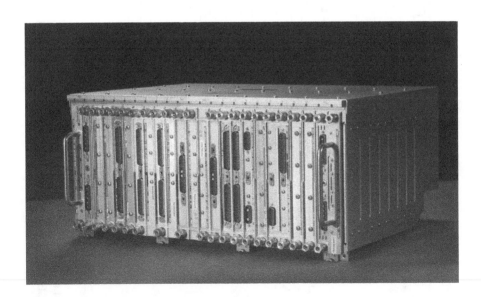

FIGURE 14.5
Primus Epic Cabinet (MAU) (Courtesy of Honeywell)

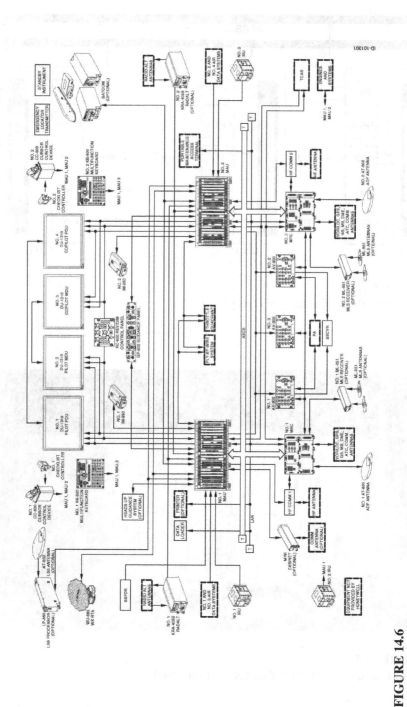

FIGURE 14.6
Primus Epic Example System Diagram (Courtesy of Honeywell)

COTS and custom tools that help in isolating software from hardware. DEOS supports reduction of re-validation and verification efforts of software that changes.

DEOS differs from earlier ARINC 653 [143, 145, 144] systems in that it uses preemptive fixed priority (PFP) scheduling, and consequently the time partitioning model includes preemption. The most well known PFP algorithm is Rate Monotic Scheduling [202], where tasks with higher rate have higher priority. Additionally, DEOS supports dynamic thread and dynamic time partition allocation. Upgrades of DEOS slack and time partitioning lead to several improvements. For example, about a threefold increase in communication throughput between a processor and a remote file server and approximately a sevenfold reduction in the amount of a priori reserved execution time required for response times of noncritical applications were reported. Also, certain display tasks were able to achieve higher average refresh update rates using slack. DEOS algorithms provide robust time partitioning capabilities, enabling the safe co-hosting of COTS software with safety critical software without decreases in the COTS software performance [33].

The concept of *slack stealing* has been a unique feature in safety-critical operating systems deployed in aerospace and is deployed in DEOS. It enables many improvements like higher processor utilization and quicker response times. Slack stealing is a preemptive processor scheduling algorithm that delays the execution of high-priority periodic tasks to improve response times of aperiodic tasks provided the periodic tasks will not miss any of their deadlines. When the set of periodic tasks is fixed, there is predictable slack inherent in the execution timeline, known as timeline slack, for threads scheduled using Preemptive Fixed Priority (PFP) scheduling. Timeline slack can be calculated offline and table lookups can be used at runtime, in combination with other quantities, to quickly determine the amount of time a periodic task can be delayed and still meet its deadline. Slack stealing also makes available reclaimed slack, or equivalently unused worst-case compute time of periodic tasks at the priority at which the compute time was initially reserved [33].

The key for the success of slack stealing in DEOS are the relatively high worst-case execution times compared to actual average execution times. Slack stealing boosts processor utilization efficiency and reduces response times significantly. As can be imagined, slack "moves" the actual execution time of tasks within a frame period. This may be a good illustration to readers why Primus Epic does not deploy sub-frame-level scheduling. As described already in Chapter 13, sub-frame-level scheduling is a concept denoting the fact that data sent on a network is used by an application with its frame-period. This may tightly couple processor and network execution time. As slack requires some "freedom" of moving actual task execution times within its frame period, Primus Epic designers have chosen *not* to leverage end-to-end worst-case latency improvements obtained by deploying sub-frame-scheduling, but rather leverage the slack of execution time to boost performance. Each approach has its advantages, but both cannot practically be deployed at the same time.

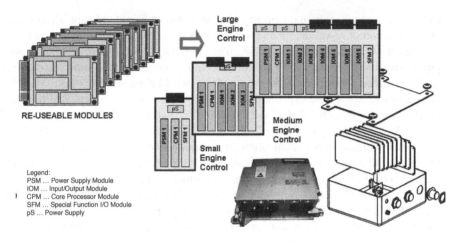

FIGURE 14.7
Concept behind Modular Aerospace Control (MAC) [345] (Courtesy of Honeywell)

14.2.3.3 Honeywell's Modular Aerospace Controller

The Honeywell Modular Aerospace Control (MAC) has been developed using TTP/C
as a backplane databus for intermodule communication. For engine control, MAC
facilitates simplified overspeed and uncommanded thrust protection. Honeywell has
created the MAC platform having re-usable modules in mind. Module functions are
scaled to the control needs of multiple applications. Furthermore, reduced development cycle times are possible due to reuse. This reduces non-recurring cost by reuse
of modules. The core architecture comprises multiple modules (i.e., cards) connected
by TTP/C scalable within a wide range of anticipated use cases. The replication level
for each module is scalable to the needs of the application for this one module. If one
compares this to the most prevalently existing dual-lane engine control dependability paradigm where everything is deployed at the same dual replication needs, this
can be more efficient. In addition, this reduces recurring cost. Figure 14.7 depicts
the concept behind MAC. Figure 14.8 compares traditional dual-lane and the MAC
distributed engine control.

MAC modules are partly pre-qualified and, hence, decrease certification cost.
Furthermore, MAC supports pro-active obsolescence management by enabling a
building block approach where obsolete modules can easily be re-developed and
exchanged due to well-defined interfaces. Furthermore, TTP/C not only should ensure proper communication, but also partitioning between modules. This supports
deterministic integration and keeps efforts low.

MAC is deployed in the FADEC of F124 used, for example, on Alenia Aermacchi's M-346. M-346 is an advanced/lead-in fighter trainer. MAC is also used for the
engine control of General Electric's F110 Turbofan deployed, for example, on newer
generations F-16 Fighting Falcon.

Traditional Dual Channel Electronic Engine Control

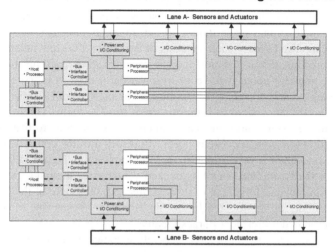

Fault-Tolerant Modular Electronic Engine Control

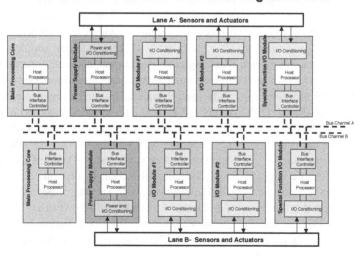

FIGURE 14.8

Engine Control Architectures–Dual-Lane Versus Distributed [345] (Courtesy of Honeywell)

14.2.3.4   TTEthernet in Orion

In the U.S. under President G.W. Bush, Orion has been planned and started to be developed under the leadership of NASA as the next-generation Crew Exploration Vehicle (CEV) replacing the Space Shuttle. Orion is part of the NASA Constellation program, which comprises several projects, among others the Ares programs, which are the rockets for lifting Orion and other space vehicles into low earth orbit to go, e.g., to the international space station, to our moon, and to Mars – at least this has been the plan. The Orion vehicle consists of two parts, the crew module (CM) and the service module (SM). The crew module provides all the facilities for housing up to six astronauts for about two weeks to lower earth orbit in initial configurations and up to six months for going to the moon. The crew module is also used for re-entry into earth atmosphere to bring the crew back to earth. The service module provides the necessary supplementary equipment for the crew to live in space and is jettisoned before re-entry. The concept of operation of the constellation program for space exploration is very similar to the Apollo program. Figure 14.9 sketches the structure of the Orion and Ares I as docked vehicles for the reader to get an idea of the configuration. At the time of writing this text, the future of the Constellation program is under discussion and further development unclear due to the new U.S. administration under President Barack Obama.

Both Ares I and Orion comprise separate avionics architectures due to operational constraints and control aspects. Orion deploys an Integrated Modular Avionics (IMA) architecture approach as this allows an open architecture, a key characteristic to NASA. With an open architecture, the system is easier to integrate and maintain. For example, the central processing unit is standardized allowing the use of commercial-off-the-shelf (COTS) hardware without any single supplier constraint and, hence, reduction of non-recurring cost for components within the system. Similarly, the use of COTS will decrease the upgrade cost since competition drives supplier upgrades to low or no cost to the project. Additionally, maintenance costs are reduced, since the COTS producers can leverage the development and maintenance tools already available. NASA's systems, such as the human-rated Orion, tend to be deployed for long durations. The space shuttle, for example, has been operational for about 30 years. An IMA architecture used in such systems benefits of more easily supporting midlife upgrades to the system and, hence, incorporating mid-life technology upgrades into the system. Furthermore, with varying mission durations, it is also important that the avionics architecture allows for different safety and criticality protections. IMA architectures support this flexibility [219]. Figure 14.10 shows the Orion/Ares interface and some avionics architecture aspects of Orion and Ares.

Orion has very specific mission requirements. It has to meet a mission requirement of up to 5000 hours (about 200 days) of continuous operation at a time. It must be a flexible vehicle that will interface to several other space elements (the International Space Station, the Altair Lunar Lander, Ground Systems and the Space Network, in addition to the Ares I interface) over the next several decades. Additionally, Orion has strict weight and power restrictions that make it desirable for its avionics system to maintain full protection against erroneous behavior even when operated in

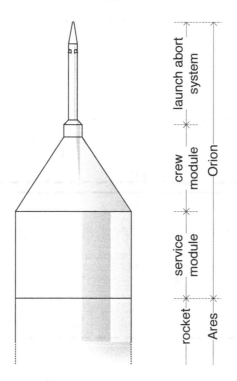

FIGURE 14.9
Sketch of Orion (Not to Scale)

a low power partial system mode, which is a mode when part of the avionics system is shut off to conserve power [219].

Early trade studies that have been conducted favored an IMA approach due to the Orion's requirements [219]:

- The Orion CEV shall meet an ascent reliability allocation of 0.9999.

- The Orion CEV shall meet an overall mission reliability allocation of 0.999.

- The Orion CEV shall be one fault-tolerant for safe return of the crew with exceptions for design for minimum risk.

- The Orion CEV shall safely recover from, and return the crew in case of loss of output and erroneous output from the vehicle flight computers due to software common cause failure.

- The Orion CEV shall allow the crew to manually control, inhibit and/or override autonomous or ground controlled critical functions.

- The Orion CEV shall be one fault-tolerant for mission completion with exceptions for design for minimum risk.

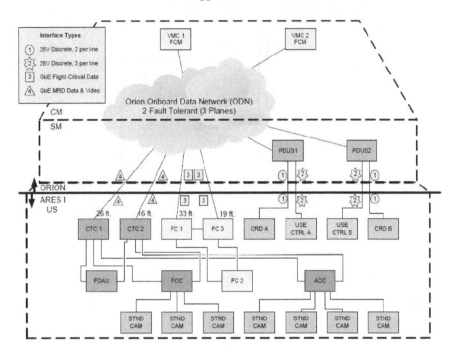

FIGURE 14.10
Orion and Ares I Interface [219]

In order to meet requirements like full error detection in low power partial system mode, a simple high-integrity, self-checking computing approach has been designed building on implementations and knowledge from avionics programs of aircraft like the Boeing 787, Boeing 777 and Boeing 777RS. Figure 14.11 provides a detailed block diagram of a typical self-checking pair computer design, which in a similar manner is deployed on Orion. Orion actually does deploy an IBM PowerPC 750 processors as indicated in Figure 14.11 with self-checking support logic such as an interface to the bus and external memory (BIPM) and backend interfaces. Yet, the actual main interface is a variant of TTEthernet, named TT-GbE, and not ARINC 659 as indicated in Figure 14.11. It shall also be noted that it is typical to perform local power and clock monitoring for a pair of processors. The presented and deployed architecture has the following advantages:

- Good error detection for bit flips in processors as well as in associated memories. In all cases, *no* inadvertent action is performed due to single event upsets with a very high probability (or in other words, the probability of undetected failures is smaller than 10^{-9}).

- No inadvertent violation of space partitioning (memory and I/O) due to features like memory management units (MMUs).

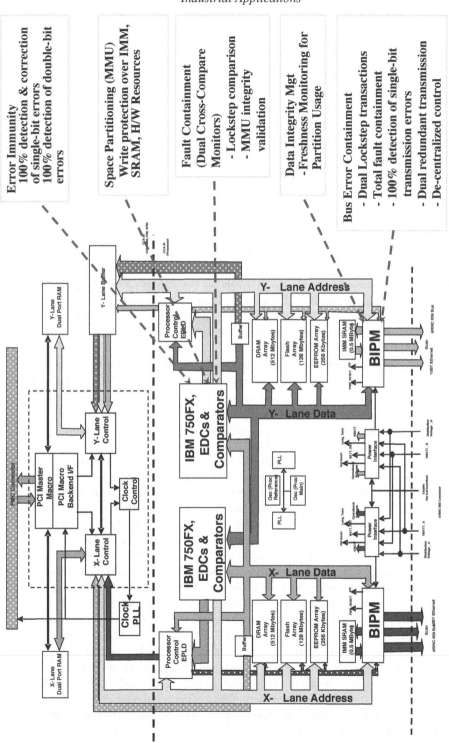

FIGURE 14.11
Self-Checking Pair Computer [103] (Courtesy of Honeywell)

- Cross compare monitors perform constant comparison of computing outputs and validate MMU integrity.

- Data integrity management (freshness monitoring for partition usage).

- Support of bus error containment (more Orion details below).

The system network is an implementation variant of TTEthernet described in detail in Chapter 8. The Orion TTEthernet deployment is called *Time-Triggered Gigabit Ethernet* (or *TT-GbE*). As the name implies, it is leveraging a standard Gigabit Ethernet physical layer and the standard TTEthernet IP core including all features and support up to ISO/OSI layer 2 (i.e., not including layer 3 according to ISO/OSI). Given the harsh natural radiation environment in space, the TTEthernet IP is embedded in a space-hardened technology together with a special space-hardened SERDES (serializer/deserializer) physical layer technology, compatible with Ethernet, and other interfaces like PCI. TT-GbE supports normal and high-integrity end systems and high-integrity switches. The high-integrity variant is based on a command/monitor configuration. Figure 14.12 shows the command/monitor configuration for the TTEthernet network for an end system. It is deployed for the switches and some end systems (such as the ones interfacing to the self-checking computers) [121, 103, 219]. Such command/monitor component configurations ensure a validated fail-silence model in that all output of a component is checked bit-for-bit against independent monitor copy and any violation forces a consistently detectable fault output. The command/monitor end system and switch and self-checking compute components enable a reduction of end-to-end integrity augmentation normally deployed in high-integrity end systems. It also helps leaving the high dependability implementation aspects out of the application and at the electronic platform system level, which simplifies platform development.

Figure 14.13 presents a very high-level overview of the Orion avionics architecture at one time during the development. The platform was under development at the time publications [103] and [219] were written and has slightly changed in the meantime. Orion in general had to adapt to weight requirements due to Ares I rocket design. Adaptation of architectures to a certain extent is a typical system design refinement. On the left side of the dotted line of the diagram labeled CM is the equipment in the crew module. On the right side is the equipment in the service module (SM) of Orion. The crew module contains the majority of the different electronic units. The component-connecting lines in Figure 14.13 (in blue in the original publication) show the major onboard data network, which is implemented using TT-GbE and labeled by the FCNet interface cards. The architecture contains – in this instance – two major computer-related cabinets called VMCs for Vehicle Management Computers. VMCs contain three different computer modules (CMs labeled FCM, LCM and RCM) with associated network cards and power supply modules. Remote Interface Units (RIUs) are cabinets with input/output interfacing cards. Orion has three display units (DUs) implementing modern glass cockpits, interface devices (keypad, RHC, THC, CCD), (external) communication means (e.g., audio I/F Unit, S-Band), absolute and relative navigation units (inertial measurement units (IMUs), star track-

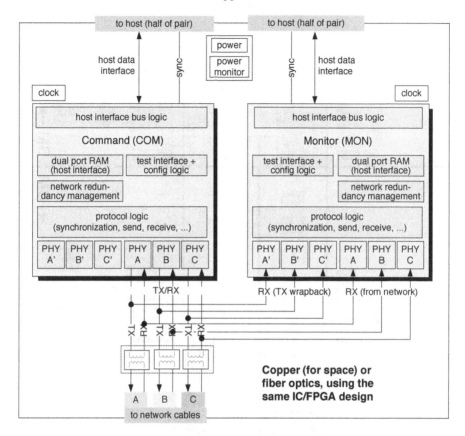

FIGURE 14.12
TTEthernet Self-Checking Pair End System [121] (Courtesy of Honeywell)

ers, GPS), environmental control units (ECLSS) and multiple vehicle power management units (MBSUs) among others.

14.3 Time-Triggered Communication in Automotive Applications

The typical automotive system is a passenger car. This system consists of several subsystems which have evolved from the typical main mechanical components of the car (e.g., the engine, the transmission, the brakes, etc.). In a modern car, nearly each mechanical component also has a dedicated node computer for controlling it. The node computers are called Electronic Control Units (ECUs) in the automotive

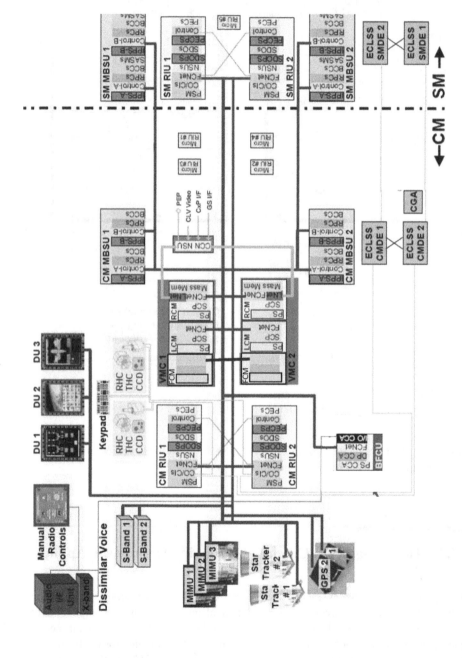

FIGURE 14.13
Orion Avionics Overview [103] (Courtesy of Honeywell)

domain and they are typically interconnected to distributed systems using different types of automotive networks such as Controller Area Network (CAN) [151], Media Orientated Systems Transport (MOST) [229] and FlexRay [226]. Figure 14.14 displays a typical automotive system with several CAN buses forming a subdivided network of interconnected subsystems.

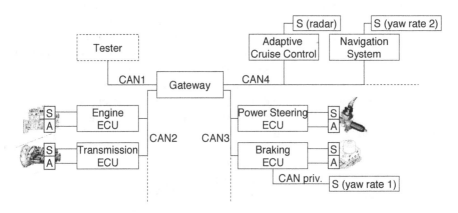

FIGURE 14.14
Typical Automotive System (CAN-Based)

Due to the continuously increasing functionality (and complexity) within a modern car, several separated CAN buses are used. This helps to overcome the bandwidth limitations of the physical layer of CAN to a certain extent. On the other hand, a significant number of CAN messages need to be forwarded from one CAN bus to another via a gateway. The shown example uses five CAN buses which are associated to certain function domains. In the displayed typical example, these domains are:

- *CAN1 – Diagnostic CAN:* Used for connection to an intelligent tester in the service station

- *CAN2 – Drivetrain CAN:* Used for the drivetrain ECUs

- *CAN3 – Chassis CAN:* Used for the ECUs controlling the chassis mechanical components

- *CAN4 – Driver assistance CAN:* Used for driver assistance functions

- *CAN priv. – Private CAN:* Used for acquisition of time-critical yaw rate sensor data by the braking ECU

Each ECU controls its dedicated mechanical component by reading the relevant sensors (S) and driving the dedicated actuators (A). Most sensors are directly connected to their dedicated ECU – using analog or digital interfaces. Even though some smart sensors have a CAN interface, they typically are still connected to a dedicated ECU via a private CAN (like the sensor "yaw rate 1"). A yaw rate sensor gives precise data of a very important portion of vehicle movement – namely the angular velocity of the vehicle turning around its vertical axis. Especially sensors providing time-critical data are not suitable for being connected to a heavily loaded CAN bus since the data will only be available for the control algorithms in the ECUs with a significant – maybe intolerable – delay and/or jitter. On top of this technical reason, quite often the simpler organization of the series production oriented development process of a subsystem leads to the private usage of a sensor by only one ECU. This may lead to the fact that certain sensors (like the yaw rate sensor) are in the vehicle more than once – not for redundancy reasons – but in order to keep the subsystems (and their development processes) independent from one another. The strongly cost-driven automotive industry has pushed the development of subsystems toward less private usage of the sensors and therefore several smart sensors with CAN inter-faces and even smart sensor clusters sending a large set of different sensor data via a FlexRay interface have evolved.

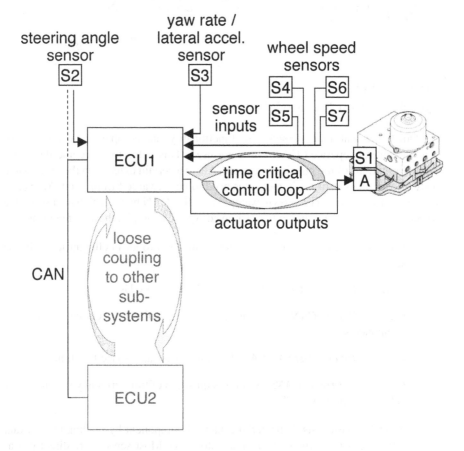

FIGURE 14.15
Typical Automotive System (CAN-Based)

Figure 14.15 gives a more detailed insight into a typical automotive subsystem within the vehicle. At the current state of today's series production cars, the interconnection between the subsystems is typically less time-critical and also less essential for fulfilling the designed function than the very close interaction of the ECU with its dedicated mechanical component. Not only sensors (like "S1") directly connected to the mechanical component but also other sensors (like "S2" – perhaps available as CAN message) may be evaluated by the ECU in order to monitor the mechanical component and the state of the vehicle and to control the actuators (A).

Let's take a modern braking ECU as an example: The wheel speed sensors of the four wheels are directly connected to the braking ECU (ECU1) so it can fulfill the time-critical 'anti-lock brake' function even without any available CAN communication. For the more complex functionality of the 'electronic stability control,' additional sensor data is required which is only available via the CAN buses (steering angle, yaw rate, lateral acceleration, etc.). The overall control loop timing for the 'electronic stability control' allows a slower control loop cycle (approximately 40 ms) compared with the time-critical 'anti-lock brake' function (approximately 5 ms). The CAN's temporal behavior on a public or perhaps a private CAN bus is acceptable for the 'electronic stability control' function – even though an improvement toward less latency and less jitter would be desirable in order to improve the responsiveness of the function. Additionally the need for synchronizing the available data in order to obtain a precise view of the vehicle movement finally led to the development of a FlexRay-based sensor cluster which provides a large set of vehicle movement data with a high update rate. It suits the needs of the 'electronic stability control' just as the needs of other driver assistance functions.

Even though taking advantage of the higher bandwidth of FlexRay will lead to further enhancements and redistributions of functions among the subsystems, the basic design principles of the described subsystems represent today's accepted state of the art in passenger cars. It is the result of an evolutionary process of engineering and reengineering to fulfill customer requirements and the cost targets of the automotive industry while continuously increasing the overall functionality of the electronically controlled subsystems. It has its roots in the independence of the former purely mechanical components. Today's electronic subsystems still try to keep their independence whenever possible. The brake subsystem is a 'fail silent' subsystem with a two-step degradation model. If the CAN communication is not available, the 'electronic stability control' function is not available anymore while the 'anti-lock brake' function remains active. If the complete braking ECU fails (is switched off) the mechanical braking performance is still available for the driver to stop the car – even though he will not receive the assistance of the higher level functions while braking.

14.3.1 Typical Design of Automotive Applications

The typical automotive application running on an ECU is focused on controlling the dedicated mechanical component in accordance with a large set of requirements (e.g., performance, efficiency, availability, robustness, safety, diagnostic capabilities, hard real-time, etc.). The toughest real-time requirements typically need to be fulfilled reliably under all operating conditions and are normally handled via interrupts which are triggered by sensors of the controlled mechanical component. Also the actuators often require very short reaction times measured from the correlating sensor interrupt event. As an example, the interrupt may be generated by an engine speed signal which is used to monitor the exact rotary movement of the engine and also to trigger the engine ECU to perform the next injection and ignition with the precise timing matching the current operating conditions. Within the application design, this environment-controlled hard real-time functionality is performed using an interrupt which typically receives the highest priority within a preemptive scheduling design. The remaining algorithms within the control unit will typically be executed at a lower priority level either as interrupts (event-triggered) or as a set of periodic tasks (time-triggered). Also the processing of CAN messages can be performed in both these ways. For stability reasons, the general approach typically is to shift as much software execution (CAN message processing and algorithms) as possible into a time-triggered 'main algorithm' which is executed according to a periodic activation scheme derived from the local timer of the CPU as shown in Figure 14.16.

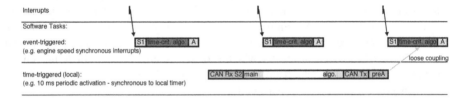

FIGURE 14.16
Typical Automotive Application Design (CAN-Based) – Timing of Software Execution

    The above figure displays a typical software execution timing situation in a simplified schematic way. Note that typically the 'time-critical algorithm' will take a smaller portion of the CPU execution time – while a 'main algorithm' in most applications will be much larger (execute longer) than displayed here. At the high priority level, the reading of sensor S1, the 'time-critical algorithm' and the output control for the actuator (A) are performed in an event-triggered manner. At the lower priority level, the reception of the CAN message containing sensor value S2, the 'main algorithm,' the CAN transmit operations and a 'prepare actuator operation' (preA) are performed according to a periodic time-triggered activation scheme. The high and the low priority software parts recur with different rates and therefore can appear with different phase shifts against each other. As all software executed at the lower priority level is not performed strictly synchronized to the S1-interrupts – and additionally can be interrupted (and delayed) by the high-priority software parts – the coupling between these software parts is quite loose. This means that S1 has a quick influence on the actuator control (A) while the latency and the jitter of S2 – which can only influence the actuator via the 'main algorithm' and the 'prepare actuator operation' – is much larger. This design is only to be used when this large latency and jitter of the sensor value S2 is acceptable for the according function.

14.3.2   Migration from CAN to FlexRay

The migration from CAN to FlexRay has already taken place for a certain set of subsystems in a limited number of series production cars today. The main reasons for this migration were the increased bandwidth and the improved determinism – resulting in an expected communication behavior leading to more precision and stability together with shorter latency and less jitter than CAN-based systems.

The new automotive subsystems shall increase the coupling between the subsystems by usage of the FlexRay network and the large amount of sensor data which can be shared among the subsystems. "Private" sensors of one ECU can become "public" when the values are sent across the FlexRay network with a sufficient frequency and are processed with short latency and jitter. This also helps to avoid equipping the vehicle with double or even triple sensors of the same kind. The intended basic subsystem design is shown in Figure 14.17.

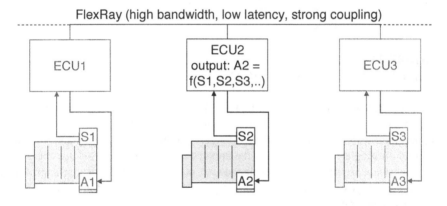

FIGURE 14.17
New Automotive Subsystems with FlexRay (CAN-to-FlexRay Migration)

The typical new automotive system now contains a mixture of CAN and FlexRay subsystems as not all CAN-based subsystems can be replaced immediately. It also is not required to migrate all subsystems from CAN to FlexRay. Today's typical solution is shown in Figure 14.18. The gateway remains the central device for interconnection between the CAN buses and also the FlexRay cluster. The gateway CPU now performs CAN-to-CAN, CAN-to-FlexRay and FlexRay-to-CAN routing. Even though today's series solutions with FlexRay typically contain only one single channel FlexRay cluster, this cluster is subdivided into four FlexRay branches on the physical layer. These four branches are interconnected by a FlexRay 'active star' which is a multiple transceiver device which forwards data from any incoming FlexRay branch to all other branches. This helps to keep the number of nodes connected to one twisted wire pair rather small and the wire lengths short which significantly reduces reflections on the physical layer and improves the robustness of the data transmission under electromagnetic disturbances. From a timing perspective the 'active star' inserts a very small – actually negligible – delay (approximately 250 ns).

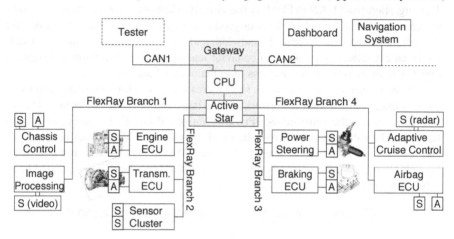

FIGURE 14.18
Typical New Automotive Systems (CAN + FlexRay)

The migration from CAN to FlexRay also has to take place in the application design. Here the key issue is the software and especially its execution timing. Even though the FlexRay network permits stronger coupling, short latency and low jitter, the main timing effects result from the application design on the ECUs. The different subsystem suppliers have developed two major application design approaches which shall be explained in the following – the event-triggered approach and the time-triggered approach.

14.3.2.1   Event-Triggered Approach – FlexRay as CAN Replacement

The event-triggered approach is based on the idea of the simplest possible migration by just replacing CAN driver software functions with the according FlexRay drivers without any other software changes. Especially application designs which were based on interrupt driven CAN communication were simply converted to FlexRay in this way. The typical adapted automotive application design is shown in Figure 14.19.

This adapted automotive application design basically inserts a third priority level in-between the event-triggered 'time-critical algorithm' and the time-triggered 'main algorithm.' This intermediate priority level performs the FlexRay receive and transmit operations synchronously to the FlexRay global time. Even though this might appear as a time-triggered solution, in most cases it actually is implemented in a purely event-triggered way. The typical automotive implementation is built by adding FlexRay–synchronous interrupts on top of the "normal" time-triggered soft-

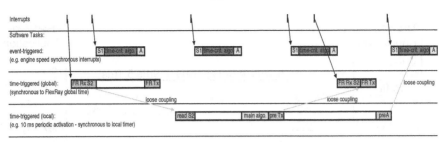

FIGURE 14.19

Adapted Automotive Application Design (Event-Triggered Approach) — Timing of Software Execution

ware tasks. As a result, the according software parts are not synchronized since they are triggered by two independent and not synchronized time sources (the FlexRay global time and the local timer). The loose coupling, which mainly results from software parts being executed on different priority levels, increases even more with this intermediate priority level in place and leads to additional delays between the actual reception of a sensor value (FR Rx S2 in Figure 14.19) and the processing by the 'main algorithm.' There are also delays between the preparation of transmit values (pre Tx in Figure 14.19) and the actual transmit operation (FR Tx) and another delay until the message is really sent over the FlexRay bus (in the pre-assigned slot). This third-priority level finally leads to less determinism of the execution timing because the worst-case delays and the maximum observed jitter increase.

Even though the event-triggered approach was intended to simply replace CAN with FlexRay, there still is one logical difference to be taken care of. Since FlexRay uses pre-assigned slots for transferring messages, the transmit operation should be performed before this slot starts. If this time constraint is not met, the according FlexRay frame will be sent anyway – but marked as containing no data (which of course is detected by the receiver). The next chance for transmitting the data is at the next slot assigned to the according message (one message period later). As shown in Figure 14.20, this leads to a certain amount of lost messages in the situation when the transmitter is "asynchronous" to FlexRay (and has a small jitter) and generates the messages at the "same" rate as they are to be sent on the FlexRay bus. This cannot be compensated by the receiver since the data really was not transmitted over the FlexRay bus.

This behavior can be observed in "bursts" which recur periodically with a quite long time interval – typically several minutes (resulting from the small clock speed deviation between the local timer and the FlexRay global time). A practical measurement of such a "burst" where every second message is not transmitted on the FlexRay bus is shown in Figure 14.21.

The receiver as the second part in the process of exchanging data can additionally miss incoming messages when its receiver application is not checking the receive buffer at the appropriate times which typically happens when the receiver is

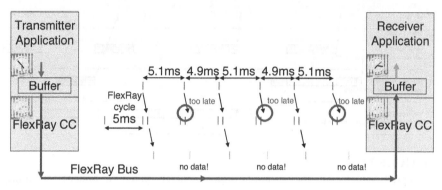

Legend:

| ... FlexRay Cycle Start
| ... Transmit Operation of Application
| ... Reserved Slot on FlexRay Bus
| ... Receive Operation of Application

FIGURE 14.20
FlexRay Communication Timing (Event-Triggered Approach) — Transmitter "Asynchronous" to FlexRay — with Lost Messages

"asynchronous" to FlexRay. In this case, the number of lost messages (actually 'non-received' messages) may even increase as the timing behavior of the transmitter and the receiver are not synchronized. As a consequence of the shortcomings described above, the synchronization of the application software of the transmitter and the receiver to the FlexRay global time is the corrective measure which shall be discussed in the following chapter of the time-triggered approach.

14.3.2.2 Time-Triggered Approach — FlexRay-Synchronous Task Execution

The time-triggered approach requires an adaptation of the software scheduling in such a way that the global time of FlexRay is used to schedule all time-triggered software parts. This reduces the number of priority levels back to two again and additionally gives the application designer the opportunity to align his main algorithm with the communication schedule of the messages. This especially helps to provide measurement data as FlexRay messages without any lost messages. This rigorous time-triggered automotive application design is shown in Figure 14.22.

This rigorous time-triggered automotive application design uses the FlexRay global time (whenever available) for scheduling the FlexRay receive operations, the 'main algorithm,' the transmit operations and the 'prepare actuator operation.' If sensor S1 does not require interrupts, the whole application design can even be simplified to run on only one priority level. The messages on the FlexRay bus (relevant for this ECU) are completely received and also sent without any gaps or jumps within the sequence of messages. The timing of the software execution needs to be synchronized to the FlexRay global time (which is typically done at the recurring cycle start

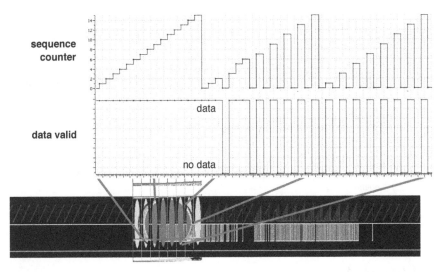

FIGURE 14.21

Practical Measurement Results with an "Asynchronous" Transmitter on FlexRay —
with Lost Messages

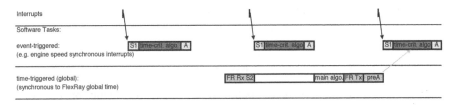

FIGURE 14.22

Rigorous Time-Triggered Automotive Application Design (Time-Triggered Approach) — Timing of Software Execution

event) as shown in Figure 14.23. With a "synchronous" transmitter and receiver, no messages are lost.

Whenever the FlexRay global time is not available (before the FlexRay startup or during a temporary loss of communication), the local timer shall be used for scheduling instead – this in return leads to the need for a resynchronization strategy whenever the FlexRay global time becomes available again. This is performed by a smooth synchronization as shown in Figure 14.24.

The smooth synchronization is performed according to the following design rules:

- The application runs on the local timer and re-synchronizes smoothly when the global time is available

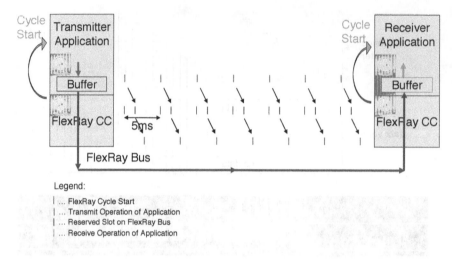

FIGURE 14.23
FlexRay Communication Timing (Time-Triggered Approach) — Transmitter "Synchronous" to FlexRay — without Lost Messages

- The maximum jump width of the resynchronization steps is limited to a specified range

- After reaching the synchronous operation mode, only tiny resynchronization adjustments are necessary to stay synchronous with the global time

- Some messages may be lost before reaching the synchronous operation

14.3.2.3 Discussion

When comparing the event-triggered and the time-triggered approach, the fundamentally simpler timing behavior is achieved with the time-triggered design which leads to "synchronously" operating application software on the CPUs within the subsystems and to a deterministic behavior of the data transfer without losing messages. This does not mean that asynchronous operation does not exist in such systems but it is limited to startup and resynchronization scenarios. The overall performance of a rigorous time-triggered system is not only more stable than of an event-triggered one but it also achieves the shorter maximum values for latency and jitter. Even though time-triggered is in fact the simpler design and has more robustness and performs in a more deterministic way, the automotive industry only progresses slowly toward this direction.

The author has identified the following main reasons for the reluctance of the developers working in the automotive industry to migrate to rigorous time-triggered designs:

- Time-triggered design is significantly different to the normal human models of

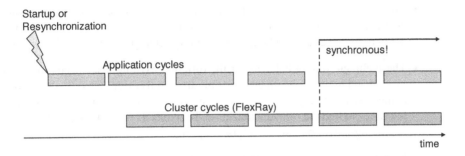

FIGURE 14.24
Smoothly Synchronizing the Local Time to the FlexRay Global Time (Time-Triggered Approach)

behavior. A human being is an event-triggered performer driven by multiple dynamic priority levels and that is also how most of our software is designed.

- The complexity of the current automotive subsystems is ranked to be "manageable without any special measures" – even though the coupling between the subsystems is becoming stronger and hidden dependencies are present in many ways but not so easy to identify.

- The timing of the software scheduling – normally controlled by an operating system – is not dared to be touched. Even though modern operating systems do provide an interface to "tune" the starting point of the time-triggered tasks, there's a certain fear of possible side effects of such a manipulation.

- The fact that FlexRay has a different temporal behavior than CAN is often denied. FlexRay is simply seen as a kind of high-bandwidth version of the well-known CAN. Even though FlexRay introduces a new notion of time (namely its own global time which is quite independent from the subsystems), the software is not redesigned to reflect this difference.

14.3.3 Practical Experience with the Time-Triggered Approach in Automotive Subsystems

Even though only a small number of automotive subsystems have been developed following the rigorous time-triggered design principles (and the author has only had detailed insight to a limited number of such application designs), the experiences with these time-triggered subsystems were quite promising. The author's personal experience from such projects is condensed into the following list:

- All projects which "dared" to switch to a time-triggered design kept up this new design (and didn't switch back) since they achieved a stable subsystem development level and had less "surprises" while testing.

- The positive effect of stable time-triggered behavior helps to find software errors during testing. For example, a deadline miss of a single task execution is not accepted as a sporadic interrupt-timing-effect and not ignored any more.

- The complete timing behavior of the different existing tasks needs to be reviewed and re-aligned with the new time-triggered design. This should be done at the very beginning of the development work and may require some "searching time" to find the right documents and persons. If you get it right at the very beginning, it will be a stable migration project.

- In order to reach synchronous operation, the time-triggered software parts not only need to be synchronized to the FlexRay cycle start, but even the whole application cycle needs to be synchronized to the relevant FlexRay cluster cycle which typically is an integer number of FlexRay cycles long.

14.4 Time-Triggered Communication Services in Railway Applications

This section will introduce the reader to requirements and solutions for applications in the railway domain with special emphasis on communication aspects. Starting with a brief sketch of the spectrum of applications needed to safely operate trains, we will outline general characteristics of safety critical applications in the railway domain and discuss requirements on communication services. Since all these applications share similar requirements with respect to fault tolerance and reliability, it is beneficial to have a common technology basis to build upon. In the following, we will briefly describe the realization of such a common architecture for applications in the railway domain.

A more in-depth discussion on the use of time triggered communication in the railway domain will be based on the example of an interlocking system. After a brief introduction to the communication needs of interlocking systems in general, a more detailed elaboration on the use of the time-triggered protocol TTP/C will be given. A discussion on further trends concludes this chapter.

14.4.1 Railway Applications

Figure 14.25 gives an overview on the landscape of safety-critical applications for train routing and train protection. The main building blocks are the interlocking system and track-side equipment such as axle counters and on-board systems. Within this ensemble, interlocking guarantees the establishment of non-conflicting routes and train movements by assigning a movement authority to each train in the realm of control. A critical input to any interlocking system is the capturing of track occupancy. In Figure 14.25 this functionality is realized with so-called axle counters

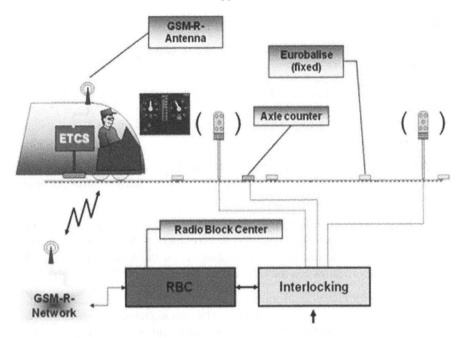

FIGURE 14.25
Train Routing and Train Protection

which count the number of axles entering and leaving a block section. In conventional signalling systems, the movement authority is indicated to the train by trackside light signals.

In turn, train protection systems are in charge of enforcing movement authorities to prevent a train from overrunning a red signal. One example for such a train protection system is ETCS, the European Train Control System, as depicted in Figure 14.25. In the most basic specification of ETCS (ETCS Level 1), the signal aspect is picked up from the signal post and transmitted to an on-board vital computer through a Euro-Balise[1] via a radio signal. The on-board unit in turn calculates the braking curve and activates the brakes when necessary. A more advanced level of ETCS involves a Radio Block Center (RBC). The RBC is fed the movement authority from the interlocking and relays this information to the train via a GSM-R[2] network. This approach eliminates the need for track-side signal poles.

---

[1]A Euro-Balise is a transponder placed within the rail tracks and informs a bypassing train about its location and/or the aspects of an attached signal.
[2]GSM-R is a variant of GSM adapted for the specific needs in the railway domain.

14.4.2   Requirements on Railway Applications

The railway applications pointed out in the last section share a great number of non-functional properties:

Product life-time: Typically a railway control product has an operating life in the order of 20–30 years, including software maintenance, function upgrades and the delivery of spare computing elements and replacement of faulty components.

Safety: The most important aspect of railway signaling applications is safety. CENELEC 50129 [57] requires less than 10^{-9} safety-critical failures per hour for interlocking systems.

Reliability and availability: High availability and reliability are essential to keep trains operating on schedule, thereby avoiding any loss of profit and reputation. As an example, for an interlocking system the Austrian Federal Railway Authority requires less than one service interruption in 10 years.

Performance: Railway applications can be seen as soft real-time applications. Timeliness constraints are determined by operational sequences, i.e., all safety conditions are fulfilled before the movement authority for a train is changed. For example, all switches are set to the correct position before the corresponding signals are set. Timing constraints typically result in deadlines by which some activities need to be completed in the order of hundred milliseconds.

Fail safe systems: In general, railway applications are fail safe systems, i.e., there exists a safe state which can be entered in case anomalies, either temporal, behavioral or in the value domain, are detected. One example for driving the system into a safe state is to set signals to red and thus stop all trains.

Certification: Railway signalling applications need to be certified by a railway authority according to CENELEC standards [54] and [57]. Four Safety Integrity Levels (SIL 1 to SIL 4) are distinguished. The CENELEC certification process includes the various product development phases, verification, validation, etc. It also covers the certification of all the hardware and software, including driver software, operating system software, compilation tool-chain and application software.

Costs: Competition in the railway control area has increased cost pressures and raised the demand for quickly realized, cost-effective and competitive solutions with low maintenance costs. Separating the complex railway-specific applications from the hardware and software technologies is a major step in the direction of guaranteeing long product life-times with reasonable maintenance costs.

Visibility of basic technology: For suppliers and customers, the value of the system lies in the applications. The use of computer technology (operating system, fault-tolerance, communication, etc) is only a means to provide the required system services. However, the product price is dominated by the delivered system services and not by the system realization.

14.4.3 Requirements on Communication Systems

In the following, we will discuss requirements on communication systems which are shared by all or most communication subsystems of a railway application. In terms of required throughput, addressing, routing capabilities, physical layer, etc., the requirements are quite heterogeneous. Communication systems range from Internet connections over closed or public networks to remotely control rural interlocking stations over local area networks within an interlocking to field-buses for reading sensor data on-board a train or connecting hardware which drives signals and switches. However, in terms of certification, safety and real time aspects, these communication systems share the same properties and requirements.

Certification

All safety-critical railway applications targeted for the European market have to be compliant with the relevant CENELEC standards describing the requirements for the approval of safety-related systems. The safety requirements for safety-related communication in the railway domain are stated in EN 50159-1 [55] and EN 50159-2 [56]. While EN 50159-1 refers to closed transmission systems, EN 50159-2 deals with open transmission systems.

These standards do not assume any specific precautions toward safety of the underlying transmission system. Upper layers have to independently guarantee the safety of the communication.

Fault Model

The CENELEC standard 50159-1 [55] lists the following protective measures to be taken:

- Detection of transmitter identifier error.

- Detection of data type error.

- Detection of data value error.

- Detection of out-dated data or data not received in due time.

- Loss of communication after a predefined delay.

As an additional protective measure, the functional independence of the safety-related transmission functions and the used layers of the non-trusted transmission system has to be ensured.

The fault detection mechanisms to cover these faults have to be implemented independently from the used layers of the underlying non-trusted transmission system.

In the context of 50159-1, the transmission system is only specified to exhibit the following properties:

- The physical characteristics of the transmission system have to be fixed

- The number of communication participants has to be fixed

- The transmission system has to be closed

No other assumptions on the transmission system are being made by the standard. The fault detection mechanisms to cover these faults have to be implemented independently from used layers of the underlying non-trusted transmission system. Technical and functional safety of the communication system are governed by the measures and procedures described in EN 50128 [54] and EN 50129 [57].

Real Time Aspects

In railway applications in general, there is a safe state the system can enter upon the detection of a fault in either the time or the value domain. To detect failures in the time domain, one approach is to implement a hierarchy of supervision systems checking the temporal behavior of the respective underlying components. If a temporal deadline has been missed, a safety reaction is initiated. Such a missed deadline causes system downtime but is not considered safety critical. In this context, it is worth noting that the standard does not require any properties with respect to real time delivery of messages.

Redundancy

To reach the required level of availability of railway applications, the system must be able to cope with all single node and network failures. Based on an in-depth failure analysis, a redundancy concept for each system component has to be elaborated. In an architecturally heterogeneous system, the redundancy concept for the inter-node communication links always depends on the redundancy architecture of the involved components. As such, communication redundancy is solved on an architectural level rather than on a link level.

14.4.4   Generic System Architecture

Important aspects for products in the railway domain include stringent technical and functional application characteristics in terms of safety, reliability and real-time aspects as well as their long service life in the order of 20–30 years. On the other hand, basic technology such as available processor boards and operating systems change at a much quicker pace. To bridge this gap and to decouple long lived railway products from fast changing technologies, a generic technology platform for the great majority of safety-critical railway applications, provided by THALES, has been developed. The objective of this platform (named TAS Control Platform[3]) is to provide

---

[3]The name "TAS Control Platform" has its origin as the former Transportation Automation Solution division of Alcatel.

a technology basis in terms of computer systems, operating system, communication system and redundancy mechanisms for safety and fault tolerance.

In particular, TAS Control platform comprises:

Operating system: Railway applications can be classified as soft real-time systems. As such, there are no stringent real-time requirements on the underlying operating system. Hence, TAS Control platform builds upon readily available POSIX compliant operating systems.

Generic safety strategy and safety case: TAS Control Platform users shall be able to build upon a generic safety case that covers fault tolerance and redundancy aspects, safe communication aspects and hardware aspects.

Fault tolerance system: TAS Control Platform provides all services needed to tolerate all single stochastic hardware faults.

Connectivity: The communication system offers a number of standard communication services, such as Internet Protocol (IP), serial lines and field-buses (controller area network (CAN), time-triggered protocol (TTP/C), multifunction vehicle bus (MVB), Profibus (PB)), as well as specific safe communication services conforming to European Committee for Electrotechnical Standardization (CENELEC) standards.

Defined hardware platform: TAS Control platform supports two classes of hardware platforms. One class is dedicated to the requirements of indoor applications, whereas the other class is dedicated to the requirements for outdoor applications located on-board or near the track-side.

TAS Control Platform is established as basic technology for the vast majority of safety-critical railway applications provided by Thales. The spectrum of railway applications supported by TAS Control Platform comprises track-side and on-board equipment such as axle counters or vital on-board systems and indoor applications such as interlocking systems, signal controlled warning systems and radio block centers.

14.4.4.1 TAS Control Platform Redundancy Architecture

The fault tolerance concept to cope with hardware faults is based on active redundancy. Figure 14.26 depicts a so called Computing Node (CN) consisting of three Computing Elements (CEs) working in active redundancy as a loosely coupled system (2-out-of-3 system). Cyclic resynchronization is done via dedicated Ethernet based synchronization links. For cost optimization, in (sub)systems with less stringent reliability and availability requirements, a Computing node (CN) can be configured as a 2-out-of-2 system.

14.4.4.2 TAS Control Platform Communication System

The key element for providing fault tolerance services is the enforcement of replica determinism [259], which means that in a fault free scenario replicas starting from

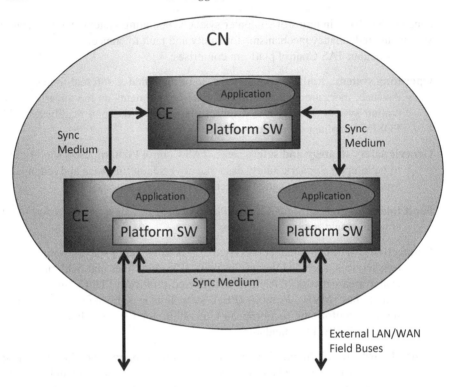

FIGURE 14.26
TAS Control Platform Redundancy Architecture

an identical initial state and consuming the same input messages in the same order, produce output messages also in the same order. In TAS Control Platform, the services needed to ensure replica determinism are provided by the TAS Control Platform Communication System.

The communication system is a run-time environment that supports replica determinism of actively redundant applications. Communication between these replica deterministic applications is realized solely as message-based. The communication system implements the following services (also see [156]):

Redundancy handling: The application knows nothing about its own redundancy or about the redundancy of its receivers or senders. Replication and fault tolerance are transparent to the application.

Location transparency: Transmission of messages is not bound to a local computer at the application level.

Authorization: To supervise communication in the system, authorization has to be explicitly defined in a static, off-line described configuration file. Message flow in the system is then checked against this configuration file.

Replica deterministic time-base: As the local clock cannot be used for replica deterministic applications, the communication system provides a replica deterministic time-base.

Application scheduling: Different application task sets (sets of closely related processes) are executed asynchronously, with pre-emptive priority scheduling allowing different interleavings of their executions on each replica. Within an application task set, run to completion scheduling is applied between pre-emption points.

Support for programming models: To preserve replica determinism of redundant applications, some semantic restrictions of the full POSIX API apply. To reduce the complexity of handling these restrictions, the communication system supports programming models to ensure the deterministic behavior of redundant applications.

14.4.4.3 TAS Control Platform Fault Tolerance Layer

Fault tolerance is achieved by comparing the message flow of actively redundant applications. All fault tolerance mechanisms are realized solely in software without the need for dedicated hardware components. To guarantee that fault-free applications perform the same operations and behave replica deterministically, the applications have to follow the programming models supported by the communication system (see [156]).

Configurable redundancy: The fault tolerance layer supports various configurable redundancy schemes at the computer and application levels. At the Computing Element level, it supports 2-out-of-3 and 2-out-of-2 configurations. In 1-out-of-1 configurations, an empty layer represents the fault tolerance layer. On the redundant computer systems, applications can execute using various redundancy schemes, ranging from 1-out-of-1 to 3- out-of-3. It is worth noting that the redundancy configuration for the applications need not be identical to that of the Computing Element. For example, it is possible to run a 2-out-of-3 application together with a 1-out-of-1 application on a 2-out-of-3 computer system.

Message-based comparison: Application data is forwarded via the communication system to the fault tolerance layer. Data is consolidated across all replicas after an interactive consistency exchange via fault tolerance dedicated synchronization links. In the next step, the fault tolerance system votes on this data according to the configured redundancy scheme and delivers the voted result to the communication system for forwarding to the receiver application.

Message comparison method: The fault tolerance system supports different comparison methods, such as majority decision, master-slave processing, byte-wise comparison and semantic comparison.

Generation of replica deterministic time-base: The fault tolerance system provides the communication system with a synchronized replica deterministic time-base.

Operational scheme: TAS Control Platform provides round based operational schemes. The start of a new execution round is driven by the progression of the synchronized time-base. In addition, external events can also initiate the beginning of a new round.

Fault management: As well as comparing redundant messages, the fault tolerance system gathers fault information from all subsystems and components, reports this information to a diagnostic system and coordinates the actions taken to resolve the problem (e.g., isolating an application, rebooting a computer).

On-line recovery: In 2-out-of-3 redundancy schemes, fault management includes the re-integration of faulty components (applications and/or computers) during operation without service interruption. To achieve this, the state of the two remaining active Computing Elements is transferred to the recovered Computing Element. Recovery units are applications and/or computers. In the event of a Computing Element failure, the entire computer and all the applications executing on it are recovered. Recovery of an application means that only the failed application is recovered, without affecting other applications running on the computer.

14.4.4.4 Connectivity

Communication to other entities outside of a computing node for system I/O is supported by a large spectrum of protocols to accommodate the communication needs of a large spectrum of safety-critical applications in the railway domain. In the following, a brief summary on the available protocols is given together with the scope of application in railway systems.

Controller Area Network: The Controller Area Network (CAN) bus is widely used in the railway domain. In particular, CAN is in use for axle counters, field element controllers, on-board units.

Time-Triggered Protocol: TAS Control Platform supports TTP/C for field element controllers.

Multifunction Vehicle Bus: The Multifunction Vehicle Bus (MVB) is a time-triggered data bus defined in IEC 61375 - 3 and mainly applied in traction units. As such, MVB is used for various on-board systems in the Thales product portfolio.

Process Field-Bus: The Process Field-bus (ProfiBus) is used for on-board systems to interface with Balise transfer modules which pick up the air gap signals from Euro-balises (see Section 14.4.1).

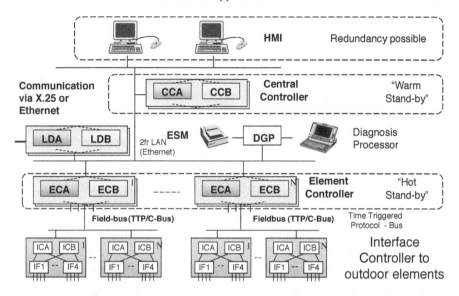

FIGURE 14.27
Architecture of the ELEKTRA Interlocking System

Ethernet-based Communication: For Ethernet-based LAN communication within an interlocking system, TAS Control Platform provides a connection oriented communication protocol which extends the concept of host-local POSIX message queues to remote message queues.

14.4.5 Application of Time-Triggered Protocols in the Railway Domain

14.4.5.1 Interlocking: Architecture (Components, Services, Interactions)

The application of TTP in the railway domain will be illustrated in the context of the TAS Control platform based interlocking system ELEKTRA. Figure 14.27 gives an overview of the components of the ELEKTRA system. The system consists of three layers, the Central Controller (CC), the HMI and the Field Element Controller (FEC) which is composed of the Element Controller (EC) and the Interface Controller (IC).

The train operator interfaces with the system via an HMI showing a schematic of the track layout of the railway station under control. The HMI indicates the status of each block, which is either occupied or free, as well as the status of the switches and signals. By selecting a starting point and an end point, the operator commands the generation of a route.

The heart of the system is the Central Controller which is responsible for generating and setting the train routes in a safe manner. Roughly speaking, a train route can be seen as an ensemble of signals and switches that have to be driven into a certain state to establish a route. After checking that the intended route does not conflict with other already existing routes, the switches are moved to the required position to establish the route. As soon as all switches have reached their end position, the switches are locked. Afterwards, the signals are set to indicate the new movement authority to the train passing the route.

14.4.5.2 Field Element Controller

The Field Element Controller (FEC) is responsible for interfacing the Central Computer layer to the track-side elements in the field. The FEC is composed of the following subsystems:

An Element Controller (EC): The Element Controller has the knowledge of the anatomy of a field element (e.g., signal) and maps the abstract notation of a signal aspect[4] commanded by the Central Controller to serial port commands for the Interface Controller (IC) subsystem, and which translates the serial scanner data generated by the IC subsystem into logical element states for the Central Controller. The EC is based on TAS Control Platform.

An Interface Controller (IC): A subsystem which transforms the serial port commands generated by the Element Controller (EC) subsystem into parallel data to control the HW interfaces, and which transforms parallel data, representing the state of HW interfaces, into the interface board specific scanner data for the Element Controller subsystem. An IC is an embedded micro-controller based computer system which consists for safety reasons of two micro-controllers (A and B) with an independent parallel IO-subsystem.

In the ELEKTRA system, time-triggered communication is used between the Element Controller and the Interface controller. The reason for this design decision was driven by the following considerations:

Quality of service: TTP provides fixed communication bandwidth in a field-bus environment. Thus the system reaction time is guaranteed even in high load situations. This is an advantage especially in case of a general scan of all field elements during the startup or a reconfiguration phase of the system.

System composability: While the number of communication partners is fixed on the HMI and Central Computer layer of the interlocking system, the number of field elements that have to be driven scales with the size of the railway station under control. This can range from a couple of signals for small stations up to several hundred signals and switches for large-scale stations.

Figure 14.28 gives an overview of the network architecture with the communication network between EC and IC.

---

[4]The term signal aspect refers to the state of a railway signal from the operational point of view, e.g.: Next block free with 40 km/h speed limit or stop.

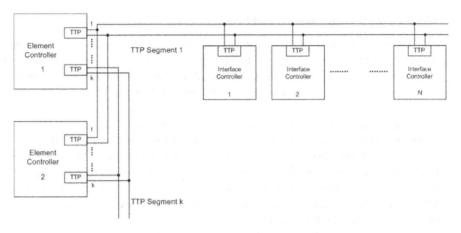

FIGURE 14.28
TTP/C Network Architecture

14.4.5.3 Availability Concept

Since Element Controllers are crucial to the availability of the overall system, the redundancy concept arranges for two redundant ECs working in a hot-hot standby mode. Interface Controllers are non-redundant, since it does not make a difference whether the field element itself fails or the corresponding IC. The failure of an Interface Controller must thus be handled by the Central Controller by entering a degraded mode depending on which field elements are no longer available for routing a train.

14.4.6 Safety Concept

The safety of the communication between Element Controller and Interface Controller is realized by the implementation of the requirements on safe communication as stated in Section 14.4.3. This means that each message is secured by a safety code on application level and holds a sequence number to detect the loss of messages. Since TTP/C can only ensure timeliness on field-bus level, the timeliness of message delivery is also supervised on application level.

14.4.6.1 Timing Requirements

From a system point of view, the following timing requirements affect the communication between the Element Controller and the Interface Controller:

- The maximum time allowed to diagnose a failure of any subsystem component is 1 second.

- The maximum time needed to drive the system into a safe state shall not exceed 1 second.

14.4.6.2 TTP-Configuration and Schedule

Each element controller can be connected to up to four TTP-segments of which each segment can be populated with up to 21 interface controllers. And the number of element controllers in each segment is limited to two. A basic design requirement to ensure composability is that for each TTP-segment the same TTP configuration is used.

Each EC needs 16 N-frames to send its data to the interface controllers and in turn each interface controller needs two TTP N-frames to send its data to the ECs.

Every TTP round consists of seven slots of which the first two slots are reserved for the element controllers. In the next two slots, the non-multiplexed interface controllers send their data while all other IC share the remaining three slots. The two non-multiplexed ICs are needed to re-integrate failed ICs even in a scenario when one element controller has failed.

The reason for multiplexing ICs is to meet the timing requirements stated in Section 14.4.6.1 while enabling 21 communication partners [5] with different message sizes in one TTP cluster cycle. These timing requirements translate to a maximum duration of a cluster cycle of 50 ms.

Table 14.3 lists the whole TTP-Cluster-Cycle. For each slot, it is indicated which node is sending data (N-Frames, normal frames) or TTP-protocol frames containing no user data (I-frames, frames with explicit controller state).

TABLE 14.3
TTP-cluster cycle.

Slot Number	0	1	2	3	4	5	6
Round 1	EC1	EC2	IC01 (I)	IC02 (I)	IC03 (I)	IC10 (I)	IC17 (I)
Round 2	EC1	EC2	IC01 (N)	IC02 (N)	IC03 (N)	IC10 (N)	IC17 (N)
Round 3	EC1	EC2	IC01 (I)	IC02 (I)	IC04 (N)	IC11 (N)	IC18 (N)
Round 4	EC1	EC2	IC01 (I)	IC02 (I)	IC05 (N)	IC12 (N)	IC19 (N)
Round 5	EC1	EC2	IC01 (I)	IC02 (I)	IC06 (N)	IC13 (N)	IC20 (N)
Round 6	EC1	EC2	IC01 (I)	IC02 (I)	IC07 (N)	IC14 (N)	IC21 (N)
Round 7	EC1	EC2	IC01 (I)	IC02 (I)	IC08 (N)	IC15 (N)	IC22 (I)
Round 8	EC1	EC2	IC01 (I)	IC02 (I)	IC09 (N)	IC16 (N)	IC23 (I)
Round 9	EC1	EC2	IC01 (I)	IC02 (I)	IC03 (I)	IC10 (I)	IC17 (I)
Round 10	EC1	EC2	IC01 (N)	IC02 (N)	IC03 (N)	IC10 (N)	IC17 (N)
Round 11	EC1	EC2	IC01 (I)	IC02 (I)	IC04 (N)	IC11 (N)	IC18 (N)
Round 12	EC1	EC2	IC01 (I)	IC02 (I)	IC05 (N)	IC12 (N)	IC19 (N)
Round 13	EC1	EC2	IC01 (I)	IC02 (I)	IC06 (N)	IC13 (N)	IC20 (N)
Round 14	EC1	EC2	IC01 (I)	IC02 (I)	IC07 (N)	IC14 (N)	IC21 (N)
Round 15	EC1	EC2	IC01 (I)	IC02 (I)	IC08 (N)	IC15 (N)	(IC22) (I)
Round 16	EC1	EC2	IC01 (I)	IC02 (I)	IC09 (N)	IC16 (N)	(IC23) (I)

---

[5]Nodes 22 and 23 do not have N-frames in the communication network interface (CNI).

14.4.7 Conclusion and Outlook

Historically, when the first electronic interlocking systems were introduced in the late 1980s, time-triggered technology has not been broadly available.

As railway applications are soft real-time applications, safety and availability targets have been met by thorough and conservative system design with respect to timing aspects, with event driven approaches based on readily available OSI layer 1 to 3 solutions.

In the railway industry, there is a trend to integrate more functionality and services on existing system architectures. This means that sharing resources gains more and more importance. Time-triggered communication is an excellent approach to guarantee bandwidth and provide a required level of quality of service.

Concerning the interfacing of field elements, there is a clear trend to move from centralized architectures to locally distributed solutions. Instead of driving distant analogue elements from controllers located at the central interlocking control room, the signal posts will be equipped with intelligent and network compatible interfaces. The adoption of time-triggered technology based on Ethernet for this kind of application is certainly interesting.

15

Development Tools

P. Pop

Technical University of Denmark

A. Goller

TTTech Computertechnik AG

T. Pop

Ericsson AB

P. Eles

Linköping University

CONTENTS

15.1 Introduction

Embedded systems are now everywhere: From medical devices to vehicles, from mobile phones to factory systems, almost all the devices we use today are controlled by embedded computers. Over 98% of microprocessors are used in embedded systems, and the number of embedded systems in use has become larger than the number of humans on the planet, and is projected to increase to 40 billion worldwide by 2020 [11, 84]. The embedded systems market size is about 100 times larger than the desktop market, with over 160 billion Euros worldwide and a growth rate of 9% [84].

The complexity of embedded systems is growing at a very high pace and their constraints in terms of performance, reliability, cost and time-to-market are getting tighter. The embedded software size is increasing 10 to 20% per year, depending on the application area. Today's cars have more than 100 million object code instructions [84], while in avionics, the size of the certified software has increased from 12 Mbytes in Airbus A340 to 80 Mbytes in A380 [11].

At the same time, high complexity, increasing density and higher operational frequencies have led to an increasing number of faults [65]. Embedded systems are increasingly used in safety-critical contexts, such as automotive applications, avionics, medical equipment, control and telecommunication devices, where any deviation from the specified functionality can have catastrophic consequences. In addition, many industries are very cost-sensitive, and thus the dependability requirements have to be met within a tight cost constraint.

Therefore, the task of designing such systems is becoming increasingly important and difficult at the same time. The difficulty of designing embedded systems is reflected by the share of the development and implementation costs from the final product price, which is 36% in the automotive area, 22% in industrial automation, 37% in the telecommunications area, 41% in consumer electronics and 33% for medical equipment [276]. This has led to a *design productivity gap*: The number of on-chip transistors is growing each year by 58% (according to Moore's law), whereas the productivity of hardware designers is only growing by 21% per year, and the software productivity is lagging even further behind [276].

Many organizations, including automotive manufacturers, are used to designing and developing their systems following some version of the "waterfall" [94] model of system development. This means that the design process starts with a specification and, based on this, several system-level design tasks are performed manually, usually in an ad-hoc fashion. Then, the hardware and software parts are developed independently, often by different teams located far away from each other. Software code is written, the hardware is synthesized and they are supposed to integrate correctly. Simulation and testing are done separately on hardware and software, respectively, with very few integration tests.

If this design approach was appropriate when used for relatively small systems produced in a well-defined production chain, it performs poorly for more complex systems, leading to an increase in the time-to-market. New approaches and tools have been proposed, which are able to: Successfully manage the complexity of embedded systems, meet the constraints imposed by the application domain, shorten the time-to-market, and reduce development and manufacturing costs. There are many development tools, and their use depends on the application area. The most important embedded systems tools are presented in [191].

In the next section, we present the typical design tasks, emphasizing the communication synthesis task, which is the focus of this chapter. We will present state-of-the-art techniques and tools for the communication scheduling and communication configuration. In Section 15.3, we will define the general problem of scheduling, discuss its complexity and the typical strategies employed. Once a schedule is generated, it can be manipulated, extended and visualized.

As we will show, communication synthesis has a strong impact at the system-level. In this context, in Section 15.4, we will discuss the integrated (holistic) scheduling of tasks and messages, and the bus schedule optimization to support the fulfillment of timing constraints. Systems are seldom built from scratch, hence, in Section 15.5 we discuss the issues related to incremental design, where a schedule has to be generated such that it is flexible, i.e., supports the addition of new functionality. Although this book is focused on time-triggered systems, using an event-triggered approach at the processor level can be the right solution under certain circumstances [205]. Hence, in Section 15.6, we present an approach to integrate event-driven tasks with a time-triggered communication infrastructure.

Once a schedule is generated, it has to be translated into a communication configuration, particular for the communication protocol used, such as TTP[1] or FlexRay. In Section 15.7 we illustrate this issue using the tool chain from TTTech. Finally, in the last section of this chapter, we discuss verification and certification aspects.

15.2 Design Tasks

The aim of a design methodology is to coordinate the design tasks such that the time-to-market is minimized, the design constraints are satisfied and various parameters are optimized. The following are the state-of-the-art methodologies in embedded systems design:

- **Function/architecture co-design:** Function/architecture co-design is a design methodology [162, 323] which addresses the design process at higher abstraction levels. Function/architecture co-design uses a top-down synthesis approach, where trade-offs are evaluated at a high level of abstraction. The main characteristic of this methodology is the use, at the same time with the top-down synthesis, of a bottom-up evaluation of design alternatives, without the need to perform a full synthesis of the design. The approach to obtain accurate evaluations is to use an accurate modeling of the behavior and architecture, and to develop analysis techniques that are able to derive estimates and to formally verify properties relative to a certain design alternative. The determined estimates and properties, together with user-specified constraints, are then used to drive the synthesis process.

  Thus, several architectures are evaluated to determine if they are suited for the specified system functionality. There are two extremes in the degrees of freedom available for choosing an architecture. At one end, the architecture is already given, and no modifications are possible. At the other end of the spectrum, no constraints are imposed on the architecture selection, and the synthesis task has to determine, from scratch, the best architecture for the required functionality. These two situations are, however, not common in practice. Often, a *hardware platform* is available, which can be *parameterized* (e.g., size of memory, speed of the buses, etc.). In this case, the synthesis task is to derive the parameters of the architecture such that the functionality of the system is successfully implemented. Once an architecture is determined and/or pa-

---

[1]Throughout this chapter, we use "TTP" instead of "TTP/C," as it is the commercial and more customary term.

rameterized, the function/architecture co-design continues with the mapping of functionality onto the instantiated architecture.

- **Platform-based design:** In order to reduce costs, especially in the case of a mass market product, the system architecture is usually reused, with some modifications, for several product lines. Such a common architecture is denoted by the term *platform*, and consequently the design tasks related to such an approach are grouped under the term platform-based design [163].

One of the most important components of any system design methodology is the definition of a *system platform*. Such a platform consists of a hardware infrastructure together with software components that will be used for several product versions, and will be shared with other product lines, in the hope to reduce costs and the time-to-market.

The authors in [163] have proposed techniques for deriving such a platform for a given family of applications. Their approach can be used within any design methodology for determining a system platform that later on can be parameterized and instantiated to a desired system architecture.

Considering a given application or family of applications, the system platform has to be instantiated, deciding on certain parameters, and lower level details, in order to suit the particular application(s). The search for an architecture instance starts from a certain platform, and a given application. The application is mapped and compiled on an architecture instance, and the performance numbers are derived, typically using simulation. If the designer is not satisfied with the performance of the instantiated architecture, the process is repeated.

- **Incremental design process:** A characteristic of the majority of approaches to the design of embedded systems is that they concentrate on the design, from scratch, of a new system optimized for a particular application. For many application areas, however, such a situation is extremely uncommon and appears only rarely in design practice. It is much more likely that one has to start from an already existing system running a certain application and the design problem is to implement new functionality (including also upgrades to the existing one) on this system. In such a context, it is very important to operate no, or as few as possible, modifications to the already running application. The main reason for this is to avoid unnecessarily large design and testing times. Performing modifications on the (potentially large) existing application increases design time and, even more, testing time (instead of only testing the newly implemented functionality, the old application, or at least a part of it, has also to be retested) [264].

However, minimizing the modification cost is not the only aspect to be considered. Such an incremental design process, in which a design is periodically upgraded with new features, is going through several iterations. Therefore, after new functionality has been introduced, the resulting system has to be implemented such that additional functionality, later to be mapped, can easily be accommodated [264].

There is a large body of literature on systems engineering that discusses various methodologies for systems development. Many methodologies employed in the development of safety-critical systems are a variant of the "V-Model" [94], named after the graphical representation in a "V" shape of the main development phases, that starts with the requirements phase, followed by hazard and risk analysis, specification, architectural design, module design, module construction and testing (at the bottom of the "V" shape), system integration and testing, system verification, system validation and, finally, certification. For example, the V-model is employed in the SETTA approach [6], which proposes system development methodologies for time-triggered systems in the automotive and aerospace domains.

The design tasks that have to be performed depend on the type of system being developed and on the design methodology employed. For safety-critical systems, the design tasks are often dictated by certification requirements, or by the development approach used. For example, the Automotive Open System Architecture (AUTOSAR) defines, besides the models for system development, the design tasks that have to be performed [18]. Regardless of the design tasks performed, *model-based design* is used throughout the development process: The interaction among design tasks is facilitated by the use of models, and the modeling is supported by graphical modeling tools. The following are the typical design tasks:

- **Functional analysis and design:** The functionality of the host system, into which the electronic system is embedded, is normally described using a formalism from that particular domain of application. For example, if the host system is a vehicle, then its functionality is described in terms of control algorithms using differential equations, which are modeling the behavior of the vehicle and its environment. At the level of the embedded real-time system which controls the host system, the functionality is typically described as a set of functions, accepting certain inputs and producing some output values.

  During the functional analysis and design stage, the desired functionality is specified, analyzed and decomposed into sub-functions based on the experience of the designer.

- **Architecture selection:** The architecture selection task decides what components to include in the hardware architecture and how these components are connected. Architecture selection relies heavily on the experience of the designer and previous product versions. If needed, new hardware components may be designed and synthesized, part of the **hardware design** task.

- **Mapping:** The mapping task has to decide what part of the functionality should be implemented on which of the selected components.

  The automotive companies integrate components from suppliers, and thus the mapping choices are often limited.

- **Software design and implementation:** This is the phase in which the software is designed and the code is written. The code for the functions is developed manually or generated automatically. The low-level software that interacts closely with the hardware is sometimes called *firmware*, and the task of designing it is hence called **firmware design**.

At this stage, the correctness of the software is analyzed through simulations, but no analysis of timing constraints is performed, which is done during the scheduling and schedulability analysis stage.

- **Scheduling and schedulability analysis:** Once the functions have been defined and the code has been written, the scheduling task is responsible for determining the execution order of the functions inside an ECU, and the transmission of messages such that the timing constraints are satisfied.

  Schedulability analysis is used to determine if an application is schedulable. A detailed discussion about scheduling and schedulability analysis is presented in the next section.

- **Integration:** In this phase, the manufacturer has to integrate the ECUs from different suppliers. The performance of the interacting functionality is analyzed using analysis tools and time-consuming simulation runs using the realistic environment of a prototype car.

  Detecting potential problems at such a late stage may lead to large delays in the time-to-market, since once a problem is identified, it takes a very long time to go through all the previous stages in order to fix it.

- **Communication synthesis:** Many real-time applications, following physical, modularity or safety constraints, are implemented using *distributed architectures.* The systems addressed in this book are composed of several different types of hardware components, interconnected in a network.

  In this context, an important design task is the communication synthesis task, which decides the scheduling of communications and the configuration parameters specific to the employed protocol. These decisions have a strong impact on the overall system properties such as predictability, performance, dependability, cost, maintainability, etc.

- **Calibration, testing, verification:** These are the final stages of the design process. If not enough analysis, testing and verification has been done in earlier stages of the design, these stages can be very time consuming, and problems identified here may lead to large delays.

15.3   Schedule Generation

According to [49], a *scheduling policy* provides two features: (i) an algorithm for ordering the use of system resources (in particular the processors, the buses, but also I/Os) and (ii) a means of predicting the worst-case behavior of the system when the scheduling algorithm is applied. The prediction, also known as *schedulability analysis*, can then be used to guarantee the temporal requirements of the application.

The aim of a schedulability analysis is to determine *sufficient* and *necessary* conditions under which an application is schedulable. An application is schedulable if there exists at least one scheduling algorithm that is able to produce a feasible schedule. A *schedule* is a particular assignment of activities to the resource (e.g., tasks to processors). A schedule is *feasible* if all tasks can be completed within the specified constraints. Before such techniques can be used, the worst-case execution times of tasks have to be determined. Tools such as aiT [98] can be used in order to determine the worst-case execution time of a piece of code on a given processor.

The analysis and optimization techniques employed depend on the scheduling policy and the model of the functionality used. The design techniques typically take as input a model of the functionality consisting of sets of interacting tasks. A *task* is a sequence of computations (corresponding to several building blocks in a programming language) which starts when all its inputs are available. When it finishes executing, the task produces its output values. Tasks can be *preemptible* or *non-preemptible*. Non-preemptible tasks are tasks that cannot be interrupted during their execution. Preemptible tasks can be interrupted during their execution. For example, a higher priority task has to be activated to service an event; in this case, the lower priority process will be temporarily preempted until the higher priority process finishes its execution. Tasks send and receive messages. Depending on the communication protocol, message transmission can be preemptible or non-preemptible. Large non-preemptible messages can be split into packets before transmission.

There are several approaches to scheduling:

- Non-preemptive *static cyclic scheduling* (SCS) algorithms are used to build, offline, a schedule table with activation times for each task (and message), such that the timing constraints of tasks (and messages) are satisfied.

- Preemptive *fixed priority scheduling* (FPS). In this scheduling approach, each task (and message) has a fixed (static) priority which is computed offline. The decision on which ready task to activate (and message to send) is taken online according to their priority.

- *Earliest deadline first* (EDF). In this case, that task will be activated (and that message will be sent) which has the nearest deadline.

For static cyclic scheduling, if building the schedule table fulfills the timing constraints, the application is schedulable. In the context of online scheduling methods, there are basically two approaches to the schedulability analysis: Utilization-based tests and response-time analysis.

- The *utilization tests* use the *utilization* of a task or message (its worst-case execution time relative to its period) in order to determine if the task sets (or messages) are schedulable.

- A *response-time analysis* has two steps. In the first step, the analysis derives the worst-case response time of each task and message (the time it takes from the moment it is ready for execution, until it has finished executing). The second step compares the worst-case response time of each task and message to its deadline and, if the response times are smaller than or equal to the deadlines, the application is schedulable.

As mentioned throughout this book, another important distinction is between two basic design approaches for real-time systems, the event-triggered and time-triggered approaches.

- **Time-Triggered:** In the time-triggered approach, activities are initiated at predetermined points in time. In a distributed time-triggered system, it is assumed that the clocks of all nodes are synchronized to provide a global notion of time. Time-triggered systems are typically implemented using *non-preemptive static cyclic scheduling*, where the task activation or message communication is done based on a schedule table built offline.

- **Event-Triggered:** In the event-triggered approach, activities happen when a significant change of state occurs. Event-triggered systems are typically implemented using *preemptive priority-based scheduling*, or *earliest deadline first*, where, as response to an event, the appropriate task is invoked to service it.

In this chapter, we are interested in time-triggered systems implemented using non-preemptive static cyclic scheduling. A static schedule is a list of activities that is repeated periodically. Each activity has an associated start time, capturing, for example, when the particular task has to be activated or the message has to be transmitted. There are several types of schedules in time-triggered systems.

- **Message schedules:** These are the schedules for the messages and frames transmitted on the bus. The message schedules are organized according to a TDMA policy: Each processor can transmit only during a predetermined time interval, the so-called TDMA slot. In such a slot, a node can send several messages packaged in a frame (TTP), or even several frames (TTEthernet). Some protocols require a fixed sequence of slots, each slot corresponding to a node, and covering all the nodes in the architecture. This sequence is called a TDMA round. Several TDMA rounds can be combined together in a cycle that is repeated periodically (cluster cycle). Other protocols (like TTEthernet) are less strict and allow a basically arbitrary pattern within a cluster cycle. However, the design of control algorithms often implies the use of TDMA rounds, and several TDMA rounds with different length may be folded into a cluster cycle. The sequence and length of slots may be required to be the same for all TDMA rounds (FlexRay). In TTP, different lengths of slots are allowed, but a fixed sequence must be maintained.

- **Task schedules:** These are the schedules for tasks running on the processors, according to a SCS policy. Such a scheduling scheme is also called "time-line scheduling," and is the most used approach to handle periodic tasks in safety-critical systems. The advantages and disadvantages of timeline scheduling (especially compared to fixed-priority preemptive scheduling) are well understood [203]. The tasks are repeated periodically, with a period called the *major cycle*. In most cases, the task periods are not identical, so the major cycle is set to the least common multiple of all periods, and is subdivided into *minor cycles*. A task with a smaller period will appear in several minor cycles, thus achieving its desired rate. The task schedules are implemented using a *cyclic executive*, typically based on a clock tick (an interrupt), which triggers the start of the minor cycle. Often, other interrupts are disabled (or severely limited) and when the tasks in the minor cycle finish executing, control is passed to a background scheduler that attends to less important activities.

- **Partition schedules:** In safety-critical systems, applications of different criticality levels are often separated from each other using spatial and temporal partitioning. Thus, with temporal partitioning, each application is allowed to run only within predefined time slots, allocated on each processor. The sequences of time slots for all applications on a processor are grouped within a *major frame*, which is repeated periodically.

- **Interrupt schedules:** While task and partition schedules mainly focus on the user application, interrupt schedules are used for middleware tasks. Certain actions, like reading and unpacking a frame, have to be executed actually for every frame received. An interrupt (or middleware task activation) therefore may occur several times within a cluster cycle or even within a TDMA round. The interrupt schedule specifies what specific actions to execute in this particular instance of an interrupt occurrence.

- **Cluster schedules:** To implement a schedule in a distributed system, a global notion of time is required. The previously mentioned schedules are typically specified at the cluster level, since clock synchronization is performed at the cluster level. A cluster schedule captures task, message and partition schedules within a cluster. Several cluster schedules can be present in a system, but they will not be synchronized with each other.

15.3.1 Requirements and Application Model

The requirements imposed on an embedded system depend on the particular application that it implements. Requirements are divided into functional requirements and non-functional requirements. The difficulty of designing embedded systems lies in the many competing non-functional requirements that have to be satisfied. Typical non-functional requirements are: Performance (in terms of latency, throughput, speedup), unit cost (the cost of manufacturing each copy of the system), non-recurring engineering cost (the one-time monetary cost of designing the system), size, power consumption, flexibility (how easy is it to change the functionality, to add new functions), time-to-prototype, time-to-market and dependability attributes such as reliability, maintainability and safety.

In a *real-time system*, the timing constraints are of utmost importance: "The correctness of the system behavior depends not only on the logical results of the computations, but also on the physical instant at which these results are produced" [169]. In *hard* real-time systems, missing a deadline can lead to a catastrophic failure. Design methodologies for these systems are based on their worst-case execution times. In *soft* real-time systems, missing a deadline does not cause catastrophic failures in the system but leads to a certain performance degradation. The following are typical constraints imposed in a hard real-time system:

- **Timing constraints**. The worst-case execution time (WCET) C_i is an upper bound on the execution times of a task τ_i, which depends on its functionality and the particular processor N_i where it runs. Tasks can have constraints on their completion or activation. Thus, a *deadline D_i* of a task τ_i is a time at which the task must complete its execution. Tasks which must be executed once every T_i units of time are called periodic tasks, and T_i is called their period. (Each execution of a periodic task is called a job.) All other tasks are called aperiodic. *Release times* restrict the start time of task activations (often to avoid resource contention). Another important timing constraint, especially in the context of control applications, is *jitter*, which captures the time-variation of a periodic event. Note that all these constraints also apply to messages.

- **Precedence constraints:** They impose an ordering in the execution of activities. The behavior of the system is often modeled as a sequence of activities. Thus, before a task can start, it has to wait for the input from another task. For example, to perform an image recognition, first the image has to be acquired. *Distance* constraints express a minimum distance between two activities, on top of a precedence constraint. The opposite of distance constraints are the *freshness* constraints, which express the maximum distance between two consecutive activities. Freshness constraints are typically placed on sensor data.

- **Resource constraints:** To perform their function, tasks have to use resources. A task may have a *locality* constraint which requires the allocation of a task to a specific processor, for example, because it has to use an actuator attached to this particular processor. When several tasks want to use the same resource (e.g., shared memory), we impose *mutual exclusion* constraints. Messages exchanged between tasks on different processors have to use the bus, thus imposing *communication* constraints.

- **Extendability constraints:** Of specific interest are changes that are considered "local." Such a local change is a new message m_{i+1} that shall be transmitted from one node A to another node B, but not to all other nodes C to Z. Ideally, the communication configuration of nodes C to Z need not be updated due to this change. A slightly different case is if message m_i, which only is transmitted between nodes A and B, gets changed in its size.

  Unfortunately, this view does not provide enough detail to decide whether this change is local or not. If it is necessary to move another message m_j due to the now bigger size of message m_i, it is obviously not simply a local change. Constraints may exist regarding the placement and alignment of messages within frames. A certain amount of bandwidth (per host) could be reserved for future extensions. Users may want to specify the layout of the frame manually, but leave the scheduling of the frames to a tool. The objective is to be able to modify and extend an existing schedule throughout the whole development and product lifetime just by local changes in order to save verification and certification efforts.

These requirements dictate the types of schedules that have to be produced, and the types of tools needed to generate the schedules. For example, the precedence constraints will capture if the interaction between components is synchronous or asynchronous. A fully synchronous application (the tasks and the communication are in phase and with the same speed) needs a more interacting design tool chain, that will produce synchronized cluster-level schedules for both tasks and messages, than an asynchronous application. There can be several setups, which will be reflected in the tools used and the tool flow employed: The time-triggered network communication and application are synchronous; the time-triggered network communication and application are asynchronous (causing oversampling and undersampling issues); the network communication is not time-triggered and the application is bound to a local clock (e.g., a control loop with CAN); and the network communication is not time-triggered and the application reacts on events.

Thus, in this section we discuss the tools needed for generating message schedules for time-triggered communication. In Section 15.4, we consider a complex setup, where tasks can be both time-triggered and event-triggered, and messages are transmitted using FlexRay, which has both static (time-triggered) and dynamic (event-triggered) segments. The assumption is that tasks and messages are synchronous. We discuss holistic scheduling: How to generate the cluster-level task and message schedules such that the timing constraints are satisfied for both time-triggered and event-triggered activities. We show how schedulability analysis has to be integrated with schedule generation to guarantee the timing constraints. In Section 15.5, we discuss how the schedules can be generated such that they are flexible, i.e., easy to extend with new functionality. Section 15.6 focuses on the interaction between event-triggered tasks, which produce event-triggered messages, and the time-triggered frames scheduled over TTP. Several approaches that schedule event-triggered messages over time-triggered frames are proposed and discussed. We propose both problem-specific heuristic algorithms and meta-heuristics for the optimization of the generated schedules. Section 15.3.2 discusses the complexity of the scheduling problem and the typical solutions employed. As we will show in the remainder of this chapter, the way the schedules are generated and optimized has a significant impact not only on the timing constraints, but also on flexibility, latency, jitter, buffer size, switching devices required and others.

15.3.1.1 Application Model

There is a lot of research in the area of system modeling and specification, and an impressive number of representations have been proposed. An overview, classification and comparison of different design representations and modeling approaches is given in [85]. The scheduling design task deals with sets of interacting tasks. Researchers have used, for example, dataflow process networks (also called task graphs, or process graphs) to describe interacting tasks, and have represented them using directed acyclic graphs, where a node is a process and the directed arcs are dependencies between processes.

In this subsection, we describe the application model assumed in the following sections. Thus, we model an application $\mathcal{A}$ as a set of directed, acyclic, polar graphs $\mathcal{G}_i(\mathcal{V}_i, \mathcal{E}_i) \in \mathcal{A}$. A node $\tau_{ij} \in \mathcal{V}_i$ represents the jth task or message in $\mathcal{G}_i$. An edge $e_{ijk} \in \mathcal{E}_i$ from τ_{ij} to τ_{ik} indicates that the output of τ_{ij} is the input of τ_{ik}. A task becomes ready after all its inputs have arrived, and it issues its outputs when it terminates. A message will become ready after its sender task has finished, and becomes available for the receiver task after its transmission has ended. The communication time between tasks mapped on the same processor is considered to be part of the task's worst-case execution time and is not modeled explicitly. Communication between tasks mapped on different processors is performed by message passing over the bus. Such message passing is modeled as a communication task inserted on the arc connecting the sender and the receiver task.

We consider that the scheduling policy for each task is known (either *SCS* or *FPS*), and we also know how the messages are transmitted. For example, for FlexRay, we would know if the message is sent in the static or dynamic segment. For a task $\tau_{ij} \in \mathcal{V}_i$, $Node_{\tau_{ij}}$ is the node to which τ_{ij} is assigned for execution. When executed on $Node_{\tau_{ij}}$, a task τ_{ij} has a known worst-case execution time $C_{\tau_{ij}}$. We also consider that the size of each message m is given, which can be directly converted into communication time C_m on the particular bus.

Tasks and messages activated based on events also have a priority, $priority_{\tau_{ij}}$. All tasks and messages belonging to a task graph G_i have the same period $T_{\tau ij} = T_{\mathcal{G}_i}$ which is the period of the task graph. A deadline $D_{\mathcal{G}_i}$ is imposed on each task graph $\mathcal{G}_i$. In addition, tasks can have associated individual release times and deadlines. If dependent tasks are of different periods, they are combined into a merged graph capturing all activations for the hyper-period (LCM of all periods) [261].

15.3.2 Scheduling Complexity and Scheduling Strategies

As mentioned earlier, a schedule defines the assignment of activities to the resources. The complexity of deriving a schedule depends on the type and quantity of resources available, the constraints imposed, and the objective function that has to be optimized. Scheduling is probably one of the most researched problems in computer science, and there is an enormous amount of results. There are several surveys available which present the scheduling problems, their complexity and the strategies used.

The following are the main findings regarding the complexity of the scheduling problems related to time-triggered systems, as reported in [300]:

- The integrated task and message scheduling problem to find the optimal schedule (the one with minimum length) is NP-complete. Thus, given a task graph model of the application, a limited number of processors interconnected by a time-triggered bus, the problem of finding a feasible schedule that minimizes the schedule length does not have a polynomial-time solution.

- The optimal task scheduling problem on a limited number of processors, but without considering the communication costs, is also NP-complete.

- The scheduling problem, considering communication costs, on an unlimited number of processors is NP-complete.

- The task scheduling problem, without the communication costs, is polynomial on an unlimited number of processors. Of course, there are never unlimited resources in a real system.

- The problem of deriving a schedule for messages, with the aim of optimizing a given design metric, is NP-complete if it can be reduced to the "knapsack" or "bin-packing" problems, which themselves are NP-complete.

These results mean that the schedules cannot be derived manually, and tool support is necessary. The scheduling problem is a very well-defined optimization problem, and has been tackled with every conceivable approach.

- **Mathematical techniques:** Researchers have proposed integer linear programming, mixed-integer programming and dynamic programming. Decomposition strategies (such as Benders-decomposition), enumerative techniques such as Branch-and-Bound and Lagrangian relaxation techniques have also been proposed. Such mathematical approaches have the advantage of producing the optimal solution. However, they are only feasible for limited problem sizes due to the prohibitive run times.

- **Artificial intelligence (AI):** AI techniques have been used for scheduling, such as expert/knowledge-based systems, distributed agents and neural networks.

- **Scheduling heuristics:** The most popular scheduling heuristics are *list scheduling* and *clustering* [300]. List scheduling (LS) is the dominant scheduling heuristic technique. LS heuristics use a sorted priority list, containing the tasks ready to be scheduled, while respecting the precedence constraints. A task is ready if all the predecessor tasks have finished executing and all the incoming messages are received. LS generates the schedule by successively scheduling each task (and message) onto the processor (bus). The start time in the schedule table is the earliest time when the resource is available to the respective task (or message). The allocation of tasks to processors has a direct influence on the communication cost. When the allocation of tasks to processors is not decided, *clustering* can be used to group tasks that interact heavily with each other, and allocate them on the same processor [300].

- **Neighborhood search:** Although very popular, the drawback of scheduling heuristics such as list scheduling is that they do not guarantee finding the optimal solution, i.e., they get stuck in a local optimum in the solution space. Neighborhood search techniques are meta-heuristics (i.e., they can be used for any optimization problem, not only scheduling) that can be used to escape from the local optimum. Neighborhood search techniques use design transformations (moves) applied to the current solution, to generate a set of neighboring solutions that can be further explored by the algorithm. Popular meta-heuristics in this category are Simulated Annealing, Tabu Search and Genetic Algorithms [46].

In the following subsections, we will use constructive heuristics such as list scheduling to generate schedules, and meta-heuristics (neighborhood search techniques) such as Simulated Annealing and Tabu Search to optimize a given schedule for a certain metric. In the next subsections, some concepts based on and extending the list-scheduling heuristic are discussed in detail. These concepts are partly implemented in the scheduler of ^{TTP}Plan [344], the cluster design tool for TTP clusters from TTTech. Lastly, we provide further details on the scheduling approach chosen for ^{TTP}Plan.

15.3.2.1 Incremental Scheduling

Once a schedule has been generated and optimized, an important aspect is the extension of a schedule. The goal is to keep the scheduled tasks or messages as they are, and to only add new tasks or messages in the free places. Incremental scheduling (a.k.a. *schedule extension*) thus means that scheduling is done in discrete steps.

Schedule Steps

Each time a schedule is made, this is called a "schedule step." These schedule "steps" do not really form a sequence of *different* steps, but the whole process is a quite iterative one: After an initial schedule has been created, some properties or objects may be changed, and a new schedule is made, which is possibly analyzed. Due to this analysis or to change requests, further modifications are done, and a new schedule is made. Each such cycle of changing and scheduling is considered a *schedule step*. It is possible to make as many schedule steps as needed, until the result is satisfactory. The concept of schedule steps fits well into the list-scheduling approach as discussed above. Furthermore, a schedule step does not imply that already placed tasks or messages are kept in their places. Any modification of the output is possible.

Freezing and Thawing

One can keep a schedule by "freezing" the current schedule step. By adding new messages (with their type, period and further attributes, such as sender and receiver) to it, and scheduling again, the "holes" in the original schedule are filled without changing the already placed parts. The inverse operation is to "thaw" a schedule step. This means to actually throw away the schedule that was computed in this very step, but keeping the schedule parts from previous schedule steps. The additions made in this step are then merged with the new additions (made after the just thawed schedule step), and together considered the change set for the current schedule step. Obviously, only the last frozen schedule step can be thawed. The concept of freezing and thawing schedule steps also nicely fits into the list-scheduling approach.

Apart from adding new messages, other possible additions after a frozen schedule step are:

- Additional hosts and subsystems

- Additional message types

- Mapping of new subsystems to hosts

In ^{TTP}Plan, only "frozen" schedule steps are stored and actually *counted* as steps. Schedule steps are numbered to identify them later on. The first schedule step is also called the "base step." It contains all information necessary to make the MEDL (*Message Descriptor List*, see Chapter 5, Section 5.3.1) for each host. In later schedule steps, additional messages can be added for transmission in previously unused portions of frames. Since the MEDL only contains information about the lengths of the frames, but not their contents, the addition of messages can be done without changing the MEDL.

^{TTX}Plan

^{TTX}Plan is the cluster design tool for FlexRay clusters. Incremental scheduling is of special interest here, as the Field-Bus Exchange Format (FIBEX) [13] is used, and FIBEX also allows us to save just parts of a cluster schedule. Furthermore, FlexRay comprises a static and a dynamic segment, but the concept of schedule steps is not applicable to the dynamic segment.

During FIBEX import, any already existing schedule information is imported first, then the *static* part of the schedule is frozen and the rest of the information is imported. With the command "Make new schedule," this remaining data, including the whole dynamic segment, is included in the schedule. The dynamic segment is *always* scheduled from scratch, regardless of any already existing schedule information. Part of the reason is that the length and the structure of dynamic frames change when messages are added.

^{TTX}Plan adds all schedule increments to its model. When the scheduler is then started to generate a new schedule, it takes into account the original schedule while computing a schedule for the "extended" model. It will not change the global FlexRay configuration, but will eventually allocate additional free slots to hosts and map additional messages to empty spaces in frames. Hosts, subsystems, messages, frames and their associations that were present in the original cluster design remain unchanged. The advantage of this concept is that hosts which are not affected by a change need not be touched. Moreover, a host may support different versions of the schedule by identifying which messages are sent.

Change Management

If, for example, only two hosts A and B need additional messages, only these two must be updated, while all other hosts can remain at the base step of the scheduling. Later, host C might be updated to use the second schedule step, too. Eventually, hosts A, D, and E might get updated to yet another schedule step with additional messages. At runtime, a cluster using incremental scheduling can thus contain hosts with differing schedule steps.

Each schedule step is an extension of the cluster's communication properties. It can place messages into unused parts of already allocated frames or assign yet unused frames to the host and put messages there. When a host has exhausted the spare capacity of its frames, or is known not to want to participate in any further schedule steps, it should be excluded from further schedule steps. The user may then still add increments to other hosts. The dynamic segment is not affected by this exclusion.

To allow for safe interoperation of hosts at various steps of an incremental schedule, each of the hosts participating in a schedule step should send one message per schedule step carrying the schedule-step checksum (e.g., computed by a design tool) which allows for online consistency checks. For a schedule step to be safely usable, the schedule-step checksum sent by the sender must be equal to the schedule-step checksum expected by the receiver.

15.3.2.2   Host Multiplexing

Host Multiplexing is a means to describe the fact that two or more hosts use the same sending slot in different rounds. Although this is a general concept, it is only available for TTP clusters.

A rather simple scenario is given in Figure 15.1. The first three slots are occupied as usual: Each slot is assigned to one node. The last slot is assigned to three nodes, where "Node 3" occupies two rounds, and "Node 4" and "Node 5" each occupy a single round in this four-round schedule.

In the following example scenario, a special kind of host has been designed to be non-periodic and still participate in the multiplexing. It is important to notice that the messages of this host are still periodic! It meets additional requirements like the following:

- One slot (in a schedule of 32 rounds) shall be shared by six hosts.

- Each host shall be assigned one round-slot every 8th round (periodic data).

- In the remaining $4 * 2$ rounds (two per multiplexing period), each host shall be assigned one additional round-slot (event data, higher-level protocols).

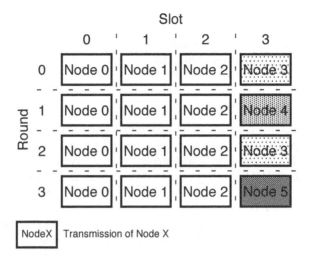

FIGURE 15.1
Multiplexed Slots

- With hosts A to F, the 32 round-slots shall be shared like this (typed in four lines, each representing 8 rounds, for better readability):

```
A B C D E F A B
A B C D E F C D
A B C D E F E F
A B C D E F ? ?
```

- The remaining two round-slots (marked "? ?") can be assigned to any multiplexing partner.

The pattern required is non-periodic in the sense that transmissions by one multiplexing host are not separated by a constant number of rounds anymore. However, it can still be modeled by assigning *multiple* periods to a single *multiplexing* host (e.g., in the above example, both "mux_periods" 8 and 32 could be assigned to the same host). This type of host is called "MUX_Ghost" (in the following, simply called "ghost") and has the following properties:

- A ghost behaves like a host in that it can run subsystems in a cluster and can thus send messages. In addition, it must be assigned a "mux_period" and a "mux_round."

- It is linked to a specific host which implements the subsystems specified for the ghost. (*Note:* A ghost must be linked to the same slot as the linked host.)

- A ghost has no "Host_in_Cluster" link in the object model.

- A ghost has no MEDL.

- The MEDL of a host contains the host's own round-slots ("R_Slot") and the round-slots of all ghosts linked to it.

15.3.2.3   Dynamic Messaging

Dynamic messaging is a concept to support the separation of concerns. One concern is the time, period and data size in which a specific host is permitted to send its data. The other concern is the actual layout and content of the frame being sent. This means that the middleware (e.g., the COM layer) needs to know both "when" and "what" to receive. Hence, it must be configured accordingly. Any time the "what" changes, it needs to be reconfigured.

The general idea — or rather: the requirement — behind dynamic messaging is that the middleware only should know the "when," and consequently only should need to be reconfigured in case of big changes, such as the timing of frames, if at all. Reconfiguration shall not be necessary if a message is added to a "hole" in an existing frame. It definitely shall not be necessary for *all* hosts in the cluster. Dynamic messaging therefore allows us to keep changes local, and to reduce certification efforts.

With dynamic messaging, every message is assigned an ID that is part of the message. It is placed at the beginning of the message, similarly to a frame header, and has a fixed length. With this ID, the embedded software or the COM layer can identify the message within a frame. The obvious disadvantage is that an additional ID per message needs to be transmitted, which requires more bandwidth. The major advantage is that a middleware layer (e.g., the COM layer) does not need any information about the location of a message within a frame. The middleware is able to pack and unpack any message without the communication configuration (MEDL) being modified, too. Allocation is statically predefined, so that overloading of frames cannot occur.

Initially, all hosts get a description of *all* possible messages that exist in the cluster, including their ID, length and other relevant properties for packing and unpacking. Once known, there is no need to update this information, regardless of whether the middleware is transferred to another host, or the message is placed at another position in the frame. Middleware configuration data only needs to be created once, and is the same for all hosts of the cluster. Having host hardware with preloaded and preconfigured middleware on stock becomes feasible, as it can be used right out of the box.

Dynamic messaging can be seen as an alternative to incremental scheduling. While for incremental scheduling, the bin-packing problem needs to be solved for placing messages in frames, and enough room must be reserved for potential future extensions, this is not relevant for dynamic messaging. The layout of the frame is determined at runtime.

15.3.2.4 Scheduling Strategies in ^{TTP}Plan

The basic input data for the message scheduler of ^{TTP}Plan consists of general cluster information (e.g., cycle durations, transmission speed, topology), information about hosts connected to this cluster and the messages sent by these hosts (e.g., size, period, redundancy).

The message scheduler of ^{TTP}Plan is an algorithm to produce a static, cyclic schedule. It is implemented as a heuristic scheduler, or more precisely, as a combination of a list scheduler, followed by an optimization step. The schedule output is basically a set of frames with a specific message allocation and a predefined transmission time instance.

In terms of programming, the message scheduler consists of five steps:

1. Initialization of the scheduler

2. Preparation for the scheduling (including checking the input object model)

3. Scheduling of the messages (including placement of the messages within a frame)

4. Write back the scheduling results to the object model

5. Finish scheduling

Preparation for Scheduling

Before the actual message scheduling takes place, various preparation steps have to be performed inside the message scheduler. This includes increasing the global cluster schedule step and figuring out the number of cluster modes. Usually, there is one user mode and one pseudo mode for TTP startup, but there might be more.

Afterwards, some messages are created that are needed for certain services. Such messages include "RPV messages" for the remote-pin-voting feature, as well as subsystem status messages. Every subsystem that was designed to send its status needs to send such a message. If the cluster allows schedule extensions, special messages carrying schedule step checksums have to be created as well.

Algorithmic Steps

In terms of algorithmic structure and complexity, only the third step from the above list is of interest. It can be broken down further into eight steps. These — basically independent — steps of the message scheduler are described in the order of invocation inside ^{TTP}Plan.

1. *Increment the schedule step.* The "scheduled" attribute of all objects is increased by one. This attribute is initially zero if no schedule step has been made so far (base step), and therefore incremented to one. If a schedule of an old, not frozen schedule step exists, this schedule is deleted. All frozen schedule information will be kept.

2. *Create the grid.* This step is only done inside the base step and is skipped for every additional schedule step. The grid is derived from the basic bus parameters like bus speed, the shortest and longest period of messages to be sent and the number of hosts in the cluster. Each cell of the grid represents a round-slot, and an "R_Slot" object is created accordingly. In this step, the number of rounds per cluster cycle is calculated, too.

3. *Schedule messages.*

   (a) Assign one slot to each host, depending on the shortest message period this host wants to use.

   (b) Assign additional slots to hosts according to the user settings regarding reserved bandwidth. With bandwith reservation, the amount of free space within a frame can be influenced, thus facilitating extensions in future schedule steps.

   (c) Determine the "difficulty" of a host by the number of messages, the replica level, and the ID of the host. (The ID is used to obtain a deterministic ordering.)

   (d) For every host, starting with the most difficult one, do the following:

      i. Determine the difficulty of a message in the following order: Channel freedom, redundancy degree, round-delta, round freedom, size and name.

      ii. Assign messages to frames starting with the most difficult message.

      iii. For each message: If there is an available R_Slot, use the R_Slot with "good" round-delta. Otherwise, try to assign a new R_Slot.

      iv. For each slot: Try to balance channels, then try to balance rounds. Slots are not balanced.

4. *Schedule messages in frame.* Place the messages in a specified position inside the frame. There are several options for this placement: The placement can be optimized for data access, leading to messages aligned with byte and word boundaries, as far as possible. It is also possible to specify that a message may be placed in fragments (i.e., not contiguously). A very simple approach is to place one message after the other, in the order they have been added to the frame.

5. *Schedule messages in message boxes.* If message boxes exist, place the messages inside the defined message box depending on alignment, size and ID.

6. *Place I-Frames.*: Place the frames necessary for synchronization of TTP wherever possible. If too few locations can be identified, a warning is issued. In this case, the user may try scheduling with different parameters, or switch over to using X-frames.

7. *Check schedule invariants.* These checks are executed to ensure the consistency of the schedule itself. If an internal error occurs, all schedule information

collected so far will be deleted again. In addition, the schedule signatures and the checksum are computed and set during this check.

15.3.3 Schedule Visualization

The more complex a communication system is, the greater the need for a means to visualize its schedule. It has been shown that increased complexity makes it more difficult to recognize design faults, simply due to a lack of overview. Thus, if the system can be visualized in terms of underlying communication *structures* instead of just pouring out all schedule details over the user, design comprehension is improved [280].

Many characteristics of time-triggered systems — such as their repetitive character (i.e., periodic transmission), predefined "active intervals," the use of state messages for data sharing and highly self-contained components — provide this kind of structure and hence support design comprehension.

For example, the points in time when events in a time-triggered system take place are well-defined. This information can be used to add to an understanding of the system, as the time axis can serve as the basis for conceptual structuring.

On the application level, strictly time-triggered systems just use interfaces based on state messages. This means that the interfaces of all components only consist of a number of state messages that must either be read or written. No other communication or coordination mechanisms are required. As time-triggered systems are of repetitive nature, a component regularly reads the same input messages and then writes the same set of output messages — usually at about equidistant points in time. Only the content of the messages changes, but not the messages themselves.

With these characteristics of a time-triggered system in mind, we can define basically three possibilities for schedule visualization: a *textual* representation (in the following called *schedule browser*), a graphical one (in the following called *schedule viewer* or *schedule editor*) and animation.

While a schedule editor may give a better overview of the whole schedule and eases real "schedule editing" (for example, manually moving frames), a schedule browser may be simpler to use when searching for specific information or wanting to compare certain properties of messages. Animation, although trendy, is not covered here, as we do not consider it a viable solution. In our opinion, it does not satisfy the user's need for interaction (editing) and customized views the way browsers and editors do. Therefore, only examples of these two types are briefly outlined in the following, as they are also implemented in TTTech's readily available cluster design and scheduling tool ^{TTP}Plan. ^{TTP}Plan can generate a cluster (i.e., message) schedule either from scratch or by extending an existing schedule (*schedule extension*), and provides both textual and graphical schedule editing. Further details can be found in [344].

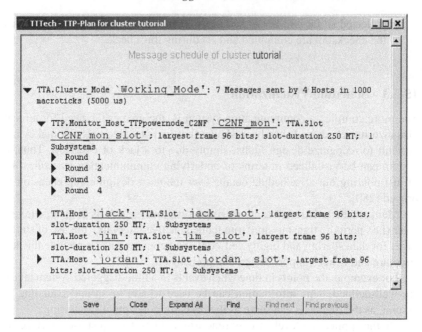

FIGURE 15.2
The Schedule Browser of ^{TTP}Plan

15.3.3.1 The Schedule Browser

The schedule browser of ^{TTP}Plan employs a hierarchical structure, similar to the well-known treeview of other browsers, listing all objects participating in the schedule (hosts, frames, transmission slots). See Figure 15.2 for a screenshot. Each object is displayed as clickable hyperlink, allowing for direct access to the corresponding *object editor*, where the object's attributes can be edited. Expanding an object node in the browser displays the actual timing information of the schedule, e.g., slot durations, frame and message sizes and transmission periods.

A shorter version of the schedule browser, the *schedule summary*, can be useful for a first quick overview. It could be automatically displayed in a design tool right after successful schedule generation, as it is done in ^{TTP}Plan. It only displays the basic data of the generated schedule (number and duration of rounds, transmission speed of messages and frames).

15.3.3.2 The Schedule Editor

In TTP and FlexRay, the communication schedule is based on *rounds* and *slots*. This fact lends itself to a grid-like representation, with the rows corresponding to rounds and the columns corresponding to slots. Each intersection of a row and a column thus represents a *round-slot*, the basic "transmission window" for scheduled data. The grid as a whole displays one cluster cycle in its entirety. Due to the periodic nature of a time-triggered schedule, where only the transmitted contents change, but not the timing behavior, this gives a perfect overview.

In TTP, each transmitting host in the cluster is assigned its own transmission slot. Consequently, the columns automatically also represent the hosts. For FlexRay, an indication which slot is used by which hosts needs to be added.

In a redundant system, i.e., with data being transmitted twice on two different communication channels, each round-slot can be split into two sections to display the frames transmitted on both channels. Vertical alignment of these sections is preferred as the structure of the frames on both channels can be compared quickly, giving an immediate understanding of whether the frames are truly redundant (i.e., have exactly the same structure), or only some messages in the frames are redundant, while others are not.

The schedule editor of ^{TTP}Plan is shown in Figure 15.3. It provides drop-down lists to select certain parts of the schedule; this is very helpful when dealing with huge and complex schedules. If a host, frame or message is selected, all occurrences of it are highlighted (as far as the schedule is displayed, that is). For example, selecting a message is useful to see in which slots or rounds it has been scheduled for transmission.

For working with large clusters, the display area of the schedule grid can be set by selecting the desired number of hosts/slots or rounds. On the one hand, this makes the frames larger, easier to see and easier to select with the mouse. On the other hand, it allows us to obtain an overview by viewing all slots and rounds at the same time and to identify "similar" patterns in the communication structure.

As the round-slot fields of the grid may not be large enough (even with a reduced number of visible slots/rounds) to display all relevant information, a "magnifier" function, like the "magnifier window" shown in Figure 15.3, allows the user to view the frames of a selected round-slot — as well as the messages contained in the frames — in a separate window area. In addition, details about the messages (size and timing) are listed below the magnifier window.

Actual *schedule editing* is best done by drag-and-drop: Drag a message from its current position (frame or round) to another and release it there. This implicitly changes the affected attributes of the message. In this way, one can optimize the current schedule and generate shorter slots, thus allowing for shorter overall rounds. Manual editing also can provide a way out in case the scheduling tool failed to find a feasible schedule.

However, certain actions are prohibited by the schedule editor because they would either violate design constraints or have to be performed prior to rescheduling, i.e., in the scheduling tool itself:

- Drop messages into rounds where their period or phase constraints would be violated

- Drop messages on I-frames (for TTP)

- Move replicated messages to a round-slot where there is not enough space on the other channel (in TTP: where there is an I-frame on one of the channels)

Object selection
list boxes   **Schedule grid**

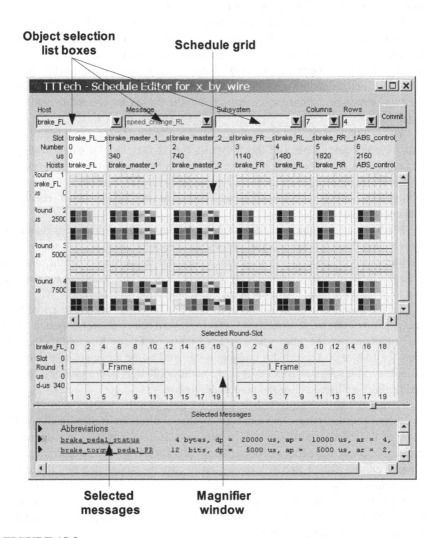

Selected   **Magnifier**
messages   **window**

FIGURE 15.3
The Schedule Editor of ᵀᵀᴾPlan

- Move messages out of their slot (in TTP) or out of the slots the sending host may use (in FlexRay). We consider it bad practice to implicitly change the communication requirements (i.e., who sends what) by editing the schedule. Editing should only refine the timing in detail.

- Move messages within the frame (there should never be a need for this).

15.3.3.3 The Round-Slot Viewer

Similar to a schedule editor, a *round-slot viewer* has a grid-like structure, with the rows representing rounds and the columns representing slots. Each intersection of a row and a column thus represents a *round-slot*. For large schedules, scrolling and limiting the number of displayed items can be useful. After the successful generation of a schedule, one might want to open the round-slot viewer to have a look at the schedule timing.

Like the schedule editor, the round-slot viewer shown in Figure 15.4 provides a magnifier window below the schedule grid. Selecting a round-slot highlights it and also shows it in the magnifier window. At the top of the magnifier window, the slot time is displayed for both channels (first channel above, second one below). The time is split into four parts that are equal for both channels (from left to right):

- **Transmission phase**: The time span needed for transmission of the frames. I-frames and N-frames are displayed in different colors. Overfull N-frames would be displayed in red to highlight them.

- **Post-receive-phase (prp)**: The time span immediately after transmission phase, during which certain services are performed.

- **Idle time**: This time is needed to stretch the durations of the slots to meet the specified round duration. This idle time is unused bandwidth.

- **Pre-send-phase (psp)**: The time span immediately before action time, during which frame transmission is prepared. The sum of prp, idle time and psp determines the inter-frame gap (IFG). It is limited by the slowest controller in the cluster.

Below the slot time, the user interrupts for both channels are displayed. The magnifier window itself displays additional information about the selected round-slot. Among this information there is the kind of each item in the round-slot, as well as a time grid showing the time from the beginning of the cluster cycle.

15.3.3.4 Visualization of Message Paths

TTEthernet communication, although time-triggered, is not as strict in its structure as TTP. It is not based on rounds and individual sending slots for each device, but rather on "communication links," i.e., physical connections between sender and receiver, that are basically independent of each other. In contrast to TTP, TTEthernet

**Object selection
list boxes**   **Schedule grid**

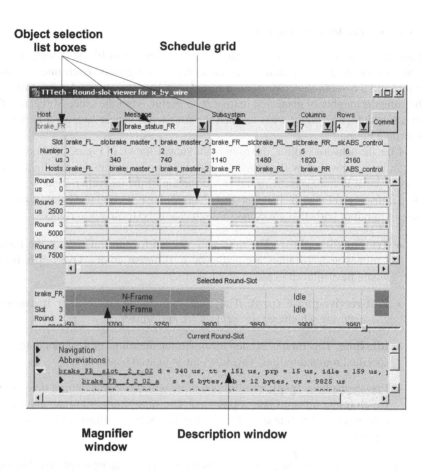

**Magnifier   Description window
window**

FIGURE 15.4
The Round-Slot Viewer of ^{TTP}Plan

allows the simultaneous reception and transmission on the same link, as well as simultaneous communication on several links. Therefore, a rigid grid like that of the schedule editor or the round-slot viewer presented above is not the optimal visualization strategy.

The approach presented here is instead based on the communication links and "message paths." A message path denotes the logical path a message takes through the network from the original sender to the last receiver or receivers, including intermediate receiving and resending by one or more switches. Figure 15.5 shows a possible schedule viewer for TTEthernet, based on such a visualization approach. Note that — for simplicity — only strictly time-triggered messages are considered here (i.e., no rate-constrained or best-effort messages).

As usual, the schedule viewer is based on a horizontal time axis. In parallel to it there are the lines representing the communication links. Each link connects two devices, whose names are stated at the left edge of the schedule, above and below the line. The colored rectangles *above* each line are the messages transmitted from the upper to the lower device, the rectangles *below* the line those in the other direction.

The example schedule in Figure 15.5 displays one cluster cycle with a duration of 1 *ms*, with the first 100 μs being reserved (by design) for special purposes, e.g., clock synchronization. For simplicity, all messages are transmitted once per cluster cycle, i.e., their periods equal the cluster cycle duration.

Following the path of the message OUT (gray rectangle) from the main controller to all other devices provides some insight into the way of interpreting the displayed schedule. Moving along the time axis, the following transmissions take place:

1. The *Main Controller* sends OUT to switch *sw1* (lower left corner of the schedule).

2. *sw1* takes some time processing OUT, hence the gap between the first and the second transmission.

3. *sw1* simultaneously sends OUT to the end system IO Node1 and the switch *sw2*.

4. *sw2* takes some time processing OUT.

5. *sw2* sends OUT to *sw3* just after the processing time.

6. *sw2* simultaneously sends OUT to IO Node2 and IO Node3. This transmission takes place later than that to *sw3* because there are some other messages scheduled for transmission in the same direction on these links.

7. *sw3* takes some time processing OUT.

8. *sw2* simultaneously sends OUT to IO Node4 and IO Node5.

In the same way, the paths of the messages INX and IN_MC_X can be traced from the end systems (IO NodeX) to the main controller (starting at the upper left corner of the schedule).

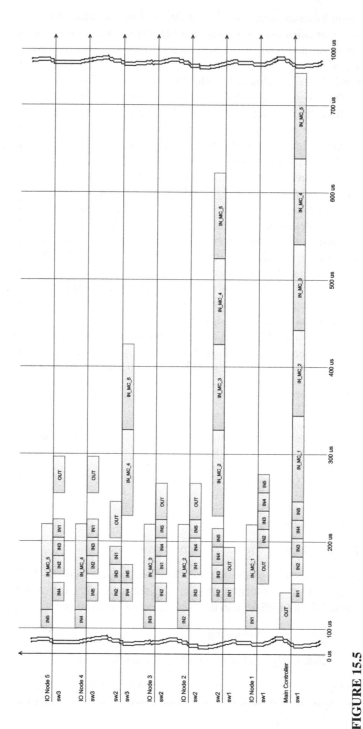

FIGURE 15.5
Illustration of Data Traffic on All Full-Duplex Connections of a TTEthernet Network

Clicking on a message not only highlights all its occurrences, but even shows its message paths as arrows indicating the intra-network communication. User interaction is not viable here, as drawing all arrows for all messages would result in a mess. However, showing the paths of a selected message on request gives the user a quick notion of where processing delays occur and which way the data flows, resulting in a good impression about the latency of this message.

Displaying messages with a transmission period of exactly one cluster cycle — as in the above example — is simple, but a system is not always designed this way. For messages with a higher frequency, i.e., that are transmitted more than once per cluster cycle, additional "depth" of the display is needed. Basically, there are two possibilities to include the periodicity in the displayed schedule:

- To draw one message instance after the other, with the schedule viewer always displaying one whole cluster cycle. The advantage is that this representation is simple. The disadvantage is that the viewing area can become very long and thus will need a lot of scrolling and zooming. By introducing a magnifier or an overview window, navigation in this "wide" schedule representation can be made more comfortable.

- To wrap each link after the shortest period (cf. the TTP schedule editor), which means that each instance of the shortest period starts "at the beginning," i.e., the left edge. The advantage is that the messages with the highest frequency are placed below each other, and messages with periods that are integer multiples of the shortest period are also nicely displayed. But there are at least two disadvantages. First, it is difficult to keep an overview with several links. Second, adding arrows to show the communication paths makes the schedule quite unreadable, as the arrows may cross message instances of interest.

15.4 Holistic Scheduling and Optimization

Applications consist of a set of interacting tasks that communicate through messages. Depending on the functionality, tasks and messages may be time-triggered or event-triggered, or, in certain situations [205], a combination of both. There are many applications where the interaction between the functions is tightly coupled, and design decisions cannot be taken in isolation, they have to be taken considering the complete system, i.e., in a holistic manner. For example, when TT tasks and TT messages are synchronized, the schedule of the tasks has to be constructed at the same time with the message schedule. Also, the worst-case end-to-end delays for ET messages may impact the worst-case response times of ET tasks, and in this case the analysis and optimization of messages has to be considered at the same time with the analysis and optimization of tasks.

In this section, we present an approach to holistic analysis and optimization of

FlexRay-based systems. Although the work here considers FlexRay, the holistic analysis and optimization principles are also valid for other protocols. FlexRay is composed of static (ST) and dynamic (DYN) segments, which are arranged to form a bus cycle that is repeated periodically. The ST segment is similar to TTP, and employs a generalized time-division multiple-access (GTDMA) scheme. The DYN segment of the FlexRay protocol is similar to Byteflight and uses a flexible TDMA (FTDMA) bus access scheme. We propose techniques for determining the timing properties of messages transmitted in the static and the dynamic segments of a FlexRay communication cycle. We first briefly present a static cyclic scheduling technique for TT messages transmitted in the ST segment. Then, we develop a worst-case response time analysis for ET messages sent using the DYN segment, thus providing predictability for messages transmitted in this segment. The analysis techniques for messages are integrated in the context of a holistic schedulability analysis algorithm that computes the worst-case response times of all the tasks and messages in the system.

Such an analysis, while being able to bound the message transmission times on both the ST and DYN segments, represents the first step toward enabling the use of this protocol in a systematic way for time-critical applications. The second step toward an efficient use of FlexRay is concerned with optimization techniques that consider the particular features of an application during the process of finding a FlexRay bus configuration that can guarantee that all time constraints are satisfied.

15.4.1   System Model

We consider architectures consisting of nodes connected by one FlexRay communication channel[2] (see Figure 15.6a). Each processing node connected to a FlexRay bus is composed of two main components: A CPU and a communication controller (see Figure 15.7a) that are interconnected through a two-way controller-host interface (CHI). The controller runs independently of the node's CPU and implements the FlexRay protocol services.

For the systems we are studying, we have made some basic assumptions about the features of a software architecture which runs on the CPU of each node. The main component of the software architecture is a real-time kernel that contains two schedulers, for static cyclic scheduling (SCS) and fixed priority scheduling (FPS), respectively[3] (see Figure 15.6b).

When several tasks are ready on a node, the task with the highest priority is activated, and preempts the other tasks. Let us consider the example in Figure 15.6b, where we have six tasks sharing the same node. Tasks τ_1 and τ_6 are scheduled using SCS, while the rest are scheduled with FPS. The priorities of the FPS tasks are indicated in the figure. The arrival time of a task is depicted with an upward pointing arrow. Under these assumptions, Figure 15.6b presents the worst-case response times of each task. SCS tasks are non-preemptable and their start time is offline fixed in the

---

[2]FlexRay is a dual-channel bus.
[3]EDF can also be added, as presented by us in [266].

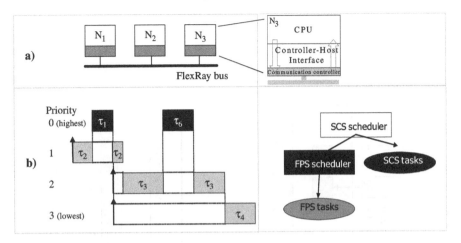

FIGURE 15.6
System Architecture Example

schedule table (they also have the highest priority, denoted with priority level "0" in the figure). FPS tasks can only be executed in the slack of the SCS schedule table.

FPS tasks are scheduled based on priorities. Thus, a higher priority task such as τ_3 preempts a lower priority task such as τ_4. SCS activities are triggered based on a local clock in each processing node. The synchronization of local clocks throughout the system is provided by the communication protocol.

15.4.2   The FlexRay Communication Protocol

In this section, we will describe how messages generated by the CPU reach the communication controller and how they are transmitted on the bus. Let us consider the example in Figure 15.7 where we have three nodes, N_1 to N_3 sending messages $m_a, m_b, \ldots, m_h$ using a FlexRay bus.

In FlexRay, the communication takes place in periodic cycles (Figure 15.7b depicts two cycles of length T_{bus}). Each cycle contains two time intervals with different bus access policies: An ST segment and a DYN segment.[4] We denote with ST_{bus} and DYN_{bus} the length of these segments. In Figure 15.7 there are three static slots for the ST segment. For details on the FlexRay communication protocol, the reader is directed to the FlexRay chapter.

In Figure 15.7, node N_1 has been allocated ST slot 2 and DYN slot 3, N_2 transmits through ST slots 1 and 3 and DYN slots 2 and 4, while node N_3 has DYN slots 1 and 5. For each of these slots, the CHI reserves a buffer that can be written by the CPU and read by the communication controller (these buffers are read by the

---

[4]The FlexRay bus cycle also contains a *symbol window* and a *network idle time*, but their size does not affect the equations in our analysis. For simplicity, they will be ignored during the examples throughout the section.

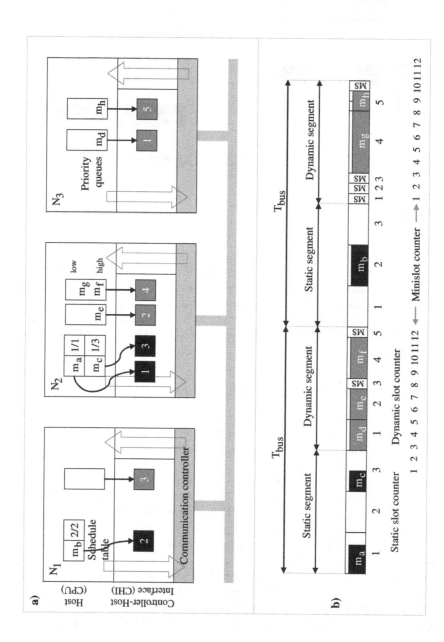

FIGURE 15.7
FlexRay Communication Cycle Example

communication controller *at the beginning* of each slot, in order to prepare the transmission of frames). The associated buffers in the CHI are depicted in Figure 15.7a. We denote with $DYNSlots_{N_p}$ the number of dynamic slots associated to a node N_p (this means that for N_2 in Figure 15.7, $DYNSlots_{N_2}$ has value 2).

We use different approaches for ST and DYN messages to decide which messages are transmitted during the allocated slots. For ST messages, we consider that the CPU in each node holds a schedule table with the transmission times. When the time comes for an ST message to be transmitted, the CPU will place that message in its associated ST buffer of the CHI. For example, ST message m_b sent from node N_1 has an entry "2/2" in the schedule table specifying that it should be sent in the second slot of the second ST cycle.

For the DYN messages, we assume that the designer specifies their *FrameID*. For example, DYN message m_e has the frame identifier "2." While nodes must use distinct *FrameID*s (and consequently distinct DYN slots) in order to avoid bus conflicts, we allow for a node to send different messages using the same DYN *FrameID*.[5] For example, messages m_g and m_f on node N_2 have both *FrameID* 4. If two or more messages with the same frame identifier are ready to be sent in the same bus cycle, a priority scheme is used to decide which message will be sent first. Each DYN message m_i has associated a priority $priority_{m_i}$. Messages with the same *FrameID* will be placed in a local output queue ordered based on their priorities. The message from the head of the priority queue is sent in the current bus cycle. For example, message m_f will be sent before m_g because it has a higher priority.

At the beginning of each communication cycle, the communication controller of a node resets the slot and minislot counters. At the beginning of each communication slot, the controller verifies if there are messages ready for transmission (present in the CHI send buffers) and packs them into frames.[6] In the example in Figure 15.7, we assume that all messages are ready for transmission before the first bus cycle.

Messages selected and packed into ST frames will be transmitted during the bus cycle that is about to start according to the schedule table. For example, in Figure 15.7, messages m_a and m_c are placed into the associated ST buffers in the CHI in order to be transmitted in the first bus cycle. However, messages selected and packed into DYN frames will be transmitted during the DYN segment of the bus cycle only if there is enough time until the end of the DYN segment. Such a situation is verified by comparing if, in the moment the DYN slot counter reaches the value of the *FrameID* for that message, the value of the minislot counter is smaller than a given value *pLatestTx*. The value *pLatestTx* is fixed for each node during the design phase, depending on the size of the largest DYN frame that node will have to send during run-time. For example, in Figure 15.7, message m_h is ready for transmission before the first bus cycle starts, but, after message m_f is transmitted, there is not enough room left in the DYN segment. This will delay the transmission of m_h for the next bus cycle.

---

[5]This assumption is not part of the FlexRay specification. If messages are not sharing *FrameID*s, this is handled implicitly as a particular case of our analysis.

[6]In this section, we do not address frame-packing [263], and thus assume that one message is sent per frame.

```
GlobalSchedulingAlgorithm()
 1  while TT_ready_list is not empty
 2     select τ_ij from TT_ready_list
 3     if τ_ij is a SCS task then
 4        schedule_TT_task(τ_ij, Node_τij)
 5     else // τ_ij is a ST message
 6        schedule_ST_msg(τ_ij, Node_τij)
 7     end if
 8     update TT_ready_list
 9  end while
end StaticScheduling
schedule_TT_task(τ_ij, Node_τij)
 10    find first available time moment t_s after ASAP_τij
       on Node_τij
 11    schedule τ_ij after t_s on Node_τij, so that holistic analysis
       produces minimal worst-case response times
       for FPS tasks and DYN messages
 12    update ASAP for all τ_ij successors
end schedule_TT_task
schedule_ST_msg(τ_ij, Node_τij)
 13    find first ST slot(Node_τij) available after ASAP_τij
 14    schedule τ_ij in that ST slot
 15    update ASAP for all τ_ij successors
end schedule_ST_msg
```

FIGURE 15.8
Global Scheduling Algorithm

15.4.3 Timing Analysis

Given a distributed system based on FlexRay, as described in the previous two sections, the tasks and messages have to be scheduled. For the SCS tasks and ST messages, this means building the schedule tables, while for the FPS tasks and DYN messages we have to determine their worst-case response times.

The problem of finding a schedulable system has to consider two aspects:

1. When performing the schedulability analysis for the FPS tasks and DYN messages, one has to take into consideration the interference from the SCS activities.

2. Among the possible correct schedules for SCS activities, it is important to build one which favors as much as possible the schedulability of FPS activities.

Figure 15.8 presents the global scheduling and analysis algorithm, in which the main loop consists of a list-scheduling based algorithm [62] that iteratively builds the static schedule table with start times for SCS tasks and ST messages.

A ready list (*TT_ready_list*) contains all SCS tasks and ST messages which are ready to be scheduled (they have no predecessors or all their predecessors have already been scheduled). From the ready list, tasks and messages are extracted one by one (Figure 15.8, line 2) to be scheduled on the processor they are mapped to (line 4), or into a static bus-slot associated to that processor on which the sender of the message is executed (line 6), respectively. The priority function which is used to select among ready tasks and messages is a critical path metric, modified by us for the particular goal of scheduling tasks mapped on distributed systems [86]. Let us consider a particular task τ_{ij} selected from the ready list to be scheduled. We consider that $ASAP_{\tau_{ij}}$ is the earliest time moment which satisfies the condition that all preceding activities (tasks or messages) of τ_{ij} are finished (line 10). With only the SCS tasks in the system, the straightforward solution would be to schedule τ_{ij} at the first time moment after $ASAP_{\tau_{ij}}$ when $Node_{\tau_{ij}}$ is free. Similarly, an ST message will be scheduled in the first available ST slot associated with the node that runs the sender task for that message.

As presented by us in [265], when scheduling SCS tasks, one has to take into account the interference they produce on FPS tasks. The function *schedule_TT_task* in Figure 15.8 places a SCS task in the static schedule in such a way that the increase of worst-case response times for FPS tasks is minimized. Such an increase is determined by comparing the worst-case response times of FPS tasks obtained with our holistic schedulability analysis before and after inserting the new SCS task in the schedule [265].

The next subsection presents our solution for computing the worst-case response times of DYN messages, while in Section 15.4.3.2 we will integrate this solution into a holistic schedulability analysis that determines the timing properties of both FPS tasks and DYN messages (which is called in line 11, of *schedule_TT_task* presented in Figure 15.8).

15.4.3.1 Schedulability Analysis of DYN Messages

The worst-case response time R_m of a DYN message m is given by the following equation:

$$R_m(t) = \sigma_m + w_m(t) + C_m, \qquad (15.1)$$

where C_m is the message communication time (see Section 15.3.1), σ_m is the longest delay suffered during one bus cycle if the message is generated by its sender task after its slot has passed, and w_m is the worst-case delay caused by the transmission of ST frames and higher priority DYN messages during a given time interval t. For example, in Figure 15.9, we consider that a message m is supposed to be transmitted in the third DYN slot of the bus cycle. The figure presents the case when message m appears during the first bus cycle after the third DYN slot has passed; therefore, the message has to wait σ_m until the next bus cycle starts. In the second bus cycle, the message has to wait for the ST segment and for the first two DYN slots to finish, delay denoted with w_m (that also contains the transmission of a message m' that uses the second DYN slot).

The communication controller decides what message is to be sent on the bus in a certain communication slot *at the beginning* of that slot. As a consequence, in the worst case, a DYN message m is generated by its sender task immediately after the slot with the $FrameID_m$ has started, forcing message m to wait until the next bus cycle starts in order to really start competing for the bus. In conclusion, in the worst

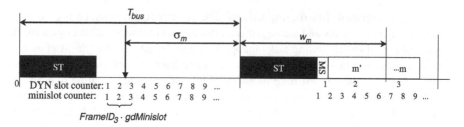

FIGURE 15.9
Response Time of a DYN Message

case, the delay σ_m has the value:

$$\sigma_m = T_{bus} - (ST_{bus} + (FrameID_m - 1) \times gdMinislot), \qquad (15.2)$$

where ST_{bus} is the length of the ST segment.

What is now left to be determined is the value w_m corresponding to the maximum amount of delay on the bus that can be produced by interference from ST frames and DYN messages. We start from the observations that the transmission of a ready DYN message m during the DYN slot $FrameID_m$ can be delayed because of the following causes:

- Local messages with higher priority, that use the same frame identifier as m. We will denote this set of *higher priority local messages* with $hp(m)$. For example, in Figure 15.7a, messages m_g and m_f share *FrameID* 4, thus $hp(m_g) = \{m_f\}$.

- Any messages in the system that can use DYN slots with lower frame identifiers than the one used by m. We will denote this set of messages having *lower frame identifiers* with $lf(m)$. In Figure 15.7a, $lf(m_g) = \{m_d, m_e\}$.

- Unused DYN slots with frame identifiers lower than the one used for sending m (though such slots are unused, each of them still delays the transmission of m for an interval of time equal with the length *gdMinislot* of one minislot); we will denote the set of such minislots with $ms(m)$. Thus, in the example in Figure 15.7b, $ms(m_g) = \{1,2,3\}$, and $ms(m_f) = \{3\}$.

Determining the interference of DYN messages in FlexRay is complicated by several factors. Let us consider the example in Figure 15.10, where we have two nodes, N_1 (with *FrameIDs* 1 and 3) and N_2 (with *FrameID* 2), and three messages m_1 to m_3. N_1 sends m_1 and m_3, and N_2 sends message m_2. Messages m_1 and m_2 have *FrameIDs* 1 and 2, respectively. We consider two situations: Figure 15.10a, where m_3 has a separate *FrameID* 3, and Figure 15.10b, where m_3 shares the same *FrameID* 1 with m_1. The values of *pLatestTx* for each node are depicted in the figure.[7]

---

[7]We use $pLatestTx_m$ to denote $pLatestTx_N$ of the node N sending message m.

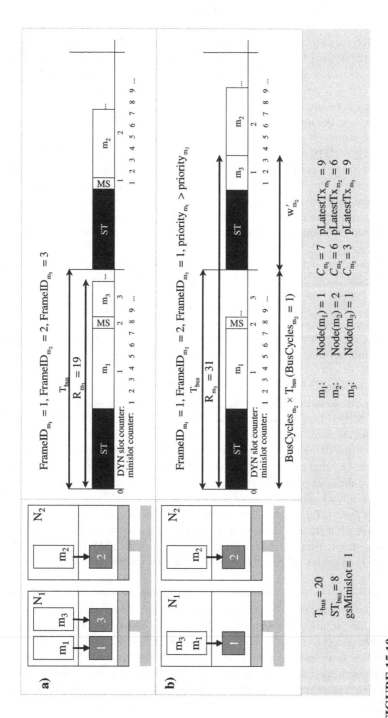

FIGURE 15.10
Transmission Scenarios for DYN Messages

In Figure 15.10a, message m_2, that has a lower FrameID than m_3, cannot be sent immediately after message m_1, because the value of the minislot counter has exceeded the value $pLatestTx_{m_2}$ when the value of the DYN slot counter becomes equal to 2 (hence, m_2 does not fit in this DYN cycle). As a consequence, the transmission of m_2 will be delayed for the next bus cycle. However, since in the moment when the DYN slot counter becomes 3 the minislot counter does not exceed the value $pLatestTx_{m_3}$, message m_3 will fit in the first bus cycle. Thus, a message (m_3 in our case) can be sent before another message with a lower $FrameID(m_2)$. Such situations must be accounted for when building the worst-case scenario.

In Figure 15.10b, message m_3 shares the same FrameID 1 with m_1 but we consider that it has a lower priority, thus $hp(m_3) = \{m_1\}$. In this case, m_3 is sent in the first DYN slot of the second bus cycle (the first slot of the first cycle is occupied with m_1) and thus will delay the transmission of m_2. In this scenario, we notice that assigning a lower frame identifier to a message does not necessarily reduce the worst-case response time of that message (compare to the situation in Figure 15.10a, where m_3 has FrameID = 3).

We next focus on determining the delay $w_m(t)$ in (15.1). The delay produced by all the elements in $hp(m)$, $lf(m)$ and $ms(m)$ can extend to one or more bus cycles:

$$w_m(t) = BusCycles_m(t) \times T_{bus} + w'_m(t), \tag{15.3}$$

where $BusCycles_m(t)$ is the number of bus periods for which the transmission of m is not possible because transmission of messages from $hp(m)$ and $lf(m)$ and because of minislots in $ms(m)$. The delay $w'_m(t)$ denotes now the time that passes, in the last bus cycle, until m is sent, and is measured from the beginning of the bus cycle in which message m is sent until the actual transmission of m starts. For example, in Figure 15.10b, $BusCycles_{m_2} = 1$ and $w'_{m_2}(t) = ST_{bus} + C_{m_3}$. Note that both these terms are functions of time, computed over an analyzed interval t. This means that when computing them we have to take into consideration all the elements in $hp(m)$, $lp(m)$ and $ms(m)$ that can appear during such a given time interval t. Thus, we will consider the multiset $hp(m,t)$ containing all the occurrences over time interval t of elements in $hp(m)$. The number of such occurrences for a message $l \in hp(m)$ is equal to: $\lceil (J_l + t)/T_l \rceil$, where T_l is the period of the message l and J_l is its worst-case jitter (such a jitter is computed as the difference between the worst-case and best-case response times of its sender task s: $J_l = R_s - R_s^b$ [245]). Similarly, $lf(m,t)$ and $ms(m,t)$ consider all the occurrences over t of elements in $lf(m)$ and $ms(m)$, respectively.

The optimal (i.e., exact) solutions for determining the values for $BusCycles_m(t)$ and $w'_m(t)$ are beyond the scope of this section, and are presented in [267]. These, can be intractable for larger problem sizes. Hence, in [267] we have proposed heuristics that quickly compute upper bounds (i.e., pessimistic) values for these terms. Once for any given time interval t we know how to obtain the values $BusCycles(t)$ and $w'_m(t)$, determining the worst-case response time for a message m becomes an iterative process that computes $R_m^k(R_m^{k-1})$, starting from $R_m^0 = C_m$ and finishing when $R_m^k = R_m^{k-1}$.

15.4.3.2 Holistic Schedulability Analysis of FPS Tasks and DYN Messages

As mentioned in Section 15.4.1, the worst-case response times of FPS tasks are influenced on one hand by higher priority FPS tasks, and on the other hand by SCS tasks. The worst-case response time R_{ij} of a FPS task τ_{ij} is determined as presented in [245], and in [265] we have shown how to take into consideration the interference on R_{ij} produced by an existing static schedule. What is important to mention is that R_{ij} depends on jitters of the higher priority tasks and predecessors of τ_{ij}. This means that for all such activities we have to compute the jitter. In the rest of this section, we will only concentrate on the situation when the jitter of a task depends on the arrival time of a message.

According to the analysis of multiprocessor and distributed systems presented in [245], the jitter for a task τ_r that starts execution only after it receives a message m depends on the values of the best-case and worst-case transmission times of that message:

$$J_{\tau_r} = R_m - R_m^b. \tag{15.4}$$

The calculation of the worst-case transmission time R_m of a DYN message m was presented in Section 15.4.3.1. For computing R_m^b we have to identify the best-case scenario of transmitting message m. Such a situation appears when the message becomes ready immediately before the DYN slot with $FrameID_m$ starts, and it is sent during that bus cycle without experiencing any delay from higher priority messages. Thus, the equation for the best-case transmission time of a message is:

$$R_m^b = C_m, \tag{15.5}$$

where C_m is the time needed to send the message m.

We notice from (15.4) that the jitters for activities in the system depend on the values of the worst-case response times, which in turn depend on the values of the jitters [266]. Such a recursive system is solved using a fixed point iteration algorithm in which the initial values for jitters are 0.

According to [245], the worst-case response time calculation of FPS tasks is of exponential complexity and the approach proposed in [245] and also used in [265] is a heuristic with a certain degree of pessimism. The pessimism of the response times calculated by our holistic analysis will, of course, also depend on the quality of the solution for the delay induced by the DYN messages transmitted over FlexRay. The calculation of this delay is our main concern in this section. Therefore, when we speak about optimal and heuristic solutions in this section we refer to the approach used for calculating the $BusCycles_m$ and w_m' (used in the worst-case response times calculation for DYN messages) and not the holistic response time analysis which is based on the heuristics in [245, 265].

For the extension of the analysis to take into account the dual-channel FlexRay bus, we direct the reader to [267].

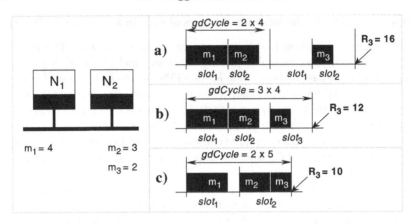

FIGURE 15.11
Optimization of the ST Segment

15.4.4 Bus Access Optimization

The design of a FlexRay bus configuration for a given system consists of a collection of solutions for the following subproblems: (1) determine the length of an ST slot, (2) the number of ST slots, and (3) their assignment to nodes; (4) determine the length of the DYN segment, (5) assign DYN slots to nodes and (6) *FrameIDs* to DYN messages.

The choice of a particular bus configuration is extremely important when designing a specific system, since its characteristics heavily influence the global timing properties of the application.

For example, notice in Figure 15.11 how the structure of the ST segment influences the response time of message m_3 (for this example, we ignored the DYN segment). The figure considers a system with two nodes, N_1 that sends message m_1 and N_2 that sends messages m_2 and m_3. The message sizes are depicted in the figure. In the first scenario, the ST segment consists of two slots, $slot_1$ used by N_1 and $slot_2$ used by N_2. In this situation, message m_3 can be scheduled only during the second bus cycle, with a response time of 16. If the ST segment consists of three slots (Figure 15.11b), with N_2 being allocated $slot_2$ and $slot_3$, then N_2 is able to send both its messages during the first bus cycle. The configuration in Figure 15.11c consists of only two slots, like in Figure 15.11a. However, in this case the slots are longer, such that several messages can be transmitted during the same frame, producing a faster response time for m_3 (one should notice, however, that by extending the size of the ST slots we delay the reception of message m_1 and m_2).

Similar optimizations can be performed with regard to the DYN segment. Let us consider the example in Figure 15.12, where we have two nodes N_1 and N_2. Node N_1 is transmitting messages m_1 and m_3, while N_2 sends m_2. Figure 15.12 depicts three configuration scenarios, a–c. Table A depicts the frame identifiers for the scenario in Figure 15.12a, while Table B corresponds to Figure 15.12b–c. The length of the

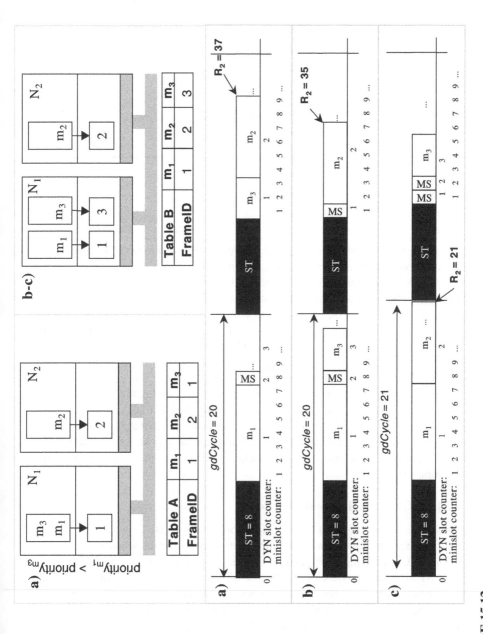

FIGURE 15.12
Optimization of the DYN Segment

```
 1  gdNumberOfStaticSlots = max(2, nodes_ST)
 2  gdStaticSlot = max(C_m), m is an ST message
 3  ST_bus = gdNumberOfStaticSlots *gdStaticSlot
 4  assign one ST slot to each node (round robin)
 5  for n = 1 to 64 do
 6    gdCycle = T_ss/n
 7    if gdCycle < 16000 μs then
 8      DYN_bus = gdCycle - ST_bus
 9      Assign FrameIDs to DYN messages
10      GlobalSchedulingAlgorithm()
11      Compute cost function Cost
12      if Cost < BestCost then save current solution
13    end if
14 end for
```

FIGURE 15.13
Basic Bus Configuration

ST slot has been set to 8. In Figure 15.12a, the length of the DYN segment is not able to accommodate both m_1 and m_2, thus m_2 will be sent during the second bus cycle, after the transmission of m_3 ends. Figure 15.12b and Figure 15.12c depict the same system but with a different allocation of DYN slots to messages (Table B). In Figure 15.12b we notice that m_3, which now does not share the same frame identifier with m_1, can be sent during the first bus cycle, thus m_2 will be transmitted earlier during the second cycle. Moreover, if we enlarge the size of the DYN segment as in Figure 15.12c, then the worst-case response time of m_2 will considerably decrease since it will be sent during the first bus cycle (notice that in this case m_3, having a greater frame identifier than that of m_2, will be sent only during the second cycle).

In order to illustrate the importance of choosing the right bus configuration, we present three approaches for optimizing the bus access such that the schedulability of the system is improved. The first approach builds a relatively straightforward, basic, bus configuration. The other two approaches perform optimization over the basic configuration.

15.4.4.1 The Basic Bus Configuration

In this section, we construct a Basic Bus Configuration (BBC) which is based on analyzing the minimal bandwidth requirements imposed by the application.

The BBC algorithm is presented in Figure 15.13 and it starts by setting the number of ST slots in a bus cycle. The length T_{bus} of the bus cycle is captured by the *gdCycle* protocol parameter. Since each node in the system that generates ST messages needs at least one ST slot, the minimum number of ST slots is $nodes_{ST}$, the number of nodes that send ST messages (line 1). The protocol specification also imposes a minimum limit on the number of ST slots; therefore, even if there are no nodes in the system that are using the ST segment, there should be at least two ST slots during a bus cycle. Next, the size of an ST slot is set so that it can accommodate the largest ST message in the system (line 2). In line 4, the configuration of the ST segment is completed by assigning in a round robin fashion one ST slot to each node that requires one (i.e., in a system with four nodes, where each node is sending in the static segment, the ST segment of the bus cycle will contain four slots; node 1 will use slot 1, node 2 will use ST slot 2, etc.).

When it comes to determining the size of the DYN segment, one has to take into consideration the fact that the period of the bus cycle (*gdCycle*) has to be an integer divisor[8] of the period of the global static schedule (T_{ss}). In addition, the FlexRay protocol specifies that each node implementing a cyclic schedule maintains in the communication controller a counter *vCycleCounter* that has values in the interval 0...63. This means that during a period of the static schedule there can be at most 64 bus cycles, which leads us to the conclusion that the value of *gdCycle* can be determined by iterating over all possible values for *vCycleCounter* (lines 5–14) and choosing the most favorable solution in terms of system schedulability (line 11). Line 7 introduces a restriction imposed by the FlexRay specification, which limits the maximum bus cycle length to 16 ms. Once the length of the bus cycle is set (line 5), knowing the length ST_{bus} of the ST segment (line 3), we can determine the length DYN_{bus} of the DYN segment (line 8).

At this point, in order to finish the design of the bus configuration, a *FrameID* has to be assigned to each of the DYN messages (and implicitly DYN slots are assigned to the nodes that generate the message). This assignment (line 9) is performed under the following guidelines:

- Each DYN message receives an unique *FrameID*; this is recommended in order to avoid large delays introduced by $hp(m)$. For example, in Figure 15.12, we notice that message m_3 has to wait for an entire *gdCycle* when it shares a frame identifier with the higher priority message m_1 (Figure 15.12a), which is not the case when it has its own *FrameID* (Figure 15.12b).

- DYN messages with a higher criticality receive smaller *FrameID*s. This is required in order to reduce, for a given message, the delays produced by $lf(m)$ and $ms(m)$. We capture the criticality of a message m as:

$$CP_m = D_m - LP_m, \tag{15.6}$$

where D_m is the deadline of the message and LP_m is the longest path in the task graph from the root to the node representing the communication of message m. A small value of CP_m (higher criticality) indicates that the message should be assigned a smaller *FrameID*.

Once we have defined the structure of the bus cycle, we can analyze the entire system (line 9) by performing the global static scheduling and analysis described in Section 15.4.3. The resulting system is then evaluated using a cost function that captures the schedulability degree of the system (line 10):

---

[8]We consider that the T_{ss} parameter is slightly adjusted, if necessary.

```
1  for gdNumberOfStaticSlots = gdNumberOfStaticSlotsmin to
   gdNumberOfStaticSlotsmax do
2    for gdStaticSlot = gdStaticSlotmin to gdStaticSlotmax step 20 *
     gdBit do
3      Assign ST slots to nodes
4      for n = 1 to 64 do
5        gdCycle = Tss/n
6        if gdCycle < 16000 µs then
7          DYNbus = gdCycle − STbus
8        do
9            Assign FrameIDs to DYN messages
10           GlobalSchedulingAlgorithm()
11           For all DYN messages, compute CPi
12           Compute cost function Cost
13           if Cost < BestCost then save current solution
14        while(BestCost unchanged for max_iterations);
15       end if
16     end for
17   end for
18 end for
```

FIGURE 15.14
Greedy Heuristic

$$Cost = \begin{cases} f_1 = \sum_{\tau_{ij}} \max(R_{ij} - D_{ij}, 0), \text{if } f_1 > 0, \\ f_2 = \sum_{\tau_{ij}} (R_{ij} - D_{ij}), \text{if } f_1 = 0 \end{cases} \qquad (15.7)$$

where R_{ij} and D_{ij} are the worst-case response times and respectively the deadlines for all the activities τ_{ij} in the system. This function is positive if at least one task or message in the system misses its deadline, and negative if the whole system is schedulable. Its value is used in line 11 when deciding whether the current configuration is the best one encountered so far.

15.4.4.2   Greedy Heuristic

The Basic Bus Configuration (BBC) generated as in the previous section can result in an unschedulable system (the cost function in (15.7) is positive). In this case, additional points in the solution space have to be explored. In Figure 15.14, we present a greedy heuristic that further explores the design space in order to find a schedulable solution.

While for the BBC the number and size of ST slots has been set to the minimum ($gdNumberOfStaticSlots_{min} = \max(2, nodes)$, $gdStaticSlot_{min} = \max(C_m)$), the heuristic explores different alternative values between these minimal values and the maxima imposed by the FlexRay protocol specification. Thus, during a bus cycle there can be at most $gdNumberOfStaticSlots_{max} = 1023$ ST slots, while the size of a ST slot can take at most $gdStaticSlot_{max} = 661$ macroticks. In addition, the payload for a FlexRay frame can increase only in 2-byte increments, which according to the FlexRay specification translates into 20 *gdBit*, where *gdBit* is the time needed for transmitting one bit over the bus (line 2).

The assignment of ST slots (line 3) to nodes is performed, like for the BBC, in a round robin fashion, with the difference that each node can have not only one but a quota of ST slots determined by the ratio of ST messages that it transmits (i.e., a node that sends more ST messages will be allocated more ST slots).

The sizes of the bus cycle and of the DYN segment are assigned in lines 4–16 in a similar way to the BBC algorithm.

However, while for the BBC the allocation of *FrameID*s to DYN messages is based on the estimated criticality (15.6), here we explore several *FrameID* assignment alternatives inside the loop in lines 8–14. We start from an initial assignment as in the BBC after which a global scheduling is performed (line 10). Using the resulted response times, in the next iteration we assign smaller *FrameID*s with priority to those DYN messages m that have a smaller value for $D_m - R_m$, where D_m is the deadline and R_m is the worst-case response time computed by the global scheduling.

15.4.4.3 Simulated Annealing-Based Approach

We have implemented a more exhaustive design space exploration than the one in Section 15.4.4.2, using a Simulated Annealing (SA) [46] approach. While relatively time consuming, this heuristic can be applied if both the BBC and the configuration produced by the greedy approach are unschedulable. Starting from the solution produced by the greedy optimization, the SA based heuristic explores the design space performing the following set of moves:

- *gdNumberOfStaticSlots* is incremented or decremented, inside the allowed limits (when an ST slot is added, it is allocated randomly to a node)

- *gdStaticSlot* is increased or decreased with $20 \times gdBit$, inside the allowed limits

- The assignment of ST slots to nodes is changed by re-assigning a randomly selected ST slot from a node N_1 to another node N_2. We also use in this context a similar transformation that switches the allocation of two ST slots, *FrameID*$_1$ and *FrameID*$_2$, used by two nodes N_1 and N_2, respectively

- The assignment of DYN slots to messages is modified by switching the slots used by two DYN messages

In Section 15.4.4.4 we used extensive, time consuming runs with the Simulated Annealing approach, in order to produce a reference point for the evaluation of our greedy heuristic.

15.4.4.4 Evaluation of Bus Optimization Heuristics

In order to evaluate our optimization algorithms, we generated seven sets of 25 applications representing systems of 2 to 7 nodes, respectively. We considered 10 tasks mapped on each node, leading to applications with a number of 20 to 70 tasks. Depending on the mapping of tasks, each such system had up to 60 additional nodes in the application task graph due to the communication tasks. The tasks were grouped

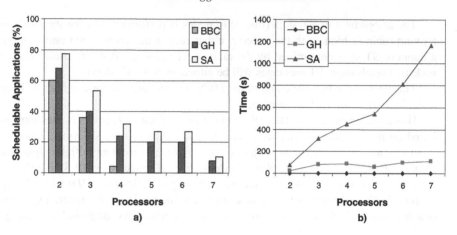

FIGURE 15.15
Evaluation of Bus Optimization Algorithms

in task graphs of five tasks each. Half of the tasks in each system were time triggered and half were event triggered. The execution times were generated in such a way that the utilization on each node was between 30% and 60% (similarly, the message transmission times were generated so that the bus utilization was between 10% and 70%). All experiments were run on an AMD Athlon 2400+ PC.

Figure 15.15 shows the results obtained after running our three algorithms proposed in Section 15.4.4 (BBC—Basic Bus Configuration, GH—Greedy Heuristic, and SA—Simulated Annealing). In Figure 15.15a, we show the percentage of schedulable applications, while in Figure 15.15b, we present the computation times required by each algorithm. One can notice that the BBC approach runs in almost zero time, but it fails to find any schedulable configurations for systems with more than four processors. On the other hand, the other two approaches continue to find schedulable solutions even for larger systems. Moreover, the percentage of schedulable solutions found by the greedy algorithm is comparable with the one obtained with the simulated annealing. Furthermore, the computation time required by the greedy heuristic is several orders of magnitude smaller than the one needed for the extensive runs of simulated annealing.[9]

15.5   Incremental Design

We have briefly introduced the issue of incremental design in Section 15.2. Incremental design has similarities with design for flexibility and scalability. The issue of

---

[9]Due to the extensive runs with SA, we can assume that the actual percentage of schedulable applications is close to that found by SA.

scalability in time-triggered systems has been investigated in [358], where the authors are interested in generating schedules which (i) allow tasks to increase their WCET without the need for rescheduling and (ii) have idle times distributed periodically to allow future expansion. Haubelt et al. [127] consider the requirement of flexibility as a parameter during design space exploration. Their goal is the generation of an architecture which, at an acceptable cost, is able to implement different applications or variants of a certain application.

In this section, we present an approach for mapping and scheduling of distributed embedded systems for hard real-time applications, aiming at a minimization of the system modification cost. We consider an incremental design process that starts from an already existing system running a set of applications. We are interested in implementing new functionality such that the timing requirements are fulfilled, and the following two requirements are also satisfied: The already running applications are disturbed as little as possible, and there is a good chance that, later, new functionality can easily be added to the resulted system. Thus, we propose a heuristic which finds the set of already running applications which have to be remapped and rescheduled at the same time with mapping and scheduling the new application, such that the disturbance on the running system (expressed as the total cost implied by the modifications) is minimized. Once this set of applications has been determined, we outline a mapping and scheduling algorithm aimed at fulfilling the requirements stated above. The approaches have been evaluated based on extensive experiments using a large number of generated benchmarks.

1. First, we consider mapping and scheduling for hard real-time embedded systems in the context of a realistic communication model based on a time division multiple access (TDMA) protocol as recommended for applications in areas like, for example, automotive electronics [180]. We accurately take into consideration overheads due to communication and consider, during the mapping and scheduling process, the particular requirements of the communication protocol.

2. Next, we have considered the design of distributed embedded systems in the context of an incremental design process as outlined above. This implies that we perform mapping and scheduling of new functionality on a given distributed embedded system, so that certain design constraints are satisfied and, in addition: (a) The already running applications are disturbed as little as possible. (b) There is a good chance that, later, new functionality can easily be mapped on the resulted system.

We propose a new heuristic, together with the corresponding design criteria, which finds the set of old applications which have to be remapped and rescheduled at the same time with mapping and scheduling the new application, such that the disturbance on the running system (expressed as the total cost implied by the modifications) is minimized. Once this set of applications has been determined, mapping and scheduling are performed according to the requirements stated above.

Supporting such a design process is of critical importance for current and future

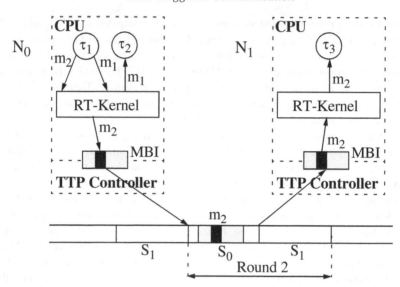

FIGURE 15.16
Message Passing Mechanism

industrial practice, as the time interval between successive generations of a product
is continuously decreasing, while the complexity due to increased sophistication of
new functionality is growing rapidly. The goal of reducing the overall cost of successive product generations has been one of the main motors behind the, currently very
popular, concept of platform-based design [163, 216]. Although, in this section, we
are not explicitly dealing with platform-based systems, most of the results are also
valid in the context of this design paradigm.

15.5.1 Preliminaries

In this section, the concepts of incremental design are investigated in the context of
TTP-based systems. However, the techniques presented here are also applicable to
other time-triggered protocols.

15.5.1.1 System Architecture

Thus, we consider architectures consisting of nodes connected by a broadcast communication channel. Every node consists of a TTP controller, processor, memory
and an I/O interface to sensors and actuators. For the details of TTP, please refer to
Chapter 5.

We assume that each node in the architecture has a real-time kernel as its main
component. Each kernel has a schedule table that contains all the information needed
to take decisions on activation of tasks and each communication controller has a
schedule table to decide the transmission of messages.

The message passing mechanism is illustrated in Figure 15.16, where we have three tasks, τ_1 to τ_3. τ_1 and τ_2 are mapped to node N_0 that transmits in slot S_0, and τ_3 is mapped to node N_1 that transmits in slot S_1. Message m_1 is transmitted between τ_1 and τ_2 that are on the same node, while message m_2 is transmitted from τ_1 to τ_3 between the two nodes. We consider that each task has its own memory locations for the messages it sends or receives and that the addresses of the memory locations are known to the kernel through the schedule table.

τ_1 is activated according to the schedule table, and when it finishes it calls the send kernel function in order to send m_1, and then m_2. Based on the schedule table, the kernel copies m_1 from the corresponding memory location in τ_1 to the memory location in τ_2. When τ_2 will be activated, it finds the message in the right location. According to our scheduling policy, whenever a receiving task needs a message, the message is already placed in the corresponding memory location. Thus, there is no overhead on the receiving side, for messages exchanged on the same node.

Message m_2 has to be sent from node N_0 to node N_1. At a certain time, known from the schedule table, the kernel transfers m_2 to the TTP controller by packaging it into a frame in the MBI. Later on, the TTP controller knows from its MEDL when it has to take the frame from the MBI, in order to broadcast it on the bus. In our example, the timing information in the schedule table of the kernel and the MEDL is determined in such a way that the broadcasting of the frame is done in the slot S_0 of *Round* 2. The TTP controller of node N_1 knows from its MEDL that it has to read a frame from slot S_0 of *Round* 2 and to transfer it into the MBI. The kernel in node N_1 will read the message m_2 from the MBI. When τ_3 will be activated based on the local schedule table of node N_1, it will already have m_2 in its right memory location.

In [260] we presented a detailed discussion concerning the overheads due to the kernel and to every system call. We also presented formulas to derive the worst-case execution delay of a task, taking into account the overhead of the timer interrupt, the worst-case overhead of the task activation and message passing functions.

15.5.1.2 Application Mapping and Scheduling

Considering a system architecture like the one presented earlier, the mapping of a task graph $\mathcal{G}(\mathcal{V}, \mathcal{E})$ is given by a function $M : \mathcal{V} \rightarrow PE$, where $PE = \{N_1, N_2, .., N_{npe}\}$ is the set of nodes (processing elements). For a task $\tau_i \in \mathcal{V}$, $M(\tau_i)$ is the node to which τ_i is assigned for execution. Each task τ_i can potentially be mapped on several nodes. Let $N_{\tau_i} \subseteq PE$ be the set of nodes to which τ_i can potentially be mapped. For each $N_i \in N_{\tau_i}$, we know the worst-case execution time $t_{\tau_i}^{N_i}$ of task τ_i, when executed on N_i. Messages transmitted between tasks mapped on different nodes are communicated through the bus, in a slot corresponding to the sending node. The maximum number of bits transferred in such a message is also known.

In order to implement an application, represented as a set of task graphs, the designer has to map the tasks to the system nodes and to derive a static cyclic schedule such that all deadlines are satisfied. We first illustrate some of the problems related to mapping and scheduling, in the context of a system based on a TDMA communi-

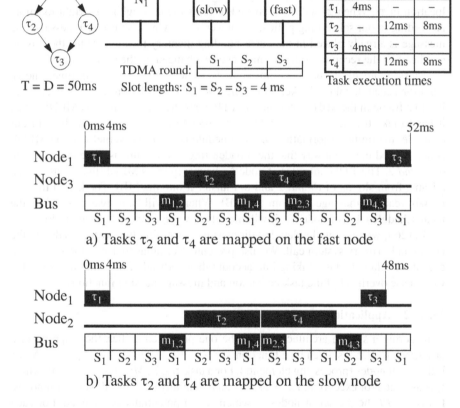

a) Tasks τ_2 and τ_4 are mapped on the fast node

b) Tasks τ_2 and τ_4 are mapped on the slow node

FIGURE 15.17
Mapping and Scheduling Example

cation protocol, before going on to explore further aspects specific to an incremental design approach.

Let us consider the example in Figure 15.17 where we want to map an application consisting of four tasks τ_1 to τ_4, with a period and deadline of 50 ms. The architecture is composed of three nodes that communicate according to a TDMA protocol, such that N_i transmits in slot S_i. For this example, we suppose that there is no other previous application running on the system. According to the specification, tasks τ_1 and τ_3 are constrained to node N_1, while τ_2 and τ_4 can be mapped on nodes N_2 or N_3, but not N_1. The worst-case execution times of tasks on each potential node and the sequence and size of TDMA slots are presented in Figure 15.17. In order to keep the example simple, we suppose that the message sizes are such that each message fits into one TDMA slot.

We consider two alternative mappings. If we map τ_2 and τ_4 on the faster processor N_3, the resulting schedule length (Figure 15.17a) will be 52 ms which does not meet the deadline. However, if we map τ_2 and τ_4 on the slower processor N_2, the schedule length (Figure 15.17b) is 48 ms, which meets the deadline. Note, that the total traffic on the bus is the same for both mappings and the initial processor load is 0 on both N_2 and N_3. This result has its explanation in the impact of the communication protocol. τ_3 cannot start before receiving messages $m_{2,3}$ and $m_{4,3}$. However, slot S_2 corresponding to node N_2 precedes in the TDMA round slot S_3 on which node N_3 communicates. Thus, the messages which τ_3 needs are available sooner in the case τ_2 and τ_4 are, counter-intuitively, mapped on the slower node.

But finding a valid schedule is not enough if we are to support an incremental design process as discussed in the introduction. In this case, starting from a valid design, we have to improve the mapping and scheduling so that not only the design constraints are satisfied, but also there is a good chance that, later, new functionality can easily be mapped on the resulted system.

To illustrate the role of mapping and scheduling in the context of an incremental design process, let us consider the example in Figure 15.18. For simplicity, we consider an architecture consisting of a single processor. The system is currently running application Ψ (Figure 15.18a). At a particular moment, application $\mathcal{A}_1$ has to be implemented on top of Ψ. Three possible implementation alternatives for $\mathcal{A}_1$ are depicted in Figure 15.18b1, 15.18c1 and 15.18d1. All three are meeting the imposed time constraint for $\mathcal{A}_1$. At a later moment, application $\mathcal{A}_2$ has to be implemented on the system running Ψ and $\mathcal{A}_1$. If $\mathcal{A}_1$ has been implemented as shown in Figure 15.18b1, there is no possibility to map application $\mathcal{A}_2$ on the given system (in particular, there is no time slack available for task τ_7). If $\mathcal{A}_1$ has been implemented as in Figure 15.18c1 or 15.18d1, $\mathcal{A}_2$ can be correctly mapped and scheduled on top of Ψ and $\mathcal{A}_1$. There are two aspects which should be highlighted based on this example:

1. If application $\mathcal{A}_1$ is implemented like in Figure 15.18c1 or 15.18d1, it is possible to implement $\mathcal{A}_2$ on top of the existing system, without performing any modifications on the implementation of previous applications. This could be the case if, during implementation of $\mathcal{A}_1$, the designers have taken into consideration the fact that, in future, an application having the characteristics of $\mathcal{A}_2$ will possibly be added to the system.

2. If $\mathcal{A}_1$ has been implemented like in Figure 15.18b1, $\mathcal{A}_2$ can be added to the system only after performing certain modifications on the implementation of $\mathcal{A}_1$ and/or Ψ. In this case, of course, it is important to perform as few as possible modifications on previous applications, in order to reduce the development costs.

15.5.2   Problem Formulation

As shown in Section 15.5.1, we capture the functionality of a system as a set of applications. An application $\mathcal{A}$ consists of a set of task graphs $\mathcal{G}_i \in \mathcal{A}$. For each task τ_i in a task graph we know the set N_{τ_i} of potential nodes on which it could be mapped and its worst-case execution time on each of these nodes. We also know the maximum number of bits to be transmitted by each message. The underlying architecture is as presented in Section 15.5.1.1. We consider a non-preemptive static cyclic scheduling policy for both tasks and message passing.

Our goal is to map and schedule an application $\mathcal{A}_{current}$ on a system that already implements a set Ψ of applications, considering the following requirements:

- Requirement a: All constraints on $\mathcal{A}_{current}$ are satisfied and minimal modifications are performed to the implementation of applications in Ψ.

- Requirement b: New applications $\mathcal{A}_{future}$ can be mapped on top of the resulting system.

We illustrate such an incremental design process in Figure 15.19. The product is implemented as a three processor system and its version $N-1$ consists of the set Ψ of two applications (the tasks belonging to these applications are represented as white and black disks, respectively). At the current moment, application $\mathcal{A}_{current}$ is to be added to the system, resulting in version N of the product. However, a new version, $N+1$, is very likely to follow and this fact is to be considered during implementation of $\mathcal{A}_{current}$.[10]

If it is not possible to map and schedule $\mathcal{A}_{current}$ without modifying the implementation of the already running applications, we have to change the scheduling and mapping of some applications in Ψ. However, even with remapping and rescheduling all applications in Ψ, it is still possible that certain constraints are not satisfied. In this case, the hardware architecture has to be changed by, for example, adding a new processor, and the mapping and scheduling procedure for $\mathcal{A}_{current}$ has to be restarted. In this section, we will not further elaborate on the aspect of adding new resources to the architecture, but will concentrate on the mapping and scheduling aspects. Thus, we consider that a possible mapping and scheduling of $\mathcal{A}_{current}$ which satisfies the imposed constraints can be found (with minimizing the modification of the already

---

[10]The design process outlined here also applies when $\mathcal{A}_{current}$ is a new version of an application $\mathcal{A}_{old} \in \Psi$. In this case, all the tasks and communications belonging to $\mathcal{A}_{old}$ are eliminated from the running system Ψ, before starting the mapping and scheduling of $\mathcal{A}_{current}$.

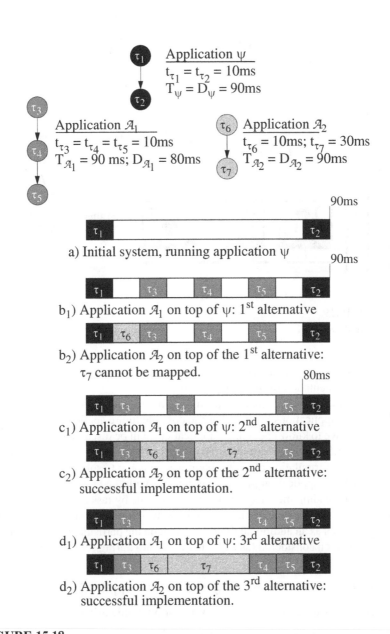

FIGURE 15.18
Application $\mathcal{A}_2$ Implemented on Top of Ψ and $\mathcal{A}_1$

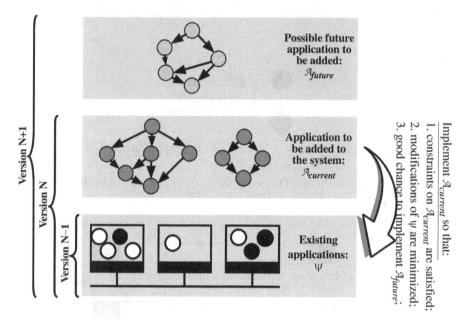

FIGURE 15.19
Incremental Design Process

running applications), and this solution has to be further improved in order to facilitate the implementation of future applications.

In order to achieve our goal, we need certain information to be available concerning the set of applications Ψ as well as the possible future applications $\mathcal{A}_{future}$. What exactly we have to know about these applications will be discussed in Section 15.5.3. In Section 15.5.4 we then introduce the quality metrics which will allow us to give a more rigorous formulation of the problem we are going to solve.

The tasks in application $\mathcal{A}_{current}$ can interact with the previously mapped applications Ψ by reading messages generated on the bus by tasks in Ψ. In this case, the reading task has to be synchronized with the arrival of the message on the bus, which is easy to model as an additional time constraint on the particular receiving task. This constraint is then considered (as any other deadline) during scheduling of $\mathcal{A}_{current}$.

15.5.3   Characterizing Existing and Future Applications

15.5.3.1   Characterizing the Already Running Applications

To perform the mapping and scheduling of $\mathcal{A}_{current}$, the minimum information needed, concerning the already running applications Ψ, consists of the local schedule tables for each processor node. Thus, we know the activation time for each task previously mapped on the respective node and its worst-case execution time. As for messages, their length as well as their place in the particular TDMA frame are known.

If the initial attempt to schedule and map $\mathcal{A}_{current}$ does not succeed, we have to modify the schedule and, possibly, the mapping of applications belonging to Ψ, in the hope of finding a valid solution for $\mathcal{A}_{current}$. The goal is to find that minimal modification to the existing system which leads to a correct implementation of $\mathcal{A}_{current}$. In our context, such a minimal modification means remapping and/or rescheduling a subset Ω of the old applications, $\Omega \subseteq \Psi$, so that the total cost of re-implementing Ω is minimized.

Remapping and/or rescheduling a certain application $\mathcal{A}_i \in \Psi$ can trigger the need to also perform modifications of one or several other applications because of, for example, the dependencies between tasks belonging to these applications. In order to capture such dependencies between the applications in Ψ, as well as their modification costs, we have introduced a representation called the *application graph*. We represent a set of applications as a directed acyclic graph $G(V, E)$, where each node $\mathcal{A}_i \in V$ represents an application. An edge $e_{ij} \in E$ from $\mathcal{A}_i$ to $\mathcal{A}_j$ indicates that any modification to $\mathcal{A}_i$ would trigger the need to also remap and/or reschedule $\mathcal{A}_j$, because of certain interactions between the applications.[11] Each application in the graph has an associated attribute specifying if that particular application is allowed to be modified or not (in which case, it is called frozen). To those nodes $\mathcal{A}_i \in V$ representing modifiable applications, the designer has associated a cost $R_{\mathcal{A}_i}$ of re-implementing $\mathcal{A}_i$. Given a subset of applications $\Omega \subseteq \Psi$, the total cost of modifying the applications in $\mathcal{A}$ is:

$$R(\Omega) = \sum_{\mathcal{A}_i \in \Omega} R_{\mathcal{A}_i}. \tag{15.8}$$

Modifications of an already running application can only be performed if the task graphs corresponding to that application, as well as the related deadlines (which have to be satisfied also after remapping and rescheduling), are available. However, this is not always the case, and in such situations that particular application has to be considered frozen.

In Figure 15.20, we present the graph corresponding to a set of 10 applications. Applications $\mathcal{A}_6$, $\mathcal{A}_8$, $\mathcal{A}_9$ and $\mathcal{A}_{10}$, depicted in black, are frozen: No modifications are possible to them. The rest of the applications have the modification cost $R_{\mathcal{A}_i}$ depicted on their left. $\mathcal{A}_7$ can be remapped/rescheduled with a cost of 20. If $\mathcal{A}_4$ is to be re-implemented, this also requires the modification of $\mathcal{A}_7$, with a total cost of 90. In the case of $\mathcal{A}_5$, although not frozen, no remapping/rescheduling is possible as it would trigger the need to modify $\mathcal{A}_6$, which is frozen.

To each application $\mathcal{A}_i \in V$ the designer has associated a cost $R_{\mathcal{A}_i}$ of re-implementing $\mathcal{A}_i$. Such a cost can typically be expressed in man-hours needed to perform retesting of $\mathcal{A}_i$ and other tasks connected to the remapping and rescheduling of the application. If an application is remapped or rescheduled, it has to be validated again. Such a validation phase is very time consuming. In the automotive industry, for example, the time-to-market in the case of the powertrain unit is 24 months. Out

---

[11]If a set of applications has a circular dependence, such that the modification of any one implies the remapping of all the others in that set, the set will be represented as a single node in the graph.

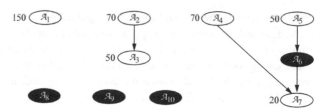

FIGURE 15.20
Characterizing the Set of Already Running Applications

of these, five months, representing more than 20%, are dedicated to validation. In the case of the telematic unit, the time to market is less than one year, while the validation time is two months [291]. However, if an application is not modified during implementation of new functionality, only a small part of the validation tasks have to be re-performed (e.g., integration testing), thus reducing significantly the time-to-market, at no additional hardware or development cost.

How to concretely perform the estimation of the modification cost related to an application is beyond the topic of this section. Several approaches to cost estimation for different phases of the software life-cycle have been elaborated and are available in the literature [75, 271]. One of the most influential software cost models is the Constructive Cost Model (COCOMO) [37]. Such estimations can be used by the designer as the cost metrics assigned to the nodes of an application graph.

In general, it can be the case that several alternative costs are associated to a certain application, depending on the particular modification performed. Thus, for example, we can have a certain cost if tasks are only rescheduled, and another one if they are also remapped on an alternative node. For different modification alternatives considered during design space exploration, the corresponding modification cost has to be selected. In order to keep the discussion reasonably simple, we present the case with one single modification cost associated to an application. However, the generalization for several alternative modification costs is straightforward.

15.5.3.2 Characterizing Future Applications

What do we suppose to know about the family $\mathcal{A}_{future}$ of applications which do not exist yet? Given a certain limited application area (e.g., automotive electronics), it is not unreasonable to assume that, based on the designers' previous experience, the nature of expected future functions to be implemented, profiling of previous applications, available incomplete designs for future versions of the product, etc., it is possible to characterize the family of applications which possibly could be added to the current implementation. This is an assumption which is basic for the concept of incremental design. Thus, we consider that, with respect to the future applications, we know the set $S_t = \{t_{min}, \dots t_i, \dots t_{max}\}$ of possible worst-case execution times for tasks, and the set $S_b = \{b_{min}, \dots b_i, \dots b_{max}\}$ of possible message sizes. We also assume that over these sets we know the distributions of probability $f_{S_t}(t)$ for $t \in S_t$ and

$f_{S_b}(b)$ for $b \in S_b$. For example, we might have predicted possible worst-case execution times of different tasks in future applications $S_t = \{50, 100, 200, 300, 500 \ ms\}$. If there is a higher probability of having tasks of 100 ms, and a very low probability of having tasks of 300 ms and 500 ms, then our distribution function $f_{S_t}(t)$ could look like this: $f_{S_t}(50) = 0.20$, $f_{S_t}(100) = 0.50$, $f_{S_t}(200) = 0.20$, $f_{S_t}(300) = 0.05$, and $f_{S_t}(500) = 0.05$.

Another piece of information is related to the period of task graphs which could be part of future applications. In particular, the smallest expected period T_{min} is assumed to be given, together with the expected necessary processor time t_{need}, and bus bandwidth b_{need}, inside such a period T_{min}. As will be shown later, this information is treated in a flexible way during the design process and is used in order to provide a fair distribution of available resources.

The execution times in S_t, as well as t_{need}, are considered relative to the slowest node in the system. All the other nodes are characterized by a speedup factor relative to this slowest node. A normalization with these factors is performed when computing the metrics C_1^τ and C_2^τ introduced in the following section.

15.5.4 Quality Metrics and Objective Function

A designer will be able to map and schedule an application $\mathcal{A}_{future}$ on top of a system implementing Ψ and $\mathcal{A}_{current}$ only if there are sufficient resources available. In our case, the resources are processor time and the bandwidth on the bus. In the context of a non-preemptive static scheduling policy, having free resources translates into having free time slots on the processors and having space left for messages in the bus slots. We call these free slots of available time on the processor or on the bus, *slack*. It is to be noted that the total quantity of computation and communication power available on our system after we have mapped and scheduled $\mathcal{A}_{current}$ on top of Ψ is the same regardless of the mapping and scheduling policies used. What depends on the mapping and scheduling strategy is the distribution of slacks along the time line and the size of the individual slacks. It is exactly this size and distribution of the slacks that characterizes the quality of a certain design alternative from the point of view of flexibility for future upgrades. In this section, we introduce two criteria in order to reflect the degree to which one design alternative meets the requirement (b) presented in Section 15.5.2. For each criterion, we provide metrics which quantify the degree to which the criterion is met. The first criterion reflects how well the resulted slack sizes fit to a future application, and the second criterion expresses how well the slack is distributed in time.

15.5.4.1 Slack Sizes (the first criterion)

The slack sizes resulted after implementation of $\mathcal{A}_{current}$ on top of Ψ should be such that they best accommodate a given family of applications $\mathcal{A}_{future}$, characterized by the sets S_t, S_b and the probability distributions f_{S_t} and f_{S_b}, as outlined in Section 15.5.3.2.

Let us go back to the example in Figure 15.18 where $\mathcal{A}_1$ is what we now call

$\mathcal{A}_{current}$, while $\mathcal{A}_2$, to be later implemented on top of Ψ and $\mathcal{A}_1$, is $\mathcal{A}_{future}$. This $\mathcal{A}_{future}$ consists of the two tasks τ_6 and τ_7. It can be observed that the best configuration from the point of view of accommodating $\mathcal{A}_{future}$, taking into consideration only slack sizes, is to have a contiguous slack after implementation of $\mathcal{A}_{current}$ (Figure 15.18d1). However, in reality, it is almost impossible to map and schedule the current application such that a contiguous slack is obtained. Not only is it impossible, but it is also undesirable from the point of view of the second design criterion, to be discussed next. However, as we can see from Figure 15.18b1, if we schedule $\mathcal{A}_{current}$ such that it fragments the slack too much, it is impossible to fit $\mathcal{A}_{future}$ because there is no slack that can accommodate task τ_7. A situation such as the one depicted in Figure 15.18c1 is desirable, where the resulted slack sizes are adapted to the characteristics of the $\mathcal{A}_{future}$ application.

In order to measure the degree to which the slack sizes in a given design alternative fit the future applications, we provide two metrics, C_1^τ and C_1^m. C_1^τ captures how much of the largest future application which theoretically could be mapped on the system can be mapped on top of the current design alternative. C_1^m is similar relative to the slacks in the bus slots.

How does the largest future application which theoretically could be mapped on the system look like? The total processor time and bus bandwidth available for this largest future application is the total slack available on the processors and bus, respectively, after implementing $\mathcal{A}_{current}$. Process and message sizes of this hypothetical largest application are determined knowing the total size of the available slack, and the characteristics of the future applications as expressed by the sets S_t and S_b, and the probability distributions f_{S_t} and f_{S_b}. Let us consider, for example, that the total slack size on the processors is 2800 ms and the set of possible worst-case execution times is $S_t = \{50, 100, 200, 300, 500 \ ms\}$. The probability distribution function f_{S_t} is defined as follows: $f_{S_t}(50) = 0.20$, $f_{S_t}(100) = 0.50$, $f_{S_t}(200) = 0.20$, $f_{S_t}(300) = 0.05$ and $f_{S_t}(500) = 0.05$. Under these circumstances, the largest hypothetical future application will consist of 20 tasks: 10 tasks (half of the total, $f_t(100) = 0.50$) with a worst-case execution time of 100 ms, 4 tasks with 50 ms, 4 with 200 ms, one with 300 and one with 500 ms.

After we have determined the number of tasks of this largest hypothetical $\mathcal{A}_{future}$ and their worst-case execution times, we apply a *bin-packing algorithm* [215] using the *best-fit policy* in which we consider tasks as the objects to be packed, and the available slacks as containers. The total execution time of tasks which are left unpacked, relative to the total execution time of the whole task set, gives the C_1^τ metric. The same is the case with the metric C_1^m, but applied to message sizes and available slacks in the bus slots.

Let us consider the example in Figure 15.18 and suppose a hypothetical $\mathcal{A}_{future}$ consisting of two tasks like those of application $\mathcal{A}_2$. For the design alternatives in Figure 15.18c1 and 15.18d1, $C_1^\tau = 0\%$ (both alternatives are perfect from the point of view of slack sizes). For the alternative in Figure 15.18b1, however, $C_1^\tau = 30/40 = 75\%$ the worst-case execution time of τ_7 (which is left unpacked) relative to the total execution time of the two tasks.

15.5.4.2 Distribution of Slacks (the second criterion)

In the previous section, we defined a metric which captures how well the sizes of the slacks fit a possible future application. A similar metric is needed to characterize the distribution of slacks over time.

Let τ_i be a task with period T_{τ_i} that belongs to a future application, and $M(\tau_i)$ the node on which τ_i will be mapped. The worst-case execution time of τ_i is $t_{\tau_i}^{M(\tau_i)}$. In order to schedule τ_i, we need a slack of size $t_{\tau_i}^{M(\tau_i)}$ that is available periodically, within a period T_{τ_i}, on processor $M(\tau_i)$. If we consider a group of tasks with period T, which are part of $\mathcal{A}_{future}$, in order to implement them, a certain amount of slack is needed which is available periodically, with a period T, on the nodes implementing the respective tasks.

During implementation of $\mathcal{A}_{current}$, we aim for a slack distribution such that the future application with the smallest expected period T_{min} and with the necessary processor time t_{need}, and bus bandwidth b_{need}, can be accommodated (see Section 15.5.3.2).

Thus, for each node, we compute the minimum periodic slack, inside a T_{min} period. By summing these minima, we obtain the slack which is available periodically to $\mathcal{A}_{future}$. This is the C_2^τ metric. The C_2^m metric characterizes the minimum periodically available bandwidth on the bus and it is computed in a similar way.

In Figure 15.21 we consider an example with $T_{min} = 120$ *ms*, $t_{need} = 90$ *ms* and $b_{need} = 65$ *ms*. The length of the schedule table of the system implementing Ψ and $\mathcal{A}_{current}$ is 360 ms (in Section 15.5.5 we will elaborate on the length of the global schedule table). Thus, we have to investigate three periods of length T_{min} each. The system consists of three nodes. Let us consider the situation in Figure 15.21a. In the first period, *Period* 0, there are 40 ms of slack available on $Node_1$, in the second period 80 ms, and in the third period no slack is available on $Node_1$. Thus, the total slack a future application of period T_{min} can use on $Node_1$ is $min(40, 80, 0) = 0$ *ms*. Neither can $Node_2$ provide slack for this application, as in *Period* 1 there is no slack available. However, on $Node_3$ there are at least 40 ms of slack available in each period. Thus, with the configuration in Figure 15.21a we have $C_2^\tau = 20$ *ms*, which is not sufficient to accommodate $t_{need} = 90$ *ms*. The available periodic slack on the bus is also insufficient: $C_2^m = 60$ *ms* $< b_{need}$. However, in the situation presented in Figure 15.21b, we have $C_2^\tau = 120$ *ms* $> t_{need}$, and $C_2^m = 90$ *ms* $> b_{need}$.

15.5.4.3 Objective Function and Exact Problem Formulation

In order to capture how well a certain design alternative meets the requirement (b) stated in Section 15.5.2, the metrics discussed before are combined in an objective function, as follows:

$$C = w_1^\tau (C_1^\tau)^2 + w_1^m (C_1^m)^2 + w_2^\tau \max(0, t_{need} - C_2^\tau) + w_2^m \max(0, b_{need} - C_2^m) \quad (15.9)$$

where the metric values introduced in the previous section are weighted by the constants w_1^τ, w_2^τ, w_1^m and w_2^m. Our mapping and scheduling strategy will try to minimize this function. The first two terms measure how well the resulted slack sizes fit to a future application (the first criterion), while the second two terms reflect the distribution of slacks (the second criterion). In order to obtain a balanced solution that favors a good fitting both on the processors and on the bus, we have used the squares of the metrics.

We call a *valid solution* that mapping and scheduling which satisfies all the design constraints (in our case the deadlines) and meets the second criterion ($C_2^\tau \geq t_{need}$ and $C_2^m \geq b_{need}$).[12]

At this point, we can give an exact formulation of our problem. Given an existing set of applications Ψ which are already mapped and scheduled, and an application $\mathcal{A}_{current}$ to be implemented on top of Ψ, we are interested in finding the subset $\Omega \subseteq \Psi$ of old applications to be remapped and rescheduled such that we produce a valid solution for $\mathcal{A}_{current} \cup \Omega$ and the total cost of modification $R(\Omega)$ is minimized. Once such a set Ω of applications is found, we are interested in optimizing the implementation of $\mathcal{A}_{current} \cup \Omega$ such that the objective function C is minimized, considering a family of future applications characterized by the sets S_t and S_b, the functions f_{S_t} and f_{S_b} as well as the parameters T_{min}, t_{need}, and b_{need}.

A mapping and scheduling strategy based on this problem formulation is presented in the following section.

15.5.5 Mapping and Scheduling Strategy

As shown in the algorithm in Figure 15.22, our mapping and scheduling strategy (MS) consists of two steps. In the first step we try to obtain a valid solution for the mapping and scheduling of $\mathcal{A}_{current} \cup \Omega$ so that the modification cost $R(\Omega)$ is minimized. Starting from such a solution, the second step iteratively improves the design in order to minimize the objective function C. In the context in which the second criterion is satisfied after the first step, improving the cost function during the second step aims at minimizing the value of $w_1^\tau(C_1^\tau)^2 + w_1^m(C_1^m)^2$.

If the first step has not succeeded in finding a solution such that the imposed time constraints are satisfied, this means that there are not sufficient resources available to implement the application $\mathcal{A}_{current}$. Thus, modifications of the system architecture have to be performed before restarting the mapping and scheduling procedure. If, however, the timing constraints are met but the second design criterion is not satisfied, a larger T_{min} (smallest expected period of a future application, see Section 15.5.3.2) or smaller values for t_{need} and/or b_{need} are suggested to the designer. This, of course, reduces the frequency of possible future applications and the amount of processor and bus resources available to them.

In the following section, we briefly discuss the basic mapping and scheduling algorithm we have used in order to generate an initial solution. The heuristic used to iteratively improve the design with regard to the first and the second design criteria is presented in Section 15.5.5.2. In Section 15.5.5.3, we describe three alternative

---

[12]This definition of a valid solution can be relaxed by imposing only the satisfaction of deadlines. In this case, the algorithm in Figure 15.22 will look after a solution which satisfies the deadlines and $R(\Omega)$ is minimized; the additional second criterion is, in this case, only considered optionally.

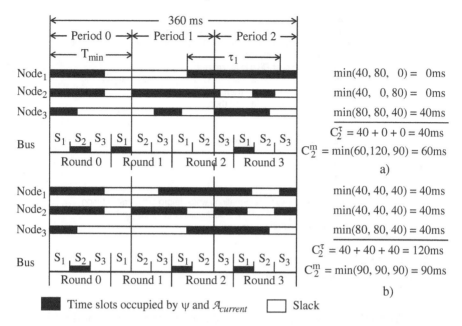

FIGURE 15.21
Example for the Second Design Criterion

heuristics which can be used during the first step in order to find the optimal subset of applications to be modified.

15.5.5.1 The Initial Mapping and Scheduling

As shown in Figure 15.23, the first step of MS consists of an iteration that tries different subsets $\Omega \subseteq \Psi$ with the intention to find that subset $\Omega = \Omega_{min}$ of old applications to be remapped and rescheduled which produces a valid solution for $\mathcal{A}_{current} \cup \Omega$ such that $R(\Omega)$ is minimized. Given a subset Ω, the InitialMappingScheduling function (IMS) constructs a mapping and a schedule for the applications $\mathcal{A}_{current} \cup \Omega$ on top of $\Psi \backslash \Omega$, that meets the deadlines, without worrying about the two criteria introduced in Section 15.5.4.

The IMS is a classical mapping and scheduling algorithm for which we have used the Heterogeneous Critical Path (HCP) algorithm [35] as a starting point. HCP is based on a list scheduling approach [62]. We have modified the HCP algorithm in three main regards:

1. We consider that mapping and scheduling does not start with an empty system but a system on which a certain number of tasks are already mapped.

2. Messages are scheduled into bus-slots according to the TDMA protocol. The TDMA-based message scheduling technique has been presented by us in [86].

MappingSchedulingStrategy
   Step 1: try to find a valid solution that minimizes $R(\Omega)$
    Find a mapping and scheduling of $\mathcal{A}_{current} \cup \Omega$ on top of $\psi \setminus \Omega$ so that:
    1. constraints are satisfied;
    2. modification cost $R(\Omega)$ is minimized;
    3. the second criterion is satisfied: $C_2^{\tau} \ge t_{need}$ and $C_2^m \ge b_{need}$
   if Step1 has not succeeded **then**
    if constraints are not satisfied **then**
      change architecture
    else
      suggest new T_{min}, t_{need} or b_{need}
    end if
    go to Step 1
   end if
   Step 2: improve the solution by minimizing objective function C
    Perform iteratively transformations which improve the first criterion
    (the metrics C_1^{τ} and C_1^m) without invalidating the second criterion.
end MappingSchedulingStrategy

FIGURE 15.22
Mapping and Scheduling Strategy (MS)

3. As a priority function for list scheduling we use, instead of the CP (critical path) priority function employed in [35], the MPCP (modified partial critical path) function introduced by us in [86]. MPCP takes into consideration the particularities of the communication protocol for calculation of communication delays.

For the example in Figure 15.17, our initial mapping and scheduling algorithm will be able to produce the optimal solution with a schedule length of 48 ms.

However, before performing the effective mapping and scheduling with IMS, two aspects have to be addressed. First, the task graphs $\mathcal{G}_i \in \mathcal{A}_{current} \cup \Omega$ have to be merged into a single graph $\mathcal{G}_{current}$ by unrolling task graphs and inserting dummy nodes as discussed in [261].

15.5.5.2 Iterative Design Transformations

Once IMS has produced a mapping and scheduling which satisfies the timing constraints, the next goal of *Step*1 is to improve the design in order to satisfy the second design criterion ($C_2^{\tau} \ge t_{need}$ and $C_2^m \ge b_{need}$). During the second step, the design is then further transformed with the goal of minimizing the value of $w_1^{\tau}(C_1^{\tau})^2 + w_1^m(C_1^m)^2$, according to the requirements of the first criterion, without invalidating the second criterion achieved in the first step. In both steps, we iteratively improve the design using a transformational approach. These successive transformations are performed inside the (innermost) `repeat` loops of the first and second step, respectively (Figure 15.23). A new design is obtained from the current one by performing a transformation called *move*. We consider the following two categories of moves:

1. Moving a task to a different slack found on the same node or on a different node

2. Moving a message to a different slack on the bus

In order to eliminate those moves that will lead to an infeasible design (that violates deadlines), we do as follows. For each task τ_i, we calculate the $ASAP(\tau_i)$ and $ALAP(\tau_i)$ times considering the resources of the given hardware architecture. $ASAP(\tau_i)$ is the earliest time τ_i can start its execution, while $ALAP(\tau_i)$ is the latest time τ_i can start its execution without causing the application to miss its deadline. When moving τ_i we will consider slacks on the target processor only inside the interval $[ASAP(\tau_i), ALAP(\tau_i)]$. The same reasoning holds for messages, with the addition that a message can only be moved to slacks belonging to a slot that corresponds to the sender node (see Section 15.5.1.1). Any violation of the data dependency constraints caused by a move is rectified by shifting tasks or messages concerned in an appropriate way. If such a shift produces a deadline violation, the move is rejected.

At each step, our heuristic tries to find those moves that have the highest potential to improve the design. For each iteration, a set of potential moves is selected by the PotentialMoveX functions. SelectMoveX then evaluates these moves with regard to the respective metrics and selects the best one to be performed. We now briefly discuss the four PotentialMoveX functions with the corresponding moves.

PotentialMoveC$_2^P$ and PotentialMoveC$_2^m$. Consider Figure 15.21a. In *Period* 2 on *Node*$_1$ there is no available slack. However, if we move task τ_1 with 40 ms to the left into *Period* 1, as depicted in Figure 15.21b, we create a slack in *Period*2 and the periodic slack on node N_1 will be $min(40, 40, 40) = 40$ *ms*, instead of 0 ms.

Potential moves aimed at improving the metric C_2^τ will be the shifting of tasks inside their $[ASAP, ALAP]$ interval in order to improve the periodic slack. The move can be performed on the same node or on less loaded nodes. The same is true for moving messages in order to improve the metric C_2^m. For the improvement of the periodic bandwidth on the bus, we also consider movement of tasks, trying to place the sender and receiver of a message on the same processor and, thus, reducing the bus load.

PotentialMoveC$_1^P$ and PotentialMoveC$_1^m$. The moves suggested by these two functions aim at improving the C_1 metric through reducing the slack fragmentation. The heuristic is to evaluate only those moves that iteratively eliminate the smallest slack in the schedule. Let us consider the example in Figure 15.24, where we have three applications mapped on a single processor: Ψ, consisting of τ_1 and τ_2, $\mathcal{A}_{current}$, having tasks τ_3, τ_4 and τ_5, and $\mathcal{A}_{future}$, with τ_6, τ_7 and τ_8. Figure 15.24 presents three possible schedules; tasks are depicted with rectangles, the width of a rectangle represents the worst-case execution time of that task. The PotentialMoveC$_1$ functions start by identifying the smallest slack in the schedule table. In Figure 15.24a, the smallest slack is the slack between τ_1 and τ_3. Once the smallest slack has been identified, potential moves are investigated which either remove or enlarge the slack. For example, the slack between τ_1 and τ_3 can be removed by attaching τ_3 to τ_1, and it can be enlarged by moving τ_3 to the right in the schedule table. Moves that remove the

Step 1: try to find a valid solution that minimizes $R(\Omega)$
$\Omega = \emptyset$
repeat
    succeeded = InitialMappingScheduling($\psi \setminus \Omega$, $\mathcal{A}_{current} \cup \Omega$)
    -- compute ASAP-ALAP intervals for all tasks
    ASAP($\mathcal{A}_{current} \cup \Omega$); ALAP($\mathcal{A}_{current} \cup \Omega$)
    if *succeeded* **then**-- if time constraints are satisfied
        -- design transformations in order to satisfy
        -- the second design criterion
        repeat
            -- find set of moves with the highest potential to
            -- maximize C_2^{τ} or C_2^m
            move_set = PotentialMoveC$_2^{\tau}$ ($\mathcal{A}_{current} \cup \Omega$) $\cup$
                PotentialMoveC$_2^m$($\mathcal{A}_{current} \cup \Omega$)
            -- select and perform move which improves most C_2
            move = SelectMoveC$_2$(*move_set*); Perform(*move*)
            succeeded =$C_2^{\tau} \geq t_{need}$ **and** $C_2^m \geq b_{need}$
        until *succeeded* **or** maximum number of iterations reached
    end if
    if *succeeded* and $R(\Omega)$ smallest so far **then**
        $\Omega_{valid} = \Omega$; *solution*$_{valid}$ = *solution*$_{current}$
    end if
            Ω=NextSubset(Ω) -- try another subset
until termination condition

Step 2: improve the solution by minimizing objective function C
solution$_{current}$ = *solution*$_{valid}$; $\Omega_{min} = \Omega_{valid}$
-- design transformations in order to satisfy the first design criterion
repeat
    -- find set of moves with highest potential to minimize C_1^{τ} or C_1^m
    move_set = PotentialMoveC$_1^{\tau}$ ($\mathcal{A}_{current} \cup \Omega_{min}$) $\cup$
        PotentialMoveC$_2^m$($\mathcal{A}_{current} \cup \Omega_{min}$)
    -- select move which improve $w_1^{\tau}(C_1^{\tau})^2 + w_1^m(C_1^m)^2$,
    -- and does not invalidate the second criterion
    move = SelectMoveC$_1$(*move_set*); Perform(*move*)
until $w_1^{\tau}(C_1^{\tau})^2 + w_1^m(C_1^m)^2$ has not changed **or**
    maximum number of iterations reached

FIGURE 15.23
Step 1 and Step 2 of the Mapping and Scheduling Strategy in Figure 15.22

slack are considered only if they do not lead to an invalidation of the second design criterion, measured by the C_2 metric improved in the previous step (see Figure 15.23, Step 1). Also, the slack can be enlarged only if it does not create, as a result, other unusable slack. A slack is unusable if it cannot hold the smallest object of the future application, in our case τ_6. In Figure 15.24a, the slack can be removed by moving τ_3 such that it starts from time 20, immediately after τ_1, and it can be enlarged by moving τ_3 so that it starts from 30, 40, or 50 (considering an increment which here was set by us to 10, the size of τ_6, the smallest object in $\mathcal{A}_{future}$). For each move, the improvement on the C_1 metric is calculated, and that move is selected by the *SelectMoveC$_1$* function to be performed, which leads to the largest improvement on C_1 (which means the smallest value). For all the previously considered moves of τ_3, we are not able to map τ_8 which represents 50% of the $\mathcal{A}_{future}$, therefore $C_1 = 50\%$. Consequently, we can perform any of the mentioned moves, and our algorithm selects the first one investigated, the move to start τ_3 from 20, thus removing the slack. As a result of this move, the new schedule table is the one in Figure 15.24b. In the next call of the *PotentialMoveC$_1$* function, the slack between τ_5 and τ_2 is identified as the smallest slack. Out of the potential moves that eliminate this slack, listed in Figure 15.24 for case b, several lead to $C_1 = 0\%$, the largest improvement. *SelectMoveC$_1$* selects moving τ_5 to start from 90, and thus we are able to map task τ_8 of the future application, leading to a successful implementation in Figure 15.24c.

The previous example has only illustrated movements of tasks. Similarly, in *PotentialMoveC$_1^m$*, we also consider moves of messages in order to improve C_1^m. However, the movement of messages is restricted by the TDMA bus access scheme, such that a message can only be moved into a slot corresponding to the node on which it is generated.

15.5.5.3 Minimizing the Total Modification Cost

The first step of our mapping and scheduling strategy, described in Figure 15.23, iterates on successive subsets Ω searching for a valid solution which also minimizes the total modification cost $R(\Omega)$. As a first attempt, the algorithm searches for a valid implementation of $\mathcal{A}_{current}$ without disturbing the existing applications ($\Omega = \emptyset$). If no valid solution is found, successive subsets Ω produced by the function NextSubsetset are considered, until a termination condition is met. The performance of the algorithm, in terms of runtime and quality of the solutions produced, is strongly influenced by the strategy employed for the function NextSubset and the termination condition. They determine how the design space is explored while testing different subsets Ω of applications. In the following, we present three alternative strategies. The first two can be considered as situated at opposite extremes: The first one is potentially very slow but produces the optimal result while the second is very fast and possibly low quality. The third alternative is a heuristic able to produce good quality results in relatively short time, as demonstrated by the experimental results presented in Section 15.5.6.

Exhaustive Search (ES). In order to find Ω_{min}, the simplest solution is to try successively all the possible subsets $\Omega \subseteq \Psi$. These subsets are generated in ascend-

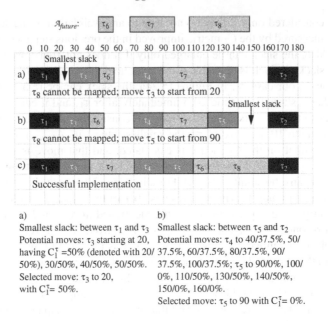

a)
Smallest slack: between τ_1 and τ_3
Potential moves: τ_3 starting at 20,
having C_1^τ =50% (denoted with 20/
50%), 30/50%, 40/50%, 50/50%.
Selected move: τ_3 to 20,
with C_1^τ= 50%.

b)
Smallest slack: between τ_5 and τ_2
Potential moves: τ_4 to 40/37.5%, 50/
37.5%, 60/37.5%, 80/37.5%, 90/
37.5%, 100/37.5%; τ_5 to 90/0%, 100/
0%, 110/50%, 130/50%, 140/50%,
150/0%, 160/0%.
Selected move: τ_5 to 90 with C_1^τ= 0%.

FIGURE 15.24
Successive Steps with Potential Moves for Improving C_1

ing order of the total modification cost, starting from $\emptyset$. The termination condition is fulfilled when the first valid solution is found or no new subsets are to be generated. Since the subsets are generated in ascending order, according to their cost, the subset Ω that first produces a valid solution is also the subset with the minimum modification cost.

The generation of subsets is performed according to the graph G that characterizes the existing applications (see Section 15.5.3.1). Finding the next subset Ω, starting from the current one, is achieved by a branch and bound algorithm that, in the worst case, grows exponentially in time with the number of applications. For the example in Figure 15.20, the call to NextSubset($\emptyset$) will generate $\Omega = \{\mathcal{A}_7\}$ which has the smallest non-zero modification cost $R(\{\mathcal{A}_7\}) = 20$. The next generated subsets, in order, together with their corresponding total modification cost are: $R(\{\mathcal{A}_3\}) = 50$, $R(\{\mathcal{A}_3, \mathcal{A}_7\}) = 70$, $R(\{\mathcal{A}_4, \mathcal{A}_7\}) = 90$ (the inclusion of $\mathcal{A}_4$ triggers the inclusion of $\mathcal{A}_7$), $R(\{\mathcal{A}_2, \mathcal{A}_3\}) = 120$, $R(\{\mathcal{A}_2, \mathcal{A}_3, \mathcal{A}_7\}) = 140$, $R(\{\mathcal{A}_3, \mathcal{A}_4, \mathcal{A}_7\}) = 140$, $R(\{\mathcal{A}_1\}) = 150$, and so on. The total number of possible subsets according to the graph G in Figure 15.20 is 16.

This approach, while finding the optimal subset Ω, requires a large amount of computation time and can be used only with a small number of applications.

Greedy Heuristic (GH). If the number of applications is larger, a possible solution could be based on a simple greedy heuristic which, starting from $\Omega = \emptyset$, progressively enlarges the subset until a valid solution is produced. The algorithm looks at all the non-frozen applications and picks that one which, together with its dependencies, has the smallest modification cost. If the new subset does not produce a valid solution, it is enlarged by including, in the same fashion, the next application with its dependencies. This greedy expansion of the subset is continued until the set is large enough to lead to a valid solution or no application is left. For the example in Figure 15.20, the call to NextSubset($\emptyset$) will produce $R(\{\mathcal{A}_7\}) = 20$, and will be successively enlarged to $R(\{\mathcal{A}_7, \mathcal{A}_3\}) = 70$, $R(\{\mathcal{A}_7, \mathcal{A}_3, \mathcal{A}_2\}) = 140$ ($\mathcal{A}_4$ could have been picked as well in this step because it has the same modification cost of 70 as $\mathcal{A}_2$ and its dependence $\mathcal{A}_7$ is already in the subset), $R(\{\mathcal{A}_7, \mathcal{A}_3, \mathcal{A}_2 \mathcal{A}_4\}) = 210$, and so on.

While this approach very quickly finds a valid solution, if one exists, it is possible that the resulted total modification cost is much higher than the optimal one.

Subset Selection Heuristic (SH). An intelligent selection heuristic should be able to identify the reasons due to which a valid solution has not been produced and to find the set of candidate applications which, if modified, could eliminate the problem. The failure to produce a valid solution can have two possible causes: An initial mapping which meets the deadlines has not been found, or the second criterion is not satisfied.

Let us investigate the first reason. If an application $\mathcal{A}_i$ is to meet its deadline D_i, all its tasks $\tau_j \in \mathcal{A}_i$ have to be scheduled inside their $[ASAP, ALAP]$ intervals. InitialMappingScheduling (IMS) fails to schedule a task inside its $[ASAP, ALAP]$ interval if there is not enough slack available on any processor, due to other tasks scheduled in the same interval. In this situation, we say that there is a *conflict* with tasks belonging to other applications. We are interested to find out which applications are responsible for conflicts encountered during the mapping and scheduling of $\mathcal{A}_{current}$, and not only that, but also which ones are *flexible* enough to be moved away in order to avoid these conflicts.

If IMS is not able to find a solution that satisfies the deadlines, it will determine a metric $\Delta_{\mathcal{A}_i}$ that characterizes both the degree of conflict and the flexibility of each application $\mathcal{A}_i \in \Psi$ in relation to $\mathcal{A}_{current}$. A set of applications Ω will be characterized, in relation to $\mathcal{A}_{current}$, by the following metric:

$$\Delta(\Omega) = \sum_{\mathcal{A}_i \in \Omega} \Delta_{\mathcal{A}_i}. \qquad (15.10)$$

This metric $\Delta(\Omega)$ will be used by our subset selection heuristic in the case IMS has failed to produce a solution which satisfies the deadlines. An application with a larger $\Delta_{\mathcal{A}_i}$ is more likely to lead to a valid schedule if included in Ω.

In Figure 15.25, we illustrate how this metric is calculated. Applications A, B and C are implemented on a system consisting of the three processors $Node_1$, $Node_2$ and $Node_3$. The current application to be implemented is D. At a certain moment, IMS comes to the point to map and schedule task $D_1 \in D$. However, it is not able to place it inside its $[ASAP, ALAP]$ interval, denoted in Figure 15.25 as I. The reason is that there is not enough slack available inside I on any of the processors, because tasks $A_1, A_2, A_3 \in A$, $B_1 \in B$ and $C_1 \in C$ are scheduled inside that interval. We are interested to determine which of the applications A, B and C are more likely to lend free slack for D_1, if remapped and rescheduled. Therefore, we calculate the slack

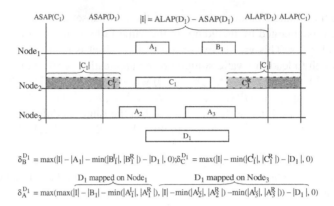

FIGURE 15.25
Metric for the Subset Selection Heuristic

resulted after we move away tasks belonging to these applications from the interval I. For example, the resulted slack available after modifying application C (moving C_1 either to the left or to the right inside its own $[ASAP, ALAP]$ interval) is of size $|I| - \min(|C_1^L|, |C_1^R|)$. With $C_1^L(C_1^R)$ we denote that slice of task C_1 which remains inside the interval I after C_1 has been moved to the extreme left (right) inside its own $[ASAP, ALAP]$ interval. $|C_1^L|$ represents the length of slice C_1^L. Thus, when considering task D_1, Δ_C will be incremented with $\delta_C^{D_1} = \max(|I| - \min(|C_1^L|, |C_1^R|) - |D_1|, 0)$. This value shows the maximum theoretical slack usable for D_1, that can be produced by modifying application C. By relating this slack to the length of D_1, the value $\delta_C^{D_1}$ also captures the amount of flexibility provided by that modification.

The increments $\delta_B^{D_1}$ and $\delta_A^{D_1}$ to be added to the values of Δ_B and Δ_A, respectively, are also presented in Figure 15.25. IMS then continues the evaluation of the metrics Δ with the other tasks belonging to the current application D (with the assumption that task D_1 has been scheduled at the beginning of interval I). Thus, as a result of the failed attempt to map and schedule application D, the metrics Δ_A, Δ_B and Δ_C will be produced.

If the initial mapping was successful, the first step of MS could fail during the attempt to satisfy the second criterion (Figure 15.23). In this case, the metric Δ_{A_i} is computed in a different way. What Δ_{A_i} will capture in this case is the potential of an application A_i to improve the metric C_2 if remapped together with $A_{current}$. Therefore, we consider a total number of moves from all the non-frozen applications. These moves are determined using the `PotentialMoveC2` functions presented in Section 15.5.5.2. Each such move will lead to a different mapping and schedule, and thus to a different C_2 value. Let us consider δ_{move} as the improvement on C_2 produced by the currently considered move. If there is no improvement, $\delta_{move} = 0$. Thus, for each move that has as subject τ_j or $m_j \in A_i$, we increment the metric Δ_{A_i} with the δ_{move} improvement on C_2.

As shown in the algorithm in Figure 15.23, MS starts by trying an implementation of $\mathcal{A}_{current}$ with $\Omega = \emptyset$. If this attempt fails for one of the two reasons mentioned above, the corresponding metrics $\Delta_{\mathcal{A}_i}$ are computed for all $\mathcal{A}_i \in \Psi$. Our heuristic SH will then start by finding the solution Ω_{GH} produced with the greedy heuristic GH (this will succeed if there exists any solution). The total modification cost corresponding to this solution is $R_{GH} = R(\Omega_{GH})$ and the value of the metric Δ is $\Delta_{GH} = \Delta(\Omega_{GH})$. SH now continues by trying to find a solution with a more favorable Ω than Ω_{GH} (a smaller total cost R). Therefore, the thresholds $R_{max} = R_{GH}$ and $\Delta_{min} = \Delta_{GH}/n$ (for our experiments we considered $n = 2$) are set. Sets of applications not fulfilling these thresholds will not be investigated by MS. For generating new subsets Ω, the function `NextSubset` now follows a similar approach like in the exhaustive search approach ES, but in a reverse direction, toward smaller subsets (starting with the set containing all non-frozen applications), and it will consider only subsets with a smaller total cost then R_{max} and a larger Δ than Δ_{min} (a small Δ means a reduced potential to eliminate the cause of the initial failure). Each time a valid solution is found, the current values of R_{max} and Δ_{min} are updated in order to further restrict the search space. The heuristic stops when no subset can be found with $\Delta > \Delta_{min}$ or a certain imposed limit has been reached (e.g., on the total number of attempts to find new subsets).

15.5.6 Experimental Results

In the following three sections, we show a series of experiments that demonstrate the effectiveness of the proposed approach and algorithms. The first set of results is related to the efficiency of our mapping and scheduling algorithm and the iterative design transformations proposed in Section 15.5.5.1 and 15.5.5.2. The second set of experiments evaluates our heuristics for minimization of the total modification cost presented in Section 15.5.5.3. As a general strategy, we have evaluated our algorithms performing experiments on a large number of test cases generated for experimental purposes. Finally, we have validated the proposed approach using a real-life example. All experiments were run on a SUN Ultra 10 workstation.

15.5.6.1 Evaluation of the IMS Algorithm and the Iterative Design Transformations

For evaluation of our approach, we used task graphs of 80, 160, 240, 320 and 400 tasks, representing the application $\mathcal{A}_{current}$, randomly generated for experimental purposes. Thirty graphs were generated for each graph dimension; thus, a total of 150 graphs were used for experimental evaluation.

We generated both graphs with random structure and graphs based on more regular structures like trees and groups of chains. We generated a random structure graph deciding for each pair of two tasks if they should be connected or not. Two tasks in the graph were connected with a certain probability (between 0.05 and 0.15, depending on the graph dimension) on the condition that the dependency would not introduce a loop in the graph. The width of the tree-like structures was controlled by the maximum number of direct successors a task can have in the tree (from 2 to 6),

TABLE 15.1
Evaluation of the initial mapping and scheduling.

Tasks	HCP			HCP		
	avg.	max.	better	avg.	max.	better
80	2.04%	31.57%	10%	0.35%	1.47%	30%
160	3.12%	48.89%	10%	1.18%	5.44%	33.33%
240	5.53%	61.27%	13.33%	1.38%	14.52%	36.66%
320	6.12%	88.57%	16.66%	2.79%	24.33%	40%
400	11.02%	120.77%	13.33%	2.78%	22.52%	36.66%

while the graphs consisting of groups of chains had 2 to 12 parallel chains of tasks. Furthermore, the regular structures were modified by adding a number of 3 to 30 random cross-connections.

Execution times and message lengths were assigned randomly using both uniform and exponential distribution within the 10 to 100 ms, and 2 to 8 bytes ranges, respectively.

We considered an architecture consisting of 10 nodes of different speeds. For the communication channel, we considered a transmission speed of 256 kbps and a length below 20 meters. The maximum length of the data field in a bus slot was 8 bytes. Throughout the experiments presented in this section, we have considered an existing set of applications Ψ consisting of 400 tasks, with a schedule table of 6s on each processor, and a slack of about 50% of the total schedule size. The mapping of the existing applications has been done using a simple heuristic that tries to balance the utilization of processors while minimizing communication. The scheduling of the applications Ψ has been performed using list scheduling, and the schedules obtained have then been stretched to their deadline by introducing slacks distributed uniformly over the schedule table.

In this section, we have also considered that no modifications of the existing set of applications Ψ are allowed when implementing a new application. We will concentrate on the aspects related to the modification of existing applications in the following section.

The first result concerns the quality of the designs produced by our initial mapping and scheduling algorithm IMS. As discussed in Section 15.5.5.1, IMS uses the MPCP priority function which considers particularities of the TDMA protocol. In our experiments, we compared the quality of designs (in terms of schedule length) produced by IMS with those generated with the original HCP algorithm proposed in [35]. Results are depicted in Table 15.1 where we have three columns for both HCP and IMS. In the columns labelled "average," we present the average percentage deviations of the schedule length produced with HCP and IMS from the length of the best schedule among the two. In the maximum column, we have the maximum percentage deviation, and the column with the heading better shows the percentage of cases in which HCP or IMS was better than the other. For example, for 240 tasks, HCP had an average percentage deviation from the best result of 5.53%, compared to 1.38% for IMS. Also, in the worst case, the schedule length obtained with HCP was 61.27% larger than the one obtained with IMS. There were four cases (13.33%) in which HCP has obtained a better result than IMS, compared to 11 cases (36.66%) where IMS has obtained a better result. For the rest of the 15 cases, the schedule lengths obtained were equal. We can observe that, in average, the deviation from the best result is 3.28 times smaller with IMS than with HCP. The average execution times for both algorithms are under half a second for graphs with 400 tasks.

For the next set of experiments, we were interested to investigate the quality of the design transformation heuristic discussed in Section 15.5.5.2, aiming at the optimization of the objective function C. In order to compare this heuristic, implemented in our mapping and scheduling approach MS, we have developed two additional heuristics:

1. A *simulated annealing strategy* (SA) [275], based on the same moves as described in Section 15.5.5.2. SA is applied on the solution produced by IMS and aims at finding the near-optimal mapping and schedule that minimizes the objective function C. The main drawback of the SA strategy is that in order to find the near-optimal solution it needs very large computation times. Such a strategy, although useful for the final stages of the system synthesis, cannot be used inside a design space exploration cycle.

2. A so-called *ad-hoc approach* (AH), which is a simple, straightforward solution to produce designs that, to a certain degree, support an incremental process. Starting from the initial valid schedule of length S obtained by IMS for a graph G with N tasks, AH uses a simple scheme to redistribute the tasks inside the $[0,D]$ interval, where D is the deadline of task graph G. AH starts by considering the first task in topological order, let it be τ_1. It introduces after τ_1 a slack of size $max(smallest\ task\ size\ of\ \mathcal{A}_{future}, (D-S)/N)$, thus shifting all descendants of τ_1 to the right (toward the end of the schedule table). The insertion of slacks is repeated for the next task, with the current, larger value of S, as long as the resulted schedule has a length $S \leq D$. Processes are moved only as long as their individual deadlines (if any) are not violated.

Our heuristic (MS), as well as SA and AH have been used to map and schedule each of the 150 task graphs on the target system. For each of the resulted designs, the objective function C has been computed. Very long and expensive runs have been performed with the SA algorithm for each graph and the best ever solution produced has been considered as the near-optimum for that graph. We have compared the objective function obtained for the 150 task graphs considering each of the three heuristics. Figure 15.26a presents the average percentage deviation of the objective function obtained with the MS and AH from the value of the objective function obtained with the near-optimal scheme (SA). We have excluded from the results in Figure 15.26a, 37 solutions obtained with AH for which the second design criterion has not been met, and thus the objective function has been strongly penalized. The average run-times of the algorithms are presented in Figure 15.26b. The SA approach performs best in terms of quality at the expense of a large execution time: The execution time can

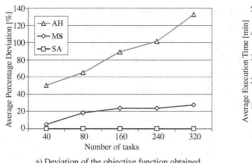
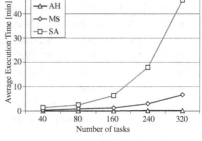

a) Deviation of the objective function obtained
with MS and AH from that obtained with SA

b) Execution times

FIGURE 15.26

Evaluation of the Design Transformation Heuristics

be up to 45 minutes for large graphs of 400 tasks. The important aspect is that MS performs very well, and is able to obtain good quality solutions, very close to those produced with SA, in a very short time. AH is, of course, very fast, but since it does not address explicitly the two design criteria presented in Section 15.5.4, it has the worst quality of solutions, as expressed by the objective function.

The most important aspect of the experiments is determining to which extent the design transformations proposed by us, and the related heuristic, really facilitate the implementation of future applications. To find this out, we have mapped graphs of 80, 160, 240 and 320 nodes representing the $\mathcal{A}_{current}$ application on top of Ψ (the same Ψ as defined for the previous set of experiments). After mapping and scheduling each of these graphs, we have tried to add a new application $\mathcal{A}_{future}$ to the resulted system. $\mathcal{A}_{future}$ consists of a task graph of 80 tasks, randomly generated according to the following specifications: $S_t = \{20, 50, 100, 150, 200 \ ms\}$, $f_t(S_t) = \{10, 25, 45, 15, 5\%\}$, $S_b = \{2, 4, 6, 8 \ bytes\}$, $f_b(S_b) = \{20, 50, 20, 10\%\}$, $T_{min} = 250 \ ms$, $t_{need} = 100$ and $b_{need} = 20 \ ms$. The experiments have been performed three times: Using MS, SA and AH for mapping $\mathcal{A}_{current}$. In all three cases, we were interested to see if it is possible to find a correct implementation for $\mathcal{A}_{future}$ on top of $\mathcal{A}_{current}$ using the initial mapping and scheduling algorithm IMS (without any modification of Ψ or $\mathcal{A}_{current}$). Figure 15.27 shows the percentage of successful implementations of $\mathcal{A}_{future}$ for each of the three cases. In the case $\mathcal{A}_{current}$ has been implemented with MS and SA (this means using the design criteria and metrics proposed in the section) we were able to find a valid schedule for 65% and 68% of the total cases, respectively. However, using AH to map $\mathcal{A}_{current}$ has led to a situation where IMS is able to find correct solutions in only 21% of the cases. Another conclusion from Figure 15.27 is that when the total slack available is large, as when $\mathcal{A}_{current}$ has only 80 tasks, it is easy for MS and, to a certain extent, even for AH to find a mapping that allows adding future applications. However, as $\mathcal{A}_{current}$ grows to 240 tasks, only MS and SA are able to find an implementation of $\mathcal{A}_{current}$ that supports an incremental design task, accommodating the future application in more than 60% of the cases. If the remaining slack is

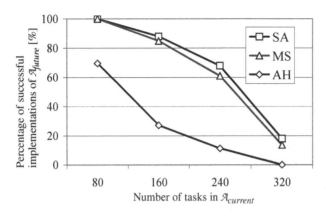

FIGURE 15.27
Percentage of Future Applications Successfully Implemented

very small, after we map an $\mathcal{A}_{current}$ of 320 tasks, it becomes practically impossible to map new applications without modifying the current system. Moreover, our mapping heuristic MH performs very well compared to the simulated annealing approach SA which aims for the near-optimal value of the objective function.

15.5.6.2 Evaluation of the Modification Cost Minimization Heuristics

For this set of experiments, we first used the same 150 task graphs as in the previous section, consisting of 80, 160, 240, 320 and 400 tasks, for the application $\mathcal{A}_{current}$. We also considered the same system architecture as presented there.

The first results concern the quality of the solution obtained with our mapping strategy MS using the search heuristic SH compared to the case when the simple greedy approach GH and the exhaustive search ES are used. For the existing applications, we have generated five different sets Ψ, consisting of different numbers of applications and tasks, as follows: 6 applications (320 tasks), 8 applications (400 tasks), 10 applications (480 tasks), 12 applications (560 tasks), 14 applications (640 tasks). The task graphs in the applications as well as their mapping and scheduling were generated as described in the introduction to Section 15.5.6.1.

After generating the applications, we have manually assigned modification costs in the range 10 to 100, depending on their size. The dependencies between applications (in the sense introduced in Section 15.5.3.1) were such that the total number of possible subsets Ω resulted for each set Ψ were 32, 128, 256, 1024 and 4096, respectively. We have considered that the future applications, $\mathcal{A}_{future}$, are characterized by the following parameters: $S_t = \{20,50,100,150,200 \; ms\}$, $f_t(S_t) = \{10,25,45,15,5\%\}$, $S_b = \{2,4,6,8 \; bytes\}$, $f_b(S_b) = \{20,50,20,10\%\}$, $T_{min} = 250 \; ms$, $t_{need} = 100 \; ms$ and $b_{need} = 20 \; ms$.

MS has been used to produce a valid solution for each of the 150 task graphs representing $\mathcal{A}_{current}$, on each of the target configurations Ψ, using the ES, GH and

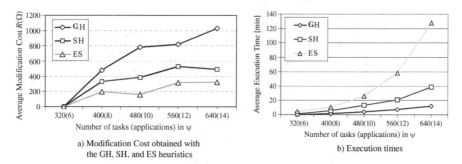

a) Modification Cost obtained with
the GH, SH, and ES heuristics

b) Execution times

FIGURE 15.28

Evaluation of the Modification Cost Minimization

SH approaches to subset selection. Figure 15.28a compares the three approaches based on the total modification cost needed in order to obtain a valid solution. The exhaustive approach ES is able to obtain valid solutions with an optimal (smallest) modification cost, while the greedy approach GH produces on average 3.12 times more costly modifications in order to obtain valid solutions. However, in order to find the optimal solution, ES needs large computation times, as shown in Figure 15.28b. For example, it can take more than two hours on average to find the smallest cost subset to be remapped that leads to a valid solution in the case of 14 applications (640 tasks). We can see that the proposed heuristic SH performs well, producing close to optimal results with a good scaling for large application sets. For the results in Figure 15.28, we have eliminated those situations in which no valid solution could be produced by MS.

Finally, we have repeated the last set of experiments discussed in the previous section (the experiments leading to the results in Figure 15.27). However, in this case, we have allowed the current system (consisting of $\Psi \cup \mathcal{A}_{current}$) to be modified when implementing $\mathcal{A}_{future}$. If the mapping and scheduling heuristic is allowed to modify the existing system, then we are able to increase the total number of successful attempts to implement application $\mathcal{A}_{future}$ from 65% to 77.5%. For the case with $\mathcal{A}_{current}$ consisting of 160 tasks (when the amount of available resources for $\mathcal{A}_{future}$ is small), the increase is from 60% to 92%. Such an increase is, of course, expected. The important aspect, however, is that it is obtained not by randomly selecting old applications to be modified, but by performing this selection such that the total modification cost is minimized.

15.6 Integration of Time-Triggered Communication with Event-Triggered Tasks

There has been a long debate in the real-time and embedded systems communities concerning the advantages of TT vs. ET approaches. Several aspects have been considered in favor of one or the other approach, such as flexibility, predictability, jitter control, processor utilization and testability. An interesting comparison of the ET and TT approaches, from a more industrial, in particular automotive, perspective, can be found in [205]. The conclusion there is that the right choice depends on the particularities of the application.

Moreover, considering preemptive priority based scheduling at the task level, with time-triggered static scheduling at the communication level, can be the right solution under certain circumstances. TT communication protocols have been classically associated with non-preemptive static scheduling of tasks, mainly for fault-tolerance reasons. A TT communication protocol, such as TTP, provides a global time-base, improves fault-tolerance and predictability. At the same time, certain particularities of the application or of the underlying real-time operating system can impose a priority based scheduling policy at the task level.

Therefore, in this section, we consider that tasks are scheduled according to a static priority preemptive policy, while messages are scheduled using a time-triggered protocol. In this section, we consider TTP-based systems, but the TT/ET integration approach is valid also for other TT protocols.

Thus, we first develop a schedulability analysis for distributed tasks with preemptive priority based scheduling considering a TTP-based communication infrastructure. Secondly, we propose four different approaches to message scheduling using static and dynamic message allocation. Finally, we show how the parameters of the communication protocol can be optimized in order to fit the communication particularities of a certain application. Thus, based on our approach, it is not only possible to determine if a certain task set implemented on a TTP-based distributed architecture is schedulable, but it is also possible to select a particular message passing strategy and also to optimize certain parameters of the communication protocol. By adapting the communication infrastructure to certain particularities of the task set, we increase the likelihood of producing an implementation which satisfies all time constraints.

15.6.1 Software Architecture

In Section 15.5.1.1, we have discussed the message passing mechanism. The organization of the message queue assembling of a frame depends on the particular approach chosen for message scheduling (see Section 15.6.3). We assume that there is a message transfer task which is activated, at certain a priori known moments, by the tick scheduler in order to perform the message transfer. Our assumption is that these activation times are stored in a message handling time table (MHTT) available to the real-time kernel in each node. Both the MEDL and the MHTT are generated off-line as a result of the schedulability analysis and optimization which will be discussed later. The MEDL imposes the times when the TTP controller of a certain node has to move frames from the MBI to the communication channel. The MHTT contains the times when messages have to be transferred by the message transfer task from the *Out* queue into the MBI, in order to be broadcasted by the TTP controller. As a result of this synchronization, the activation times in the MHTT are directly related to those in the MEDL and the first table results directly from the second one.

It is easy to observe that we have the most favorable situation when, at a certain activation, the message transfer task finds in the *Out* queue all the "expected" messages which then can be packed into the immediate following frame to be sent by the TTP controller. However, application tasks are not statically scheduled and availability of messages in the *Out* queue cannot be guaranteed at fixed times. Worst-case situations have to be considered, as will be shown in Section 15.6.3.

Let us consider Figure 15.16. There we assumed a context in which the broadcasting of the frame containing message m_2 is done in the slot S_0 of *Round* 2. The TTP controller of node N_1 knows from its MEDL that it has to read a frame from slot S_0 of *Round* 2 and to transfer it into its MBI. In order to synchronize with the TTP controller and to read the frame from the MBI, the tick scheduler on node N_1 will activate, based on its local MHTT, a so-called delivery task D. The delivery task takes the frame from the MBI and extracts the messages from it. For the case when a message is split into several packets, sent over several TDMA rounds, we consider that a message has arrived at the destination node after all its constituent packets have arrived. When m_2 has arrived, the delivery task copies it to task τ_3 which will be activated. Activation times for the delivery task are fixed in the MHTT just as explained earlier for the message transfer task.

The number of activations of the message transfer and delivery tasks depends on the number of frames transferred, and it is taken into account in our analysis, as also is the delay implied by the propagation on the communication bus.

15.6.2   Optimization Problem

We model an application as a set of tasks. Each task τ_i is allocated to a certain processor, and has a known worst-case execution time C_i, a period T_i, a deadline D_i and a uniquely assigned priority. We consider a preemptive execution environment, which means that higher priority tasks can interrupt the execution of lower priority tasks. A lower priority task can block a higher priority task (e.g., it is in its critical section), and the blocking time is computed according to the priority ceiling protocol. Tasks exchange messages, and for each message m_i we know its size S_{m_i}. A message is sent once in every n_m invocations of the sending task, with a period $T_m = n_m T_i$ inherited from the sender task τ_i, and has a unique destination task. Each task is allocated to a node of the distributed system and messages are transmitted according to the TTP.

We are interested to synthesize the MEDL of the TTP controllers (and, as a direct consequence, also the MHTTs) so that the task set is schedulable on an as cheap (slow) as possible processor set.

The next section presents the schedulability analysis for each of the four approaches considered for message scheduling, under the assumptions outlined above. In Section 15.6.4, the response times calculated using this schedulability analysis are combined in a cost function that measures the "degree of schedulability" of a given design alternative. This "degree of schedulability" is then used to drive the optimization and synthesis of the MEDL and the MHTTs.

15.6.3 Schedulability Analysis

Under the assumptions presented in the previous section, [330] integrate processor and communication scheduling and provide a "holistic" schedulability analysis in the context of distributed real-time systems with communication based on a simple TDMA protocol. The validity of this analysis has been later confirmed in [244]. The analysis belongs to the class of response time analyses, where the schedulability test is whether the worst-case response time of each task is smaller than or equal to its deadline. In the case of a distributed system, this response time also depends on the communication delay due to messages. In [330] the analysis for messages is done in a similar way as for tasks: A message is seen as an unpreemptable task that is "running" on a bus.

The basic idea in [330] is that the release jitter of a destination task depends on the communication delay between sending and receiving a message. The release jitter of a task is the worst-case delay between the arrival of the task and its release (when it is placed in the run-queue for the processor). The communication delay is the worst-case time spent between sending a message and the message arriving at the destination task.

Thus, for a task $d(m)$ that receives a message m from a sender task $s(m)$, the release jitter is

$$J_{d(m)} = r_{s(m)} + a_m + r_{deliver} + T_{tick} \qquad (15.11)$$

where $r_{s(m)}$ is the response time of the task sending the message, a_m (worst-case arrival time) is the worst-case time needed for message m to arrive at the communication controller of the destination node, $r_{deliver}$ is the response time of the delivery task (see Section 15.6.1) and T_{tick} is the jitter due to the operation of the tick scheduler. The communication delay for a message m (also referred to as the "response time" of message m) is

$$r_m = a_m + r_{deliver} \qquad (15.12)$$

where a_m itself is the sum of the access delay Y_m and the propagation delay X_m. The access delay is the time a message queued at the sending processor spends waiting for the use of the communication channel. In a_m, we also account for the execution time of the message transfer task (see Section 15.6.1). The propagation delay is the time taken for the message to reach the destination processor once physically sent by the corresponding TTP controller. The analysis assumes that the period T_m of any message m is longer than or equal to the length of a TDMA round, $T_m \geq T_{TDMA}$ (see Figure 15.29).

The pessimism of this analysis can be reduced by using the notion of offset in order to model precedence relations between tasks [328]. The basic idea is to exclude certain scenarios which are impossible due to precedence constraints. By considering dynamic offsets, the tightness of the analysis can be further improved [117, 118]. In the present section, our attention is concentrated on the analysis of network communication delays and on optimization of message passing strategies. In order to keep the discussion focused, we present our analysis starting from the results in [330]. All the conclusions of this research apply as well to the developments addressing precedence relations proposed, for example, in [117, 118].

Although there are many similarities with the general TDMA protocol, the analysis in the case of TTP is different in several aspects and also differs to a large degree depending on the policy chosen for message scheduling.

Before going into details for each of the message scheduling approaches proposed by us, we analyze the propagation delay and the message transfer and delivery tasks, as they do not depend on the particular message scheduling policy chosen. The propagation delay X_m of a message m sent as part of a slot S, with the TTP protocol, is equal to the time needed for the slot S to be transferred on the bus (this is the slot size expressed in time units; see Figure 15.29). This time depends on the number of bits which can be packed into the slot and on the features of the underlying bus.

The overhead produced by the communication activities must be accounted not only as part of the access delay for a message, but also through its influence on the response time of tasks running on the same processor. We consider this influence during the schedulability analysis of processes on each processor. We assume that the worst-case computation time of the transfer task (T in Figure 15.16) is known, and that it is different for each of the four message scheduling approaches. Based on the respective MHTT, the transfer task is activated for each frame sent. Its worst-case period is derived from the minimum time between successive frames.

The response time of the delivery task (D in Figure 15.16), $r_{deliver}$, is part of the communication delay (Equation 15.12). The influence due to the delivery task must also be included when analyzing the response time of the tasks running on the respective processor. We consider the delivery task during the schedulability analysis in the same way as the message transfer task.

The response times of the communication and delivery tasks are calculated, as for all other tasks, using the arbitrary deadline analysis from [330].

The four approaches we propose for scheduling of messages using TTP differ in the way the messages are allocated to the communication channel (either statically or dynamically) and whether they are split or not into packets for transmission. The next subsections present the analysis for each approach as well as the degrees of liberty a designer has, in each of the cases, for optimizing the MEDL.

15.6.3.1 Static Single Message Allocation (SM)

The first approach for scheduling messages using TTP is to statically (offline) schedule each of the messages into a slot of the TDMA cycle, corresponding to the node sending the message. This means that for each message we decide offline to allocate

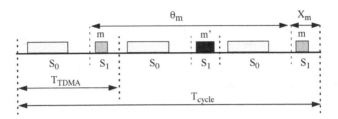

FIGURE 15.29
Worst-Case Arrival Time for SM

space in one or more frames, space that can only be used by that particular message. In Figure 15.29, the frames are denoted by rectangles. In this particular example, it has been decided to allocate space for message m in slot S_1 of the first and third rounds. Since the messages are dynamically produced by the tasks, the exact moment a certain message is generated cannot be predicted. Thus, it can happen that certain frames will be left empty during execution. For example, if there is no message m in the *Out* queue (see Figure 15.29) when the slot S_1 of the first round in Figure 15.29 starts, that frame will carry no information. A message m produced immediately after slot S_1 has left, could then be carried by the frame scheduled in the slot S_1 of the third round.

In the SM approach, we consider that each slot can hold a maximum of one single message. This approach is well suited for application areas, like safety-critical automotive electronics, where the messages are typically short and the ability to easily diagnose the system (fewer messages in a frame are easier to observe) is critical. In the automotive electronics area, messages are typically a couple of bytes, encoding signals like vehicle speed. However, for applications using larger messages, the SM approach leads to overheads due to the inefficient utilization of slot space when transmitting smaller size messages.

As each slot carries only one fixed, predetermined message, there is no interference among messages. If a message m misses its allocated frame, it has to wait for the following slot assigned to m. The worst-case access delay Y_m for a message m in this approach is the maximum time between consecutive slots of the same node carrying the message m. We denote this time by θ_m, illustrated in Figure 15.29, where we have a system cycle of length T_{cycle}, consisting of three TDMA rounds.

In this case, the worst-case arrival time a_m of a message m becomes $\theta_m + X_m$. Therefore, the main aspect influencing schedulability of the messages is the way they are statically allocated to slots, which determines the values of θ_m. θ_m, as well as X_m, depend on the slot sizes which in the case of SM are determined by the size of the largest message sent from the corresponding node plus the bits for control and CRC, as imposed by the protocol.

As mentioned before, the analysis in [330], done for a simple TDMA protocol, assumes that $T_m \geq T_{TDMA}$. In the case of static message allocation with TTP (the SM and MM approaches), this translates to the condition $T_m \geq \theta_m$.

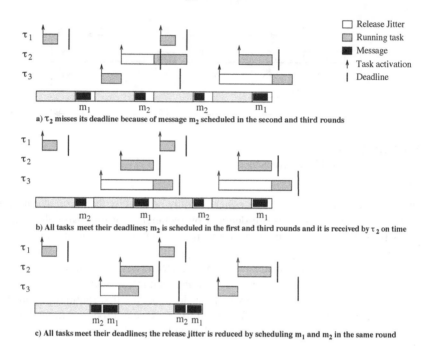

a) τ_2 misses its deadline because of message m_2 scheduled in the second and third rounds

b) All tasks meet their deadlines; m_2 is scheduled in the first and third rounds and it is received by τ_2 on time

c) All tasks meet their deadlines; the release jitter is reduced by scheduling m_1 and m_2 in the same round

FIGURE 15.30
Optimizing the MEDL for SM and MM

During the synthesis of the MEDL, the designer has to allocate the messages to slots in such a way that the task set is schedulable. Since the schedulability of the task set can be influenced by the synthesis of the MEDL only through the θ_m parameters, these are the parameters which have to be optimized.

Let us consider the simple example depicted in Figure 15.30, where we have three tasks, τ_1, τ_2 and τ_3 each running on a different processor. When task τ_1 finishes executing, it sends message m_1 to task τ_3 and message m_2 to task τ_2. In the TDMA configurations presented in Figure 15.30, only the slot corresponding to the CPU running τ_1 is important for our discussion and the other slots are represented with light gray. With the configuration in Figure 15.31a, where the message m_1 is allocated to the rounds 1 and 4 and the message m_2 is allocated to rounds 2 and 3, task τ_2 misses its deadline because of the release jitter due to the message m_2 in *Round* 2. However, if we have the TDMA configuration depicted in Figure 15.30b, where m_1 is allocated to rounds 2 and 4 and m_2 is allocated to rounds 1 and 3, all the tasks meet their deadlines.

15.6.3.2   Static Multiple Message Allocation (MM)

This second approach is an extension of the first one. In this approach, we allow more than one message to be statically assigned to a slot and all the messages transmitted in the same slot are packaged together in a frame. As for the SM approach, there is no interference among messages, so the worst-case access delay for a message m is the maximum time between consecutive slots of the same node carrying the message m, θ_m. It is also assumed that $T_m \geq \theta_m$.

However, this approach offers more freedom during the synthesis of the MEDL. We have now to decide also on how many and which messages should be put in a slot. This allows more flexibility in optimizing the θ_m parameter. To illustrate this, let us consider the same example depicted in Figure 15.31. With the MM approach, the TDMA configuration can be arranged as depicted in Figure 15.30c, where the messages m_1 and m_2 are put together in the same slot in the rounds 1 and 2. Thus, the deadline is met and the release jitter is further reduced compared to the case presented in Figure 15.31b where task τ_3 was experiencing a large release jitter.

15.6.3.3 Dynamic Message Allocation (DM)

The previous two approaches have statically allocated one or more messages to their corresponding slots. This third approach considers that the messages are dynamically allocated to frames, as they are produced.

Thus, when a message is produced by a sender task, it is placed in the *Out* queue (Figure 15.16). Messages are ordered according to their priority. At its activation, the message transfer task takes a certain number of messages from the head of the *Out* queue and constructs the frame. The number of messages accepted is decided so that their total size does not exceed the length of the data field of the frame. This length is limited by the size of the slot corresponding to the respective processor. Since the messages are sent dynamically, we have to identify them in a certain way so that they are recognized when the frame arrives at the delivery task. We consider that each message has several identifier bits appended at the beginning of the message.

Since we dynamically package messages into frames in the order they are sorted in the queue, the access delay to the communication channel for a message m depends on the number of messages queued ahead of it.

The analysis in [330] bounds the number of queued ahead *packets* of messages of higher priority than message m, as in their case it is considered that a message can be split into packets before it is transmitted on the communication channel. We use the same analysis but we have to apply it for the number of *messages* instead of packets. We have to consider that messages can be of different sizes, as opposed to packets which are always of the same size.

Therefore, the total *size* of higher priority messages queued ahead of a message m, in the worst case, is:

$$I_m = \sum_{\forall j \in hp(m)} \left\lceil \frac{r_{s(j)}}{T_j} \right\rceil S_j \tag{15.13}$$

where S_j is the size of the message m_j, $r_{s(j)}$ is the response time of the task sending message m_j and T_j is the period of the message m_j.

Further, we calculate the worst-case time that a message m spends in the *Out* queue. The number of TDMA rounds needed, in the worst case, for a message m

placed in the queue to be removed from the queue for transmission is

$$\left\lceil \frac{S_m + I_m}{S_s} \right\rceil \tag{15.14}$$

where S_m is the size of the message m and S_s is the size of the slot transmitting m (we assume, in the case of DM, that for any message x, $S_x \leq S_S$). This means that the worst-case time a message m spends in the *Out* queue is given by

$$Y_m = \left\lceil \frac{S_m + I_m}{S_s} \right\rceil T_{TDMA} \tag{15.15}$$

where T_{TDMA} is the time taken for a TDMA round.

Since the size of the messages is given with the application, the parameter that will be optimized during the synthesis of the MEDL is the slot size. To illustrate how the slot size influences schedulability, let us consider the example in Figure 15.31 where we have the same setting as for the example in Figure 15.30. The difference is that we consider message m_1 having a higher priority than message m_2 and we schedule the messages dynamically as they are produced. With the configuration in Figure 15.31a, message m_1 will be dynamically scheduled first in the slot of the first round, while message m_2 will wait in the *Out* queue until the next round comes, thus causing task τ_2 to miss its deadline. However, if we enlarge the slot so that it can accommodate both messages, message m_2 does not have to wait in the queue and it is transmitted in the same slot as m_1. Therefore, τ_2 will meet its deadline as presented in Figure 15.31b. However, in general, increasing the length of slots does not necessarily improve schedulability, as it delays the communication of messages generated by other nodes.

15.6.3.4 Dynamic Packet Allocation (DP)

This approach is an extension of the previous one, as we allow the messages to be split into packets before they are transmitted on the communication channel. We consider that each slot has a size that accommodates a frame with the data field being a multiple of the packet size. This approach is well suited for the application areas that typically have large message sizes. By splitting messages into packets, we can obtain a higher utilization of the bus and reduce the release jitter. However, since each packet has to be identified as belonging to a message, and messages have to be split at the sender and reconstructed at the destination, the overhead becomes higher than in the previous approaches.

The worst-case time a message m spends in the *Out* queue is given by the analysis in [330] which is based on similar assumptions as those for this approach:

$$Y_m = \left\lceil \frac{P_m + I_m}{S_p} \right\rceil T_{TDMA} \tag{15.16}$$

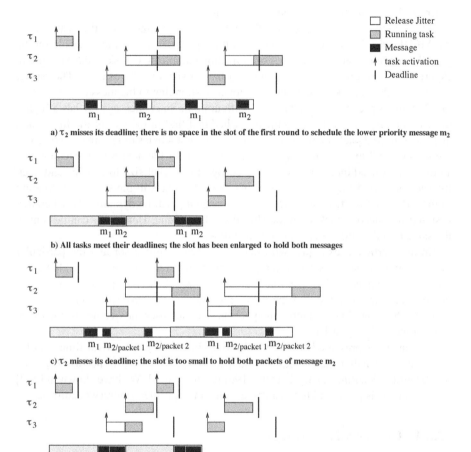

a) τ_2 misses its deadline; there is no space in the slot of the first round to schedule the lower priority message m_2

b) All tasks meet their deadlines; the slot has been enlarged to hold both messages

c) τ_2 misses its deadline; the slot is too small to hold both packets of message m_2

b) All tasks meet their deadlines; the slot has been enlarged to hold 4 packets instead of 3

FIGURE 15.31
Optimizing the MEDL for DM and DP

where p_m is the number of packets of message m, S_p is the size of the slot (in number of packets) corresponding to m and

$$I_m = \sum_{\forall j \in hp(m)} \left\lceil \frac{r_{s(j)}}{T_j} \right\rceil p_j \qquad (15.17)$$

where p_j is the number of packets of a message m_j.

In the previous approach (DM), the optimization parameter for the synthesis of the MEDL was the size of the slots. With this approach, we can also decide on the packet size which becomes another optimization parameter. Consider the example in Figure 15.31c where messages m_1 and m_2 have a size of 6 bytes each. The packet size is considered to be 4 bytes and the slot corresponding to the messages has a size of 12 bytes (3 packets) in the TDMA configuration. Since message m_1 has a higher priority than m_2, it will be dynamically scheduled first in the slot of the first round and it will need 2 packets. In the third packet, the first 4 bytes of m_2 are placed. Thus, the remaining 2 bytes of message m_2 have to wait for the next round, causing task τ_2 to miss its deadline. However, if we change the packet size to 3 bytes and keep the same size of 12 bytes for the slot, we have 4 packets in the slot corresponding to the CPU running τ_1 (Figure 15.31d). Message m_1 will be dynamically scheduled first and will need 2 packets in the slot of the first round. Hence, m_2 can be sent in the same round so that τ_2 can meet its deadline.

In this particular example, with one single sender processor and the particular message and slot sizes as given, the problem seems to be simple. This is, however, not the case in general. For example, the packet size which fits a particular node can be unsuitable in the context of the messages and slot size corresponding to another node. At the same time, reducing the packets size increases the overheads due to the transfer and delivery tasks.

The analysis presented so far is valid only in the case the arrival time a_m of a message m is smaller than or equal to its period T_m. However, in the case $a_m > T_m$ the "arbitrary deadline" analysis from [196] has to be used. We have shown in [262] how the analysis presented here can be extended to consider arbitrary deadlines.

15.6.4 Optimization Strategy

Our problem is to analyze the schedulability of a given task set and to synthesize the MEDL of the TTP controllers (and consequently the MHTTs) so that the task set is schedulable on an as cheap as possible architecture. The optimization is performed on the parameters which have been identified for each of the four approaches to message scheduling discussed before. In order to guide the optimization task, we need a cost function that captures the "degree of schedulability" for a certain MEDL implementation. Our cost function is similar to that in [329] in the case an application is not schedulable (f_1). However, in order to distinguish between several schedulable applications, we have introduced the second expression, f_2, which measures, for a feasible design alternative, the total difference between the response times and the deadlines:

$$cost(optimization\ parameters) = \begin{cases} f_1 = \sum\limits_{i=1}^{n} \max(0, R_i - D_i), \text{if } f_1 > 0 \\ f_2 = \sum\limits_{i=1}^{n} R_i - D_i), \text{if } f_1 = 0 \end{cases} \tag{15.18}$$

where n is the number of tasks in the application, R_i is the response time of a task τ_i and D_i is the deadline of a task τ_i. If the task set is not schedulable, there exists at least one R_i that is greater than the deadline D_i; therefore, the term f_1 of the function will be positive. In this case, the cost function is equal to f_1. However, if the task set is schedulable, then all R_i are smaller than the corresponding deadlines D_i. In this case, $f_1 = 0$ and we use f_2 as the cost function, as it is able to differentiate between two alternatives, both leading to a schedulable task set. For a given set of optimization parameters leading to a schedulable task set, a smaller f_2 means that we have improved the response times of the tasks, so the application can be potentially implemented on a cheaper hardware architecture (with slower processors and/or bus, but without increasing the number of processors or buses).

The response time R_i is calculated according to the arbitrary deadline analysis [330] based on the release jitter of the tasks (see Section 15.6.3), its worst-case execution time, the blocking time, and the interference time due to higher priority tasks. They form a set of mutually dependent equations which can be solved iteratively. As shown in [330], a solution can be found if the processor utilization is less than 100%.

For a given application, we are interested to synthesize a MEDL such that the cost function is minimized. We are also interested to evaluate in different contexts the four approaches to message scheduling, thus offering the designer a decision support for choosing the approach that best fits his application.

The MEDL synthesis problem belongs to the class of exponential complexity problems; therefore, we are interested to develop heuristics that are able to find accurate results in a reasonable time. We have developed optimization algorithms corresponding to each of the four approaches to message scheduling. A first set of algorithms presented in Section 15.6.4.1 is based on simple and fast greedy heuristics. In Section 15.6.4.2, we introduce a second class of heuristics which aims at finding near-optimal solutions using the simulated annealing (SA) algorithm.

15.6.4.1 Greedy Heuristics

We have developed greedy heuristics for each of the four approaches to message scheduling. The main idea of the heuristics is to minimize the cost function by incrementally trying to reduce the communication delay of messages and, by this, the release jitter of the tasks.

The only way to reduce the release jitter in the SM and MM approaches is through the optimization of the θ_m parameters. This is achieved by a proper placement of messages into slots (see Figure 15.30).

The `OptimizeSM` algorithm presented in Figure 15.32 starts by deciding on a size ($size_{S_i}$) for each of the slots. The slot sizes are set to the minimum size that

OptimizeSM
```
01      -- set the slot sizes
02      for each node Nᵢ do
03          sizeₛᵢ = max(size of messages mⱼ sent by node Nᵢ)
04      end for
05      -- find the min. no. of rounds that can hold all the messages
06      for each node Nᵢ do
07          nmᵢ = number of messages sent from Nᵢ
08      end for
09      MinRounds = max (nmᵢ)
10      -- create a minimal complete MEDL
11      for each message mᵢ
12          find round in [1..MinRounds] that has an empty slot for mᵢ
13          place mᵢ into its slot in round
14      end for
15      for each RoundsNo in [MinRounds...MaxRounds] do
16          -- insert messages in such a way that the cost is minimized
17          repeat
18              for each task Pⱼ that receives a message mᵢ do
19                  if Dⱼ - Rⱼ is the smallest so far then m = mₚⱼ end if
20              end for
21              for each round in [1..RoundsNo] do
22                  place m into its corresponding slot in round
23                  calculate the CostFunction
24                  if the CostFunction is smallest so far then
25                      BestRound = round
26                  end if
27                  remove m from its slot in round
28              end for
29              place m into its slot in BestRound if one was identified
30          until the CostFunction is not improved
31      end for
    end OptimizeSM
```

FIGURE 15.32
Greedy Heuristic for SM

can accommodate the largest message sent by the corresponding node (lines 1–4 in Figure 15.32). In this approach, a slot can carry at most one message; thus, slot sizes larger than this size would lead to larger response times.

Then, the algorithm has to decide on the number of rounds, thus determining the size of the MEDL. Since the size of the MEDL is physically limited, there is a limit to the number of rounds (e.g., 2, 4, 8, 16 depending on the particular TTP controller implementation). However, there is a minimum number of rounds *MinRounds* that is necessary for a certain application, which depends on the number of messages transmitted (lines 5–9). For example, if the tasks mapped on node N_0 send in total seven messages then we have to decide on at least seven rounds in order to accommodate all of them (in the SM approach there is at most one message per slot). Several numbers of rounds, *RoundsNo*, are tried out by the algorithm starting from *MinRounds* up to *MaxRounds* (lines 15–31).

For a given number of rounds (that determine the size of the MEDL), the initially empty MEDL has to be populated with messages in such a way that the cost function is minimized. In order to apply the schedulability analysis that is the basis for the cost function, a *complete* MEDL has to be provided. A complete MEDL contains at least one instance of every message that has to be transmitted between the tasks on different processors. A *minimal complete* MEDL is constructed from an empty MEDL by placing one instance of every message m_i into its corresponding empty slot of a round (lines 10–14). In Figure 15.30a, for example, we have a MEDL composed of four rounds. We get a minimal complete MEDL, for example, by assigning m_2 and m_1 to the slots in rounds 3 and 4, and leaving the slots in rounds 1 and 2 empty. However, such a MEDL might not lead to a schedulable system. The "degree of schedulability" can be improved by inserting instances of messages into the available places in the MEDL, thus minimizing the θ_m parameters. For example, in Figure 15.30a inserting another instance of the message m_1 in the first round and m_2 in the second round leads to τ_2 missing its deadline, while in Figure 15.30b inserting m_1 into the second round and m_2 into the first round leads to a schedulable system.

Our algorithm repeatedly adds a new instance of a message to the current MEDL in the hope that the cost function will be improved (lines 16–30). In order to decide an instance of which message should be added to the current MEDL, a simple heuristic is used. We identify the task τ_i which has the most "critical" situation, meaning that the difference between its deadline and response time, $D_i - R_i$, is minimal compared with all other tasks. The message to be added to the MEDL is the message $m = m_{P_i}$ received by the task τ_i (lines 18–20). Message m will be placed into that round (*BestRound*) which corresponds to the smallest value of the cost function (lines 21–28). The algorithm stops if the cost function cannot be further improved by adding more messages to the MEDL.

The `OptimizeMM` algorithm is similar to `OptimizeSM`. The main difference is that in the MM approach several messages can be placed into a slot (which also decides its size), while in the SM approach there can be at most one message per slot. Also, in the case of MM, we have to take additional care that the slots do not exceed the maximum allowed size for a slot.

OptimizeDM

```
01    for each node N_i do
02        MinSize_Si = max(size of messages m_j sent by node N_i)
03    end for
04    -- identifies the size that minimizes the cost function
05    for each slot S_i
06        BestSize_Si = MinSize_Si
07        for each SlotSize in [MinSize_Si...MaxSize] do
08            calculate the CostFunction
09            if the CostFunction is best so far then
10                BestSize_Si = SlotSize_Si
11            end if
12        end for
13        size_Si = BestSize_Si
14    end for
  end OptimizeDM
```

FIGURE 15.33
Greedy Heuristic for DM

The situation is simpler for the dynamic approaches, namely DM and DP, since we only have to decide on the slot sizes and, in the case of DP, on the packet size. For these two approaches, the placement of messages is dynamic and has no influence on the cost function. The `OptimizeDM` algorithm (see Figure 15.33) starts with the first slot $S_i = S_0$ of the TDMA round and tries to find that size ($BestSize_{S_i}$) which corresponds to the smallest $CostFunction$ (lines 4–14 in Figure 15.33). This slot size has to be large enough ($S_i \geq MinSize_{S_i}$) to hold the largest message to be transmitted in this slot, and within bounds determined by the particular TTP controller implementation (e.g., from 2 bits up to $MaxSize = 32$ bytes). Once the size of the first slot has been determined, the algorithm continues in the same manner with the next slots (lines 7–12).

The `OptimizeDP` algorithm has also to determine the proper packet size. This is done by trying all the possible packet sizes given the particular TTP controller. For example, it can start from 2 bits and increment with the "smallest data unit" (typically 2 bits) up to 32 bytes. In the case of the `OptimizeDP` algorithm, the slot size has to be determined as a multiple of the packet size and within certain bounds depending on the TTP controller.

15.6.4.2 Simulated Annealing Strategy

We have also developed an optimization procedure based on a simulated annealing (SA) strategy. The main characteristic of such a strategy is that it tries to find the global optimum by randomly selecting a new solution from the neighbors of the current solution. The new solution is accepted if it is an improved one. However, a worse

SimulatedAnnealing

```
01      construct an initial TDMA round x^now
02      temperature = initial temperature TI
03      repeat
04          for i = 1 to temperature length TL
05              generate randomly a neighboring solution x' of x^now
06              delta = CostFunction(x') - CostFunction(x^now)
07              if delta < 0 then x^now = x'
08              else
09                  generate q = random (0, 1)
10                  if q < e^{-delta / temperature} then x^now = x' end if
11              end if
12          end for
13          temperature = α * temperature
14      until stopping criterion is met
15      return solution corresponding to the best CostFunction
        end SimulatedAnnealing
```

FIGURE 15.34
The Simulated Annealing Strategy

solution can also be accepted with a certain probability that depends on the deterioration of the cost function and on a control parameter called temperature [275].

In Figure 15.34, we give a short description of this algorithm. An essential component of the algorithm is the generation of a new solution x starting from the current one x^{now} (line 5 in Figure 15.34). The neighbors of the current solution x^{now} are obtained depending on the chosen message scheduling approach. For SM, x is obtained from x^{now} by inserting or removing a message in one of its corresponding slots. In the case of MM, we have to take additional care that the slots do not exceed the maximum allowed size (which depends on the controller implementation), as we can allocate several messages to a slot. For these two static approaches, we also decide on the number of rounds in a cycle (e.g., 2, 4, 8, 16; limited by the size of the memory implementing the MEDL). In the case of DM, the neighboring solution is obtained by increasing or decreasing the slot size within the bounds allowed by the particular TTP controller implementation, while in the DP approach we also increase or decrease the packet size.

For the implementation of this algorithm, the parameters TI (initial temperature), TL (temperature length), α (cooling ratio) and the stopping criterion have to be determined. They define the so called cooling schedule and have a strong impact on the quality of the solutions and the CPU time consumed. We were interested to obtain values for TI, TL and α that will guarantee the finding of good quality solutions in a short time. In order to tune the parameters, we have first performed very long and expensive runs on selected large examples and the best ever solution, for each example, has been considered as the near-optimum. Based on further experiments, we have determined the parameters of the SA algorithm, for different sizes of examples, so that the optimization time is reduced as much as possible but the near-optimal result is still produced. These parameters have then been used for the large-scale experiments presented in the following section. For example, for the graphs with 320 nodes, TI is 300, TL is 500 and α is 0.95. The algorithm stops if for three consecutive temperatures no new solution has been accepted.

15.6.5   Experimental Results

For evaluation of our approaches, we first used sets of tasks generated for experimental purposes. We considered architectures consisting of 2, 4, 6, 8 and 10 nodes. Forty tasks were assigned to each node, resulting in sets of 80, 160, 240, 320 and 400 tasks. Thirty tasks sets were generated for each of the five dimensions. Thus, a total of 150 sets of tasks were used for experimental evaluation. Worst-case computation times, periods, deadlines and message lengths were assigned randomly within certain intervals. For the communication channel, we considered a transmission speed of 256 kbps. The maximum length of the data field in a slot was 32 bytes and the frequency of the TTP controller was chosen to be 20 MHz. All experiments were run on a Sun Ultra 10 workstation.

For each of the 150 generated examples and each of the four message scheduling approaches, we have obtained the near-optimal values for the cost function (Equation 15.18) as produced by our SA based algorithm (see Section 15.6.4.2). For a given example, these values might differ from one message passing approach to another, as they depend on the optimization parameters and the schedulability analysis which are particular for each approach. Figure 15.35 presents a comparison based on the average percentage deviation of the cost function obtained for each of the four approaches, from the minimal value among them. The percentage deviation is calculated according to the formula:

$$deviation = \frac{cost_{approach} - cost_{best}}{cost_{best}} \times 100. \qquad (15.19)$$

The DP approach is, generally, able to achieve the highest degree of schedulability, which in Figure 15.35 translates in the smallest deviation. In the case the packet size is properly selected, by scheduling messages dynamically we are able to efficiently use the available space in the slots, and thus reduce the release jitter. However, by using the MM approach we can obtain almost the same result if the messages are carefully allocated to slots as does our optimization strategy.

Moreover, in the case of larger task sets, the static approaches suffer significantly less overhead than the dynamic approaches. In the SM and MM approaches, the messages are uniquely identified by their position in the MEDL. However, for the dynamic approaches we have to somehow identify the dynamically transmitted messages and packets. Thus, for the DM approach we consider that each message has several identifier bits appended at the beginning of the message, while for the DP approach the identification bits are appended to each packet. Not only do the identifier bits add to the overhead, but in the DP approach, the transfer and delivery tasks

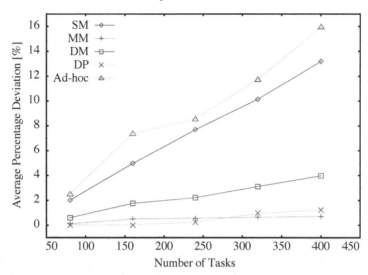

FIGURE 15.35
Comparison of the Four Approaches to Message Scheduling

(see Figure 15.16) have to be activated at each sending and receiving of a packet, and thus interfere with the other tasks. Thus, for larger applications (e.g., task sets of 400 tasks), MM outperforms DP, as DP suffers from large overhead due to its dynamic nature. DM performs worse than DP because it does not split the messages into packets, and this results in a mismatch between the size of the messages dynamically queued and the slot size, leading to unused slot space that increases the jitter. SM performs the worst as it does not permit much room for improvement, leading to large amounts of unused slot space. Also, DP has produced a MEDL that resulted in schedulable task sets for 1.33 times more cases than the MM and DM. MM, in its turn, produced two times more schedulable results than the SM approach.

Together with the four approaches to message scheduling, a so-called ad-hoc approach is presented. The ad-hoc approach performs scheduling of messages without trying to optimize the access to the communication channel. The ad-hoc solutions are based on the MM approach and consider a design with the TDMA configuration consisting of a simple, straightforward allocation of messages to slots. The lengths of the slots were selected to accommodate the largest message sent from the respective node. Figure 15.35 shows that the ad-hoc alternative is constantly outperformed by any of the optimized solutions. This demonstrates that significant gains can be obtained by optimization of the parameters defining the access to the communication channel.

Next, we have compared the four approaches with respect to the number of messages exchanged between different nodes and the maximum message size allowed. For the results depicted in Figures 15.36 and 15.37, we have assumed sets of 80 tasks allocated to four nodes. Figure 15.36 shows that, as the number of messages in-

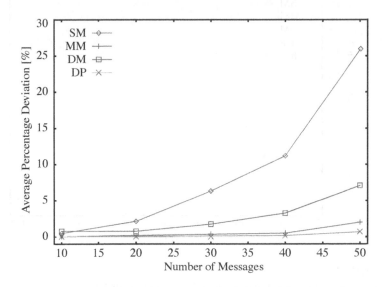

FIGURE 15.36
Four Approaches to Message Scheduling: The Influence of the Number of Messages

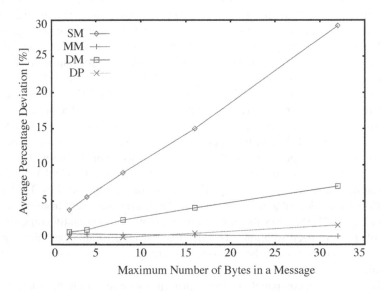

FIGURE 15.37
Four Approaches to Message Scheduling: The Influence of the Message Sizes

TABLE 15.2
Percentage deviations for the greedy heuristics compared to SA.

		80 tasks	160 tasks	240 tasks	320 tasks	400 tasks
SM	avg.	0.12%	0.19%	0.50%	1.06%	1.63%
	max.	0.81%	2.28%	8.31%	31.05%	18.00%
MM	avg.	0.05%	0.04%	0.08%	0.23%	0.36%
	max.	0.23%	0.55%	1.03%	8.15%	6.63%
DM	avg.	0.02%	0.03%	0.05%	0.06%	0.07%
	max.	0.05%	0.22%	0.81%	1.67%	1.01%
DP	avg.	0.01%	0.01%	0.05%	0.04%	0.03%
	max.	0.05%	0.13%	0.61%	1.42%	0.54%

creases, the difference between the approaches grows while the ranking among them remains the same. The same holds for the case when we increase the maximum allowed message size (Figure 15.37), with a notable exception: For large message sizes MM becomes better than DP, since DP suffers from the overhead due to its dynamic nature.

We were also interested in the quality of our greedy heuristics. Thus, we have run all the examples presented above using the greedy heuristics and compared the results with those produced by the SA based algorithm. Table 15.2 shows the average and maximum percentage deviations of the cost function values produced by the greedy heuristics from those generated with SA, for each of the graph dimensions. All four greedy heuristics perform very well, with less than 2% loss in quality compared to the results produced by the SA algorithms. The execution times for the greedy heuristics were more than two orders of magnitude smaller than those with SA. Although the greedy heuristics can potentially find solutions not found by SA, for our experiments, the extensive runs performed with SA have led to a design space exploration that has included all the solutions produced by the greedy heuristics.

The above comparison between the four message scheduling alternatives is mainly based on the issue of schedulability. However, when choosing among the different policies, several other parameters can be of importance. Thus, a static allocation of messages can be beneficial from the point of view of testing and debugging and has the advantage of simplicity. Similar considerations can lead to the decision not to split messages. In any case, however, optimization of the bus access scheme is highly desirable.

15.7   Configuration and Code Generation

Once the schedule has been created as described in Section 15.3, it is necessary to transform this schedule information into a device-specific configuration, so that the dedicated communication controller of the device knows what to do when. In TTP, this configuration is called Message Descriptor List (MEDL); different terms are used in other protocols. For brevity, we call it *communication configuration* throughout this section. The creation of such a communication configuration is described below in Section 15.7.1.

While the communication configuration is the most obvious configuration item, other parts of the system also need to be configured to be able to *process* it:

1. Middleware

   - COM layer
   - Potentially other layers, in case of a multilayer system (e.g., the AUTOSAR Basic Software Stack [19])

2. Application

3. Operating system (OS), if applicable

The creation of middleware configurations is described in Section 15.7.2. In addition, it is also possible (and often advantageous) to even generate the complete code of the middleware itself. This approach is discussed in Section 15.7.3. The application also needs some knowledge of the transmitted data, its structures and timing, and therefore requires a dedicated configuration for this specific purpose. If an operating system (OS) exists, it is also involved in the communication, and consequently also needs a configuration for its specific tasks. For brevity, all configurations needed in addition to the aforementioned communication and middleware configuration are called *third-party configurations* throughout this section. The creation of third-party configurations is described in Section 15.7.4.

15.7.1   Communication Configuration

The specific format and content of a communication configuration is hardware dependent. Each communication controller provides some specific features, and these features need to be configured correctly in order to bring the communication controller to work and interact with the other communication controllers on the network.

But not only differences in the hardware — or, more precisely, in the communication controllers — make it necessary to adapt a communication configuration on a per-node basis. Often, hardware buffers in the communication controller are very limited, but user requirements exist to provide the received frame at least for a certain amount of time (validity time span). One solution might be to copy all received frames from the hardware buffer to another location (e.g., an external RAM). But this solution is inefficient regarding execution time and resource usage. A better way is to only put those frames into buffers that are really needed by the specific host.

15.7.1.1   TTP — Personalized MEDLs

The cluster design defines the layout of rounds and cluster cycles, cluster modes, and the parameters required for clock synchronization, i.e., who *sends* what at what time.

It does not contain node-local information about the application data storage in the CNI of individual nodes. Each communication controller must have a personalized MEDL, which is derived from the cluster design. It contains node-local information and may contain special setup data required for internal purposes of specific communication controllers [340].

To optimize the CNI layout, a tool that has the node-local information, in particular the information about which messages a node *receives*, can customize the "abstract" MEDL and thus save execution time and buffer space: Only those messages really needed by the node are processed, stored and provided to upper layers and the application. Personalized MEDLs not only imply less processing work for the CPU that accesses the communication controller, but they also allow for a less strict timing of the tasks on that CPU. In addition, personalized MEDLs are usually smaller than "abstract" ones.

15.7.1.2 Monitor MEDL for TTP

However, one special node-level MEDL is created whenever MEDLs are made by the cluster design tool ^{TTP}Plan: The Monitor MEDL. This MEDL is generated automatically right after scheduling, and is loaded into the communication controller of the Monitoring Node used for monitoring a TTP network. The Monitor MEDL has a special CNI message area layout that is required by the host software operating within the Monitoring Node. The node-level information of the Monitor MEDL does not interfere with node-level designs of the cluster; however, changes to the cluster design render the Monitor MEDL invalid.

15.7.1.3 Buffer Configuration for FlexRay

FlexRay controllers have configurable hardware buffers where data is written to and read from. In the AUTOSAR stack, this concept is abstracted toward the upper layers of the system: The FlexRay driver translates the hardware-specific (i.e., controller-related) information into the more abstract data of the upper software layers. For example, the FlexRay driver maps the information "which frame shall be received" to the corresponding registers of the FlexRay controller. In contrast to the CNI, which is available in TTP controllers, this buffer interface requires that the communication configuration is personalized, i.e., optimized with node-level information.

One part of the driver is the buffer configuration, which places each frame into its hardware buffer. Configuring the FlexRay driver thus generates the meta-level specification of what happens in the cluster. This requires the introduction of logical buffers, which are also known as "L-PDUs" in AUTOSAR. Such a buffer contains one — but not necessarily always the same — frame at any point in time. In FlexRay, there can be several configurations for a buffer, and even reconfiguration during runtime is possible. AUTOSAR, however, supports only one configuration per buffer. Depending on the type of controller, one such buffer corresponds to one or more hardware buffers (mapping in the generated code).

In FlexRay and AUTOSAR, PDUs (*Protocol Data Units*) are the central elements of data transmission. A PDU is a payload of information to be exchanged between different software layers on the node. In AUTOSAR, signals are not placed directly in frames, but in PDUs, which are handled by the PDU Router, see below.

15.7.2 Middleware Configuration

Once the hardware is configured, it is also necessary to configure the "upper" layers of the communication stack. While there may be other parts of middleware software which do not belong to the communication stack, in most systems the communication stack forms the largest and also most complex part. For example, in AUTOSAR [20], the communication stack consists of at least four, but up to seven, layers for a communication based on FlexRay:

- FlexRay Driver

- FlexRay Interface (FrIf)

- PDU Router

- COM Layer

- FlexRay NM (Network Management)

- FlexRay Transport Layer

- RTE (Run-Time Environment)

While some layers do not have many configuration parameters and thus are rather straightforward to configure, other layers — like the FlexRay Interface (FrIf) layer — imply the scheduling of send and receive tasks with respect to the timing and the validity span of the messages sent and received. As a representative of a rather complex layer, the FrIf layer is described in more detail in Section 15.7.2.2 below.

Another example are the communication layers for TTP. They directly access the TTP controller and provide an interface to the application. Figure 15.38 shows their architectural differences. Table 15.3 lists the main similarities and differences between these communication layers.

In contrast to the other layers listed there, the *fault-tolerant COM layer* (FT-COM) is completely generated by the ^{TTP}Build design tool in order to optimize execution time and resource consumption. It operates closely together with TTTech's operating system ^{TTP}OS. It supports packing and unpacking, reintegration (history state handling), byte order (endianness) handling, message agreement functions and handling of replicated redundant message instances. The FT-COM layer is described in more detail in Section 15.7.3.

The *table-driven COM layer* (TD-COM), the *hardware COM layer* (HW-COM) and the *high-speed COM layer* (HS-COM) are reusable engines that execute configuration tables generated by the design tool. These configuration tables define the messages that are sent and received by a specific node, and how to process them.

Both the HW-COM and the HS-COM layer decouple the TTP communication from the application functions, also in the time domain. They provide convenient,

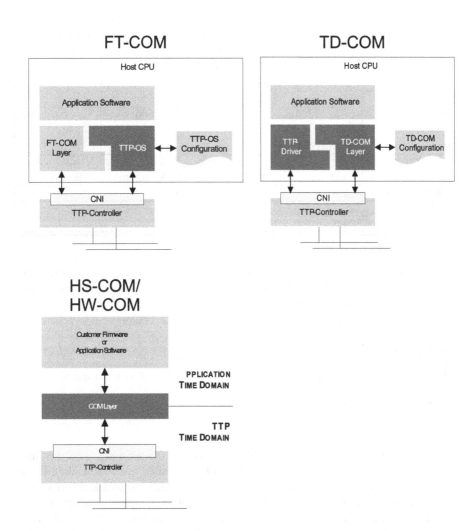

FIGURE 15.38
Examples of Different COM Layers

TABLE 15.3
COM layer properties compared.

Layer	FT-COM	TD-COM	HW-COM	HS-COM
Performance	++	+	+++	+++
Certification	none	DO-178B, level A certification for engine, verification tool for tables	DO-254 certification for IP model	DO-254 certification for IP model
Message sizes	1 to 32 bit, arrays, structured types	1 to 32 bit	32 bit only	32, 64, and 128 bit
Implementation	generated C code	C code, table-driven	VHDL code, table-driven	VHDL code, table-driven
Replication	yes	no	no	no
CPU Load	yes	yes	no	no
Asynchronous Access		yes	yes	yes

buffer-based interfaces to the application software. Their buffer interface allows for an easy mapping of ARINC 429 [10]. In addition, they are rather limited in their functionality as compared to the other layers presented. As a representative, the HS-COM layer is described in more detail in Section 15.7.2.3 below.

15.7.2.1 Configuration Format

Basically, there are two approaches to creating a middleware configuration:

- Source code, usually in C

- A binary block (memory area)

The C code actually comprises a big data structure, either a `struct` or simply an array, or any combination thereof. It might be generated just as a header file that is included in the main application code. In this case, it is automatically employed whenever the application is built. Otherwise it must be compiled and linked to the application in a separate step. As compilations are mostly done based on a Makefile, an additional file to be compiled is acceptable. The header file, which declares the data types used for the configuration structure in the C file, can be kept rather short.

An example is shown in Figure 15.41, representing a configuration for the HS-COM. Apart from the usual content of a C header file, it contains the declaration of the length of the configuration array and the array itself. The HS-COM configuration consists of 32-bit values only because they exactly match the size of an internal data access. This contributes to the high performance of the HS-COM. The comments in the table show the table index of the respective entry for easier navigation. More elaborate comments could be added if found beneficial, e.g., briefly describing each configuration parameter.

The advantages of the C code approach include the better readability and the fact that — due to prior compilation — only one file is present at runtime, which simplifies configuration management. If the configuration is not analyzed by a verification tool (see Section 15.8), good readability and means for easy navigation inside the (sometimes quite big) data structure can reduce certification efforts dramatically.

A binary block contains the configuration data in a structured form, so that the middleware directly and efficiently can access the individual parameters. It is interpreted by the middleware at runtime. Actually, the result of a compiled C code and a binary block may not differ at all for a certain configuration.

The advantages of a binary block include that it can be loaded separately from the application. If the development lifecycles of the application and the communication system are very different, or decoupling these two development tasks is advantageous for other reasons, the configuration can be generated and integrated into the system independently. A binary block needs to be loaded by the application and handed over to the middlware layer during the initialization phase.

15.7.2.2 FlexRay Interface Configuration

The FlexRay Interface (FrIf) layer is the part of the AUTOSAR communication stack that provides access to the FlexRay bus and its timing via the FlexRay Driver layer. Above the FrIf layer, there are the upper layers: PDU-Router (PduR) and FlexRay Transport Protocol (FrTp). The FrIf layer performs its actions according to the generated configuration. It is responsible for two basic tasks:

- It collects PDUs from the upper layers, packs the PDUs into frames and forwards the frames to the driver layer for sending on the FlexRay bus.

- It collects frames from the driver layer, unpacks the PDUs from the frames and forwards the PDUs to the corresponding upper layers (PduR or FrTp).

As can be seen from these characteristics, the FrIf appears PDU-based to the upper layers, but accesses the FlexRay bus in a frame-based fashion.

FrIf Actions

Receiving a frame starts when the FrIf receives the frame from the driver. The PDUs in the frame are unpacked, and the PDU data is passed to the corresponding upper layer (PduR or FrTp). This is done by calling the upper layer's respective API function, called *RxIndication* (receive indication). With this function, the PDU data is passed to the upper layer. After all PDUs have been processed, the frame reception is finished. Sending a frame starts with an upper layer (wanting to send a PDU) issuing a transmit request to the FrIf by calling the *FrIf_Transmit* API function. The FrIf stores every transmission request. It is important to note that a transmission request can occur at any point in the cluster cycle, unless the application is programmed to run synchronously with the FlexRay bus.

Later, when a frame is about to be transmitted, the FrIf checks each PDU in the frame, to see if its transmission has been requested. This point in time is determined during scheduling and can be influenced through the use of some of the advanced scheduling features described later in this chapter. For each PDU, the FrIf gets the PDU data that should be sent, packs the data into the frame and then sends the frame on to the FlexRay bus.

At some even later point in time, the FrIf confirms to the upper layer the transmission of each PDU by calling the *TxConfirmation* function. Again, this point in time is determined during scheduling. Through the use of this function, the upper layer can determine that a PDU has been sent.

For brevity, the receiving, sending and confirmation of a frame by the FrIf will in the following be referred to as *Actions*.

FrIf Job Handling

The sending and receiving of frames has to take place at predefined points in time as FlexRay is a time-triggered communication system. The timing is important for the following reasons:

- A received frame is only available for a limited time at the driver layer. If the FrIf misses the time window for getting the frame from the driver, the data of the frame might already have been overwritten and the frame data is lost. Note that the exact behavior in this situation is subject to the configuration, usage and number of the available buffers.

- If a frame is sent too late by the FrIf, the reserved bandwidth slot of the frame has already been transmitted by the driver, thus the current frame data cannot be sent. Depending on the setting of the corresponding parameter, the FlexRay controller sends either a Null frame or the current data from the frame buffer (which might be outdated).

The handling of actions at predefined points in time is implemented in the TTX-AUTOSAR FlexRay Stack by a hardware timer of the FlexRay module, which generates an interrupt each time a list of actions should be processed. A design tool with FrIf scheduling capability is responsible for calculating the timing of the actions. The output of the FrIf scheduler is called the FrIf schedule; it controls when an interrupt should occur, and which actions should be handled in a particular interrupt invocation. By accessing the compiled schedule, the FrIf layer coordinates its actions.

The main part of the schedule is the *JobList*, which is a collection of *Jobs*. There is only one JobList in the schedule. Each *Job* in turn is a collection of *Actions*; an action has an *action_type*, which can be either "*rx_frame*," "*tx_frame*" or "*tx_confirm*." The actions have already been described in the previous section.

A job stands for an invocation of the FlexRay interrupt on the target hardware. On the invocation of a particular interrupt, all the actions of the associated job are processed by the FrIf layer. The *job_activation_time* describes when the job's associated interrupt has to occur. The processing of jobs is done in the *FrIf_JobListExec*

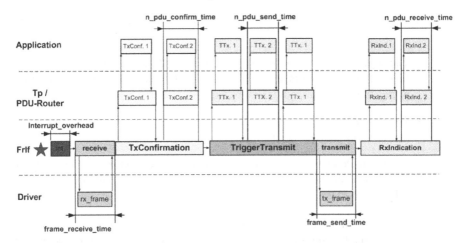

FIGURE 15.39
Sending and Receiving on FrIf Level

API function. This function has to be called in the interrupt service routine of the FlexRay interrupt. Figure 15.39 shows an example of a job and its actions.

Interrupt Overhead

The *activation_time* of the job is marked by a star in Figure 15.39. The delay between the *activation_time* and the actual processing of the first action (*rx_frame* in this case) is the *interrupt_overhead*.

The *interrupt_overhead* results from the fact that it takes some time for the CPU to get from the interrupt event into the *FrIf_JobListExec* function for the processing of the first action. Usually this time is very short. However, this is not always the case. Assume that an application needs to disable interrupts for a certain length of time, let's say $10\mu s$. If a FrIf interrupt occurs during this phase, the *FrIf_JobListExec* function is in the worst case processed after $10\mu s$ at the earliest, thus the *interrupt_overhead* needs to be configured accordingly.

Frame and Application Times

In order to put the FrIf actions into the FrIf jobs and to calculate the point in time for the interrupts, the time needed by each action must be known:

- The *frame_receive_time* is the time it takes the FrIf to receive a frame from the driver. It is defined as the time difference from the calling of the *frame_receive* function until this function returns.

- The *n_pdu_receive_time* is the time the FrIf needs to call the *RxIndication* function of the upper layer. The *RxIndication* function then passes the data to the upper layer.

- The *frame_send_time* is the time the FrIf needs to send one frame to the driver layer.

- The *n_pdu_send_time* is the time the FrIf needs to call the *TriggerTransmit* function of the upper layer. With the *TriggerTransmit* function, the upper layer passes the PDU data to be transmitted to the FrIf.

- The *n_pdu_confirm_time* is the time the FrIf needs to call the *TxConfirmation* function of the upper layer.

- The *frame_confirm_time* is not shown in the figure; it can be used to account for a constant overhead, which occurs during the processing of the *TxConfirmation* calls for all PDUs in the frame.

Using these definitions, a scheduler can calculate exactly how long the processing of a job will take (by summing up the *action times* for all *actions*). In the example from Figure 15.39, the execution time of the job can be computed with the following formula:

$$
\begin{aligned}
Duration \quad = \quad & interrupt\_overhead \\
+ \quad & frame\_receive\_time \\
+ \quad & (2 * n\_pdu\_receive\_time) \\
+ \quad & frame\_send\_time \quad\quad\quad (15.20) \\
+ \quad & (3 * n\_pdu\_send\_time) \\
+ \quad & frame\_confirm\_time \\
+ \quad & (2 * n\_pdu\_confirm\_time)
\end{aligned}
$$

Please note that this is the *worst-case execution time (WCET)* of the job. It may happen that the actual execution time for some invocations of this job on the hardware target is shorter; for example, if a received frame contains some PDUs which were not updated by the sending ECU. Then the FrIf does not need to call the *RxIndication* function for these PDUs, which results in a shorter runtime for this particular job invocation.

Figure 15.40 shows parts of a configuration for the FrIf layer. The major parts of the FrIf configuration are the definition of the PDUs as shown in the upper part of the figure, and the definition of the actions as shown in the lower part. The list of actions in this example contains 33 entries. Each entry specifies the type of the action, a reference to the frame, and a reference to the PDU. Further parts of the FrIf configuration (not shown) are the frame definitions, the definitions of the FrIf Jobs and JobLists, the action timing and the definitions of all used constants.

FrIf Schedulers

A FrIf scheduler may provide the user with advanced configuration options, such as "black-list" and "white-list" scheduling.

Black-list scheduling allows the user to specify *reserved_intervals* where no FrIf jobs may be scheduled. The intervals to be excluded from FrIf activity can be represented as a comma-separated list of ranges in microseconds. For example, setting

PDU Definitions

```
const ttx_frame_to_pdu_t _ttx_frame_to_frif_pdu_v_frame_0002_s [1] =
  { { PDU_ID_FRIF_fl_pdu_measure /* pdu_idx  */
    , 0  /* pdu_offset */
    , 8  /* pdu_len */
    , 1  /* use_update */
    , 17 /* updbit_bytepos */
    , 7  /* updbit_bitpos */
    , 0  /* is_tp_pdu */
    , PDU_ID_ROUTER_fl_pdu_measure /* destination_pdu_id */
    } /* [0] */
  };
```

FrIf Action Definitions

```
const ttx_frame_action_t _ttx_frame_action [33] =
  { { TTX_RX_AND_INDICATE /* action */
    , 1 /* frame_idx */
    , 18 /* mb_tutorial_web_018_a_r */ /* fr_pdu_id  */
    } /* [0] */
  , { TTX_TX_FRAME /* action */
    , 0 /* frame_idx */
    , 1 /* mb_tutorial_web_001_a_t */ /* fr_pdu_id */
    } /* [1] */
...
```

FIGURE 15.40
FrIf Configuration as C code — An Example

the *reserved_intervals* to 0:100,5000:5100,10000:10100,15000:15100 means that FrIf jobs may not be scheduled during the first $100\mu s$ of the first four communication cycles, assuming a cluster cycle of 20*ms*.

White-list scheduling provides the possibility to manually configure time intervals where actions for individual PDUs may be scheduled. The PDU-specific actions to be scheduled within a given time interval can be represented as a series of semicolon separated values according to the following format: "PDU/action/from:to." For example, setting the *whitelist_spec* to pdu_1/S/0:100;pdu_2/R/101:201 means that the send action for the PDU "pdu_1" can only be scheduled in the interval $0 - 100\mu s$ and the receive action of the PDU "pdu_2" can only be scheduled in the interval $101 - 201\mu s$.

Each interval of a white-list can be as large as the valid interval range or as small as the interval of the FrIf job in which the PDU action is to be scheduled, but not smaller. If the white-list spans more than one FrIf job, the user is in fact letting the scheduler choose which FrIf job to use for the processing of the action defined in the white-list. Furthermore, should the phase of the FrIf jobs vary between communication cycles, an analysis of this variation must be performed in order to ensure a large enough interval of the white-list to encompass suitable FrIf jobs in all communication cycles. More details on the configuration of the AUTOSAR communication stack for FlexRay, and especially of the FrIf, can be found in [341].

15.7.2.3   HS-COM Configuration

The HS-COM layer itself is a VHDL module that is part of an FPGA and provides the following features:

- Communication support for the AS8202NF TTP controller attached to an FPGA.

- Runtime and memory efficient packing and unpacking of messages to and from the TTP frames.

- Asynchronous access to the TTP data (buffering).

- Support for 128-bit event messages (i.e., queued best-effort transmission).

For optimization reasons, the HS-COM layer supports messages with a size of 32, 64 and 128 bits. It is further limited to the handling of message boxes whose size is an integer multiple of either 64 or 128 bits. A message box is a container that may hold one or several messages, but all messages in a message box must have the same size. The HS-COM can be executed in a highly efficient way as all these message types are aligned with the internal layout of the data registers.

Depending on the messages defined, the HS-COM acts differently:

- 64-byte messages will be assumed to be simple state messages, and the HS-COM will access the message box in 64-bit chunks.

- *Received* message boxes containing 32-bit messages will be accessed in 32-bit chunks and an additional 32-bit frame status (i.e., information whether the

Header file

```
#ifndef _HS_COM_h_
#define _HS_COM_h_ 1
#include "ptypes.h"
extern const ubyte4 hscom_config_len;
extern const ubyte4 hscom_config [];
```

C file containing the configuration

```
#include "ptypes.h"
const ubyte4 hscom_config_len = 32;
const ubyte4 hscom_config [32] =
  { 0x1          /* [0] */
  , 0xc          /* [1] */
  , 0x0          /* [2] */
  , 0x0          /* [3] */
  ...
  , 0x80000805L /* [29] */
  , 0x21         /* [30] */
  , 0x816        /* [31] */
  };
```

FIGURE 15.41
Communication Configuration as C code — An Example (HS-COM)

frame was received correctly) will be added to each message. With this feature, the content *and* the validity of a message can be retrieved in one action.

- Messages of type '128-bit' are *always* treated as event messages and are accessed in 128-bit chunks. The queue depth for event messages is 32 FIFO entries each for sending and receiving, with 128 bits (i.e., one message) per entry.

The HS-COM layer performs a so-called *destructive read* when sending data on the bus, i.e., it sets the value of a read message to 0xFFFF...FFFF. If the data in the send memory is not updated within a cluster cycle after reading, this value will be transmitted and tells the receiver that something went wrong, either with the transmission or with the send memory's update. This mechanism prevents "old" data from being transmitted and mistaken for new.

Figure 15.41 shows an example configuration for the HS-COM layer. The HS-COM configuration consists of 32-bit values only for performance reasons. It comprises entries for the Register Area, the Pointer Area and the Command Area. In the shown example, the Register Area indicates the "SyncMode" to be 0x1, and 12 lines to be used for the Pointer Area. The "host activity timeout" is set to 0. The last three shown lines represent commands (from the Command Area). Each command contains a parity bit in bit-position 31. Therefore, the first command starts with 0x8000, and the two others start with 0. Bits 0 to 3 of each command specify the type of the command. For example, `0x80000805L` means to read one 128-bit event message starting from index 0. More details on the configuration of the HS-COM layer can be found in [342].

15.7.3   Code Generation

Middleware could be written by hand, and configured as discussed in Section 15.7.2. However, it is also possible to create the entire middleware layer with a design tool. Such automated creation can be exactly tailored to the communication needs of the schedule and the application, resulting in a highly optimized code. In the following, the *fault-tolerant communication (FT-COM) layer* for TTP is described in more detail as an example.

The FT-COM layer constitutes an interface between the communication services of the hardware, the operating system and the application software. According to Time-Triggered Architecture (TTA), each node executes an appropriate part of the distributed application, handling not only the data communication, but also the fault tolerance mechanisms designed for the system. As the FT-COM layer can be generated automatically by a design tool, the application code gets decoupled from the specific communication layer and fault tolerance mechanisms. This fact allows the application programmer to write source code that is highly reusable, easy to maintain, and transparent to many changes in the communication and fault tolerance design of the system. The FT-COM layer is generated as C source code for the node CPU, compiled and linked with the application code and executed on the same hardware as the application itself.

15.7.3.1   Feature Configuration

The FT-COM layer has several features that need configuration. A selection of these features is presented here, and relevant aspects regarding configuration and automatic code generation are discussed.

Subsystem Replication

A subsystem can be regarded as a set of tasks that take some input and produce some output. Each task is part of exactly one subsystem, but each subsystem may contain as many tasks as necessary. Several subsystems may be executed — independently of each other — on one host. A subsystem may also be executed simultaneously on more than one host (*replicated subsystem*). The first step toward fault tolerance can thus be achieved by replicating functionality, i.e., by replicating a subsystem.

The FT-COM needs to know how often a subsystem is replicated, and on which hosts these replicated subsystems run. It is expected that the FT-COM layer delivers a consistent view of the entire cluster regarding the value of a message, and provides diagnostic data to assess the "quality" of the provided data.

The Replica-Deterministic Agreement (RDA) Function

The receiver of a message m that is sent by a subsystem F, which is replicated with a replication degree of n, will in fact receive several message instances or *raw values* m_i of that message — one from each F_i that is active. But what is really wanted is the "correct" or "agreed" value. Therefore, the receiver needs to take the incoming instances m_i, run a function on them, and generate a single value m that will then be used for the application:

$$m = rda(m_1, m_2, \ldots m_r) \tag{15.21}$$

The upper limit for r is the replication degree n, which applies when all replicas of F are active, the lower limit is zero. *rda*, the *agreement function*, must therefore be able to consistently handle an input vector of any length from zero to n. It must also be deterministic [258]. Several RDA functions exist and are selectable for the FT-COM, depending on the type of the subsystem (fail-safe or fail-consistent) [343]. Instead of encoding an algorithm that works for any n, it might yield a better performance to insert different implementations of the same algorithm into the FT-COM code, depending on n.

Application code that accesses the message m should never need to access the individual instances m_i, and can therefore be "ignorant" of the replication degree of the sender of m. A change of this replication degree only requires an update of the FT-COM layer, but not of the application itself.

Reintegration with H-State

Each (application) task generally takes some input, performs some function on it and produces a result as output; both input and output are messages. Furthermore, the task can contain static internal data that influences the computation and hence the output. The set of this internal data is called *h-state*.

For fast reintegration and enhanced robustness of the whole system, it might be necessary for a replicated instance of a subsystem to know this h-state of its partner instances. The network designer has to define a global message ("h-state message") that contains this information. Now the output can be considered solely a function of the input, no "hidden" data is involved anymore. For performance reasons, these h-state messages should only be received and processed when *no* valid h-state is currently present. The generated FT-COM layer needs to monitor the h-state, and to provide it when necessary.

Receiver Status

From the RDA mechanism, the *number of correctly received message copies* can immediately be derived by setting up a counter that is initialized with zero at the beginning of the message transmission interval, and increased by one for each message copy that is received correctly, finally giving r. This counter is called the *receiver status* of a message m. The receiver status is useful for several RDA functions. For example, the application software can use the receiver status to derive confidence information on how "good" m is. Another example is averaging: All valid m_i are summed up, and the result is divided by the receiver status. It would be incorrect to divide the sum by n, because in case of a failure of one or more replicas of F, the sum would contain less than n components.

In a programming language that treats the number zero as the Boolean equivalent of "false" and any number other (or at least greater) than zero as "true," the receiver status can also be queried like a Boolean flag that yields "true" if the message is present, meaning that it was received correctly at least once and the RDA has yielded a result, and "false" if the message was not received correctly in this message transmission interval.

Sender Status

If an ECU hosts several subsystems, and one (fail-safe) subsystem fails, the others still should be able to send their data. Turning off the entire ECU is thus not an option. But as the communication controller works, it sends all messages, and potentially incorrect values for messages produced by the failed subsystem.

One classic strategy to handle this problem is to define an "invalid" value. This is unfavorable because it introduces a hidden information channel; if some application program fails to check for this special value in the right way, the system becomes inconsistent. Also, the "invalid" value might fall into the range of valid values after a software extension or upgrade. Any RDA function calculated in the FT-COM layer must take this into account.

The sender status of a message is part of the message itself, and therefore part of the input vector to the RDA function. The RDA will then consider a message that was correctly received, but has a sender status of "invalid," to be non-present. Clearly, the FT-COM code performs better if the sender status is only considered for those messages actually having one, and no such code or if-statement exists for messages that have no sender status.

The receiver status of a message is generated at the receiver and is always available. Therefore, it can always be used for checking the availability of a message. But it does not carry the same amount of information that the sender status delivers: This information is generated by the sender, exists only if the system design requires it and allows the sender to explicitly *invalidate* the message contents while still sending the message; this can be necessary for a more complex node design where more than one subsystem is executed on the node.

The sender status implies additional effort for the sender, i.e., the FT-COM code generated for the sender. It must be updated, and additional bandwidth (even if only a single bit) is needed on the communication bus. Furthermore, the receiver must explicitly check this sender status, in addition to the receiver status that is always processed.

Message Timing

The message timing should not be done by the application software because, besides becoming unnecessarily complex, this could raise timing problems due to programming errors or faults during execution. Based on the separation of concerns, message timing should be handled by the FT-COM layer, which takes full responsibility and can be reused across different applications. The FT-COM code generator needs to respect all these timing constraints and "schedule" its tasks so that all messages are processed in time.

Message Buffer Handling

A replicated subsystem F that sends a message m may also want to receive this message. This sounds trivial, but requires some effort when replication is used, because in this case it is not correct to simply access the message in the local RAM.

Say N is a node where one of the replicas of F is executed, and assume that another subsystem G, which also runs on N, uses m as input. Receiving a message from a replicated subsystem requires an RDA (this is valid even for the subsystem that sends this message). Therefore, m exists twice on N: One instance is the value which is sent by F, to be entered in the RDA at all receivers (including N), the other one is the result of the RDA at N. Usually these will be equal, but if, for example, m is a sensor reading with an agreement function that computes the average, the local sensor may produce a slightly different result than the other redundant sensors in the system, and the value m that is actually used by the receivers (including G) is an average of all m_i that were transmitted in the previous round.

It follows that several message buffers can be required for a message, depending on whether the message is replicated or not:

- A *transmit buffer* for the message instance that is sent to all the receivers; this buffer is required for any message.

- *Receive buffers* for each of the m_i

- A *result buffer* for the result of the RDA; this buffer is only required for replicated messages

Each of these buffers has the size (i.e., RAM requirements) of the message itself. A generated FT-COM may only provide all these buffers for messages where it is really needed, and save RAM if a message is not consumed by F or if the RDA function allows to directly use the sent value (e.g., "one-valid").

Packing of Bit Messages

Due to the CPU architecture of a node, the C variables containing the message values often use more memory than their data representation requires. The most common representatives of such a message type are Boolean messages, which have a data content of one bit, but are usually stored in a byte or even an `int`, depending on the CPU architecture and compiler.

However, since transmission bandwidth is rather expensive, the available net data rate should be optimally utilized. For this purpose, a Boolean message should be packed into a single bit, because it wastes a lot of space if it requires 16 or more bits for transmission. Similarly, a message which can take only one of 20 different values should not require 8 bits of transmission capacity, because the data content fits into

5 bits. Likewise, sensor data from an A/D unit that has a significant range of 10 bits does not need to be transmitted in a 16 bit word — but the packing algorithm needs to know which of the 16 bits are the 10 relevant ones.

At the receiver, the message needs to be expanded into a variable that is again easy to handle, like an `int`. The algorithm needs to be the exact inverse of the packing one, but must take into account several architectural properties that may differ between sender and receiver — the most prominent of all being the byte order.

On the other hand, the effort to efficiently (in terms of computation and code size) pack bit messages must be minimized, and there are much more efficient ways to achieve this than to simply consider the transmission buffer (frame) a long bit field and store all messages sequentially in this bit field. This is even true for standard messages of a size of 1, 2 or 4 bytes, and proper alignment can result in considerable performance gains.

However, manually programming such packing and unpacking routines for each bit message, and changing them consistently if something changes in the system specification (like a 10-bit A/D result being upgraded to 12 bits), is highly error-prone. Therefore, a layer that provides packing at the sender and unpacking at the receiver needs to be configured or created automatically, and must be supported by proper tools.

An automatically generated FT-COM layer may be optimized so that it only contains code that is really necessary for this particular platform, and that as many branches as possible are eliminated from the final code.

15.7.3.2 Implementation

The FT-COM layer must handle three major operations, specifically:

- Updating of the lifesign of the communication controller

- Packing of the messages into the proper frame buffers

- Unpacking and agreement calculation of all messages used by the application

All these operations are performed by special tasks (FT-tasks). One fundamental configuration option is the location of the frame buffers. If there exists a fast access to the CNI of the TTP controller, all packing and unpacking operations can be performed directly there. If not, it is more efficient to create local copies of the frames and to perform all operations locally. By setting this configuration option, the entire code can be created as it is best suited for the actual hardware.

Depending on the kind of the FT-task, it has to run either before or after an application task. The scheduler then has to ensure that the deadlines of the application tasks are met in any case. For all operations, the design tool needs to determine an interval within which the specific operation must be performed. To reduce task switching overhead, the design tool also should try to merge as many overlapping intervals as possible and to generate one FT-task for each of the resulting intervals; this leads to a minimal number of tasks.

The following sections describe how the schedule interval is determined for specific tasks.

Life-sign Update

The life-sign of the communication controller must be updated (by the host) at least once every round. In the pre-send-phase (the phase before the actual sending slot, d_{psp}), the controller checks if an update has been performed. Let $T_{s(n)}$ be the start of the controller's own sending slot of round n. The interval for the update of the life-sign in round n then is:

$$[T_{s(n-1)} \ldots T_{s(n)} - d_{psp}] \tag{15.22}$$

If the controller notices that the life-sign has not been updated, it goes into a passive state because there does not seem to be an application. Appropriate code for updating the life-sign has to be created and inserted into an FT-task that is scheduled for execution within this time interval.

Sender Tasks

The packing of messages into the proper frames in the CNI is done by sender tasks. The scheduling interval of these sender tasks must meet the following requirements:

- The latest possible *finish time* T_f for the packing of messages is the start of the pre-send-phase of the slot (i.e., start time of the slot minus the pre-send-phase).

- The earliest *task activation time* T_a is the time when the message is stable. This time is determined by the activation time of the application task plus its deadline. If at this point in time the message is not stable (i.e., the application task violates its deadline), the sender task must not start.

  If the task has a period that is different from the period of the message transmission on the network (defined in the cluster schedule), the activation instance leading to the shortest interval shall be considered, so that the latest value produced by the application is being sent over the network.

The interval $T_a \ldots T_f$ is computed for all messages sent by the application. All overlapping intervals should then be merged and a single FT-task should be generated, considering the runtime necessary for processing the messages and for updating the frames. To further reduce the number of required tasks, the sender tasks can also be merged with the lifesign tasks, if their intervals overlap and there is still enough runtime left for the lifesign updating.

Receiver Tasks

The receiver tasks must perform two operations; first the unpacking of the message instances (these instances will be used for the agreement), and then the computing of the specified agreement function. There are two different approaches to this:

- **Store and Process:** Unpack all message instances, store them in temporary buffers and perform the agreement function using the temporary buffers.

  The advantage of this approach is that it works with any kind of agreement (including majority voting) and also allows access to the individual raw values of the message.

  The disadvantage is increased memory demand: Every single message instance has to be stored.

- **Incremental:** Unpack only one message instance, perform the agreement on this instance, unpack the next message instance, ... After all message instances have been agreed, the finalization of the agreement (e.g., divide the result by the number of values added to achieve the average) can be performed.

  The advantage is the lower memory consumption and often faster execution.

  The disadvantage is that it cannot be applied to all kinds of agreements, only to those which can be done sequentially. It must also be noted that in this case the raw values are not available to a diagnosis function at the receiver (usually not required).

The design tool can select the appropriate computation strategy for the selected agreement function, and then only insert this code into the receiver task. Dead or temporarily unused code can thus be avoided.

When it comes to optimization, it is not sufficient to just look at messages and message instances, but also their temporal distribution needs to be considered. Each time a periodically sent message is transmitted, this is called a message *generation*, not to be confused with a message *instance*. Consider a sender application that sends a specific message every 10*ms*; further assume that this message is transmitted on two channels every 2*ms*. This means that each message value generated by the sender is actually received 10 times at the receiver: Five different generations are received (one every 2*ms*), and each generation contains two instances of the message.

In order to minimize the amount of global memory required by the FT-COM layer, it is necessary to perform the complete agreement for one message generation in a single FT-task. However, it is not always possible to pack all the steps of the complete agreement into a single task, since the individual (replicated) instances of a message generation may be spread throughout a whole round, and thus may have different and potentially non-overlapping validity spans. For the incremental approach, only some intermediate results need to be allocated globally if the agreement cannot be performed in a single task. Consequently, a good default is to use the incremental approach wherever possible.

For the receiver task generation, the validity interval of a message instance may be used as a possible scheduling interval. All overlapping intervals should then be merged and a single FT-task generated, considering the runtime necessary for the unpacking of the messages and for computing the agreement function.

To further optimize the memory footprint, the required RAM, and the execution time of the FT-COM layer, the design tool that creates the FT-COM code may filter out all message generations that are not used by application tasks. This can be done by comparing the activation times of the application tasks receiving the message with the validity intervals of the message generations. Only this reduced set of message generations will be retrieved from the network and provided to the application.

Code Generation

The TTP design tool ^{TTP}Build, which is available from TTTech, is able to automatically create FT-COM layer C code. ^{TTP}Build creates three files for each node (the names of these files are defaults and can be changed to any desired filename by the user):

- The message definition file `ttpc_msg.h` contains macro and variable declarations for the message buffers of incoming and outgoing messages on the specific node. This file, when included into application code, provides access to the message buffers, which are the only interface between the application program and the FT-COM layer. Function calls are not provided as they are not necessary for communication purposes.

  Some function-like C macros are offered to increase the readability of the generated code; for example,

  ```
  tt_Message_Status (temperature)
  ```

  is provided as a macro (looking like a function) to access the sender status of a message named `temperature`. In fact, the macro simply expands to the name of a variable, which is the message buffer containing the sender status of `temperature`, but the macro call improves the clarity of the statement. It will continue to work even if the implementation of the sender status should change in the future.

- The FT-COM layer C code is written to `ttpc_ftl.c` and comprises individual tasks called by the operating system (OS).

  The generated code is documented (the comments are also generated automatically, of course) to provide some insight into the workings of the FT-COM layer, but should never be changed manually. All changes will be lost when the code is generated again.

- `ttpos_conf.c` contains the configuration tables of the operating system, which tell the OS about the activation times and deadlines of all tasks on the node (application and FT-COM tasks alike). Although these tables are not part of the FT-COM layer, they are crucial for its proper operation, and are therefore also automatically generated by ^{TTP}Build.

  The contents of this file, although correct C code, are not intended to be human-readable, because they represent binary configuration data rather than program code (see option (a) in Section 15.7.2). As different operating systems require different formats, this file needs to be generated differently for each operating system that is supported by the design tool.

Additionally, a personalized MEDL can be generated to be loaded into the controllers of the host. This enables the definition of host-specific user interrupts and an optimized CNI layout.

15.7.4 Configuration of Third-Party Software

The operating system (OS), if one is present, and the application itself need to be configured, too. Design tools specifically designed to create communication configuration also need to interact with the development environment and configuration interfaces already available for the particular OS, the application or any other third-party software, e.g., a diagnostic module.

Typically, third-party software that interacts with the middleware, and in specific with the part that handles the communication, the communication stack, needs to know about a couple of things:

- **Layout and position of the messages:** The application must know by some means where the messages it reads and writes are located and how big they are. The most practical way is to have a memory-mapped interface. In this case, a C header file is required which contains #define statements. The application can refer to a certain message by name, and, based on the definitions in the header file, this name is mapped to a location in the memory. Of course, it is mandatory that the communication stack also has the same knowledge, but this is part of the communication configuration. Another possibility is to have a function call interface. Here, too, it is advantageous to have a mapping of message names (e.g., as defined in the design tool where all messages are specified) to certain IDs.

- **Interrupts:** The design of the communication stack may require the configuration of an interrupt for internal use in the communication stack. It might be helpful if any time a frame has been received by the hardware, a distinct interrupt is raised to indicate the arrival of the frame. Usually, this is a very high-priority interrupt. In the Interrupt Service Routine (ISR), the respective functions from the communication stack are called to handle this frame. These function calls must be registered beforehand, and the OS needs to know which interrupt to look at and to propagate. In addition, it may be possible to specify the interrupt priority level, the required stack size of the ISR and the vector to the service request register of the CPU.

- **Task properties and activation times:** If the communication stack is not interrupt-driven, it might need the activation of certain functions or tasks at certain times. Especially in a fully time-triggered environment, where the application and the OS are also synchronized to the communication network, this approach is favorable. As the OS dispatches tasks, it needs to know which communication task to start when, and with which assigned resources. In a real-time and time-triggered environment, the OS also needs to know the deadline of this communication task. Usually, one task is created for every message that has to be received or sent. For performance reasons, such tasks may be put together, forming task chains. For task chains, the OS needs to know similar properties as for tasks, in order to correctly interact with the communication stack.

- **Timer configuration:** If a timer is needed by the middleware, it has to be configured. All relevant details of this timer configuration also need to be part of the data that is shared between the middleware and the OS.

Development tools may generate parts of an OS configuration in the standardized *OSEK Implementation Language (OIL)* [242] format. As OIL comes in many vendor-specific flavors, it is very important to precisely determine the OIL version as well as the vendor-specific variant the generated file should have. The data can then be transferred to the OS by an OS configuration tool. In contrast to a complete OS configuration, a development tool for the communication stack may only provide the basic information necessary to run the various layers of the communication stack.

Development tools may also provide the relevant information as discussed above in other formats, e.g., in an XML-style fashion. Many operating systems come along with their own — and sometimes very specific — definition of the structure and possible content of OS configuration files. In such cases, either the development tool for the communication stack can be extended to write these files, or an additional conversion step needs to be introduced. A special-purpose tool or a self-written script may do the conversion job, too.

If the integration of the development tool and the OS is very good, the tool creates C files that fit the application. One C file may contain the message declaration and the type definition for every message that is sent or received by the application tasks. Another file may contain the configuration tables for the OS and comprise basic information on the respective node and task schedule. Ideally, the configuration tables for the OS are read, extended and then written back so that a configuration of a different origin is preserved. These configuration files are compiled and linked to the respective application to ensure the proper dependencies.

15.8 Verification

Society and law often request evidence that a particular system is fit for use and will not fail (or only in very rare cases), especially where the safety of humans is concerned. *Certification* by an accepted authority provides this kind of evidence; hence, most systems need to get certified for a particular use. For example, without permission granted by the Federal Aviation Administration (FAA), a commercial aircraft is not allowed to be operated in the US. Similar legal directives apply in other countries.

There exists a variety of certification standards, most prominently DO-178B [269] for software in aerospace, ISO 26262 [153] for automotive and

IEC 61508 [147] for industrial applications. For example, the FAA applies DO-178B for guidance to determine if the software will perform safely and reliably in an airborne environment [97].

Verification is one means listed in said standards to provide evidence for safe and reliable operation. To get a whole system certified, verification of certain artifacts that are part of the final system is hence necessary. As the schedule and the configuration items as described in Sections 15.3 and 15.7 are part of the final system, this need for verification applies to them. Details of the verification process and the area where verification is applicable are described in the respective standard.

To actually conduct verification, the use of tools is allowed and well established. Such tools are called *verification tools*. They need to be developed according to certain processes, also described in the standards mentioned above, and need to be *qualified* to be considered fit for their purpose. Tool qualification of verification tools is thus a crucial process on the way to getting a system certified.

In this section, we will discuss the impact of the different stages of verification on the software development process and the software itself, with the focus on the benefit of verification tools and their qualification.

The requirements for the verification of configuration items, imposed by certification standards, are discussed in Section 15.8.1. Various means to reduce cost during the verification process and related activities are presented in Section 15.8.2. A very prominent way is to use verification tools instead of manual verification performed in reviews. The verification of configuration items as well as the approach to use verification tools to assist the certification is presented in Section 15.8.3. Such verification tools must have a certain quality that can be reached by performing a tool qualification process. Details of this process and the implications posed on the development of the verification tools and the structure of the configuration items are also discussed there.

15.8.1 Process Requirements

In the aerospace industry, highly integrated safety-critical systems have been developed for decades. The FAA and other authorities have thus developed stringent certification requirements to meet the needs of the industry. Safety has always been the main focus of the system development. The regulations driving the safety of an aircraft are reflected in the Federal Aviation Regulations (FAR) 25 Paragraph 1309 (for the US) or — internationally spoken — in the Joint Aviation Regulations (JAR). For the methods of compliance with the FAR and JAR 25 requirements for a new system design, five methodologies are generally adopted, some of which are described in more detail in ARP 4754 [302] and ARP 4761 [303]:

1. Analysis including engineering analysis, stress analysis, system modeling and similarity modeling.

2. Failure analysis including FMEA (Failure Mode and Effects Analysis), FTA (Fault Tree Analysis) and safety analysis (including Functional Hazard Assess-

ment (FHA), (Preliminary) System Safety Assessment ((P)SSA) and Common Cause Analysis (CCA)).

3. Laboratory tests including component tests, qualification tests, system tests and tests on an integrated systems test rig.

4. Ground tests — On-aircraft ground tests.

5. Flight tests — On-aircraft flight tests.

Nowadays, the aerospace environment is strongly influenced by software certification authorities. The rapid increase in the use of software in airborne systems in the early 1980s resulted in a need for industry-accepted guidance for satisfying airworthiness requirements. DO-178, and subsequent revisions, have been written to satisfy this need and provide guidance for system software development. These certification requirements are illustrated with an overview of the DO-178B development process below.

The emergence of safety-critical x-by-wire systems in the automotive industry now leads to similar certification bodies and standards. Safety-related recommendations are already published, such as the MISRA guidelines [230] and recently the ISO 26262 standard. The latter has been derived from the Functional Safety standard IEC 61508 to better suit the needs in automotive electric and electronic systems. However, a mandatory certification authority for the hardware and software of automotive control units is not yet established. We believe that much benefit can be gained from the aerospace industry's certification experiences and recent activities to reduce certification costs of safety-critical systems [128].

Common to all safety standards is the ALARP principle [129]. ALARP stands for "as low as reasonably practicable" and means that the residual risk shall be as low as reasonably practicable. For a risk to be ALARP, it must be possible to demonstrate that the cost involved in reducing the risk further would be grossly disproportionate to the benefit gained. Adherence to state-of-the-art standards is widely accepted to be reasonably practicable.

15.8.1.1 DO-178B

DO-178 [269] was first published by the RTCA in 1980. It is intended to be used as a guideline for the software development and verification of airborne software systems. Since its first publication, the standard has been revised twice (DO-178A in 1985, DO-178B in 1992) and a third revision is ongoing (DO-178C).

DO-178B classifies software according to five assurance levels, rated by the criticality of the software functionality. Level A, the highest criticality level, is required for software whose anomalous behavior causes a catastrophic failure condition. Level E, the lowest level, is required for software whose anomalous behavior has no effect on the system's operational capacity. For each of the classification levels, DO-178B prescribes guidelines for the planning, development, verification, configuration management, software quality assurance, certification and maintenance of the system software.

DO-178B is a process oriented document; however, it does not prescribe the use of a particular lifecycle or structured methodology. This decision is left to the practitioner; however, the guidelines do require both the lifecycle model (with transition criteria) and the development methodology to be formally identified in the software plans and agreed with the certification authority, e.g., the FAA via a Designated Engineering Representative (DER).

DO-178B implies a requirements-driven development process. System requirements are decomposed into top level software requirements, which are in turn decomposed further into lower level requirements. This decomposition continues until module-level code can be directly implemented from the lowest level of requirements definition. In addition to design requirements driven by software requirements, derived design requirements are created to facilitate completeness of the software design. It follows that each element of the code base is traceable to a system-driven requirement or derived design requirement. Source code not directly traceable to requirements is strongly discouraged by the DO-178B guidelines. Such code is termed "dead code" and must be removed before certification. Deactivated code, that is, code utilized by the control unit but not exercised in application environment (e.g., manufacturing related code), is permitted, but only when the method of deactivation is proven and verified.

The verification activities recommended by DO-178B are also requirements-driven. The level of verification effort prescribed is once again proportional to the assigned software criticality level. Level A defines the most stringent verification process. Level E requires no verification of code or configuration items at all.

Level A development requires a full independent review of all of the verification artifacts, which consist of test cases and procedures. It also mandates that full structural coverage, including modified condition decision coverage (MC/DC), is achieved for all of the software. The generation of suitable test cases and expected results to yield such coverage drives much of the cost of level A development. Even outside the aerospace industry, testing and verification can account for as much as 40% to 70% of the total development effort [29, 111].

DO-178B also requires the adherence to strict configuration management practices. These practices require the practitioner to configure the entire software life cycle environment such that it can be reconstructed upon request. It also requires that software artifacts can be reproduced in their entirety from the configured data.

15.8.1.2 IEC 61508

IEC 61508 [147] is an international standard of rules applied in industry. It is titled "Functional safety of electrical/electronic/programmable electronic safety-related systems." The goal of functional safety is to use suitable methods to reduce the probability of dangerous errors to an acceptable level.

The safety categorization of a system is determined by the quantitatively defined probability of errors (see Figure 15.42). There, the categorization as seen by the other standards mentioned here is also shown, giving a comparison of the various levels.

The residual error rate of the data communication should not rise above the ac-

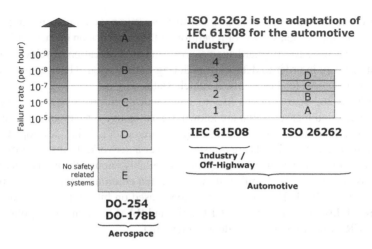

FIGURE 15.42

Comparison of Assurance Levels in Different Certification Standards

ceptable limit of 1% of the total errors of the system. For example, the following is valid for safety-relevant network signals in a SIL3 system according to IEC 61508: the probability of undetected corruption of such signals, which can lead to dangerous system errors, must be less than 10^{-9} per hour of operation (1% of the system error rate of $10^{-7}/h$). When a transmission error is detected, a corresponding system response must be triggered. In the case of a fail-safe system with only one communication bus, this means switching to a safe state; a fail-operational system must be able to transmit the data using an alternative transmission path.

IEC 61508 is less focused on requirements and, consequently, on verification of code and configuration items to satisfy these requirements. It is more concerned with functional safety: It is sufficient to show that the error rate is as low as requested, and that the system goes to a safe state in case of an error. Verification and thus the usage of verification tools is less commonly used, but may increase in the future.

15.8.1.3 ISO 26262

ISO 26262 [153] is an emerging norm for safety-relevant electrical and electronic systems in automobiles. It defines a process framework and process model together with required activities and work products, as well as applicable methods. The implementation of the norm is meant to guarantee the functional safety of an electrical/electronic system in a motor vehicle. The norm is derived from IEC 61508 specifically with regard to the domain of automobiles; compliance with this norm will tentatively start being mandatory in mid 2011 for all safety-relevant functions in motor vehicles.

In distributed control and regulating systems, data signals are transmitted over a network. The transmission route of such signals encompasses a sending device, one or more data buses (possibly including gateways), and one or more devices that receive and process the signals. Each of these components can, in case of errors, cause corruption, delay, loss, or repetition of the transmission or make it incorrect in some other way. To safeguard against errors in communication in vehicles, measures must be taken in order to detect errors during transmission. These errors must under no circumstances lead to a critical vehicle state.

ISO 26262 also contains lists of communication error classes that need to be dealt with, and measures that are known to be effective for recognizing these errors. In case a distributed system with safety requirements is being developed, in which safety-relevant data signals are transmitted over a network, it must be proven that all of these communication errors are detected reliably enough through effective mechanisms, so that the probability of an undetected communication error is below the required threshold. The calculation is done based on bit error rates of the network, the reliability of hardware (e.g., CRC units and RAM cells) and the applied methods (e.g., CRC polynomials and code word lengths).

15.8.2 Verification Best Practices

Verification is widely known as a time-consuming and costly activity. With the help of verification tools, costs can be reduced dramatically. But it is also necessary to obey a couple of best practices, so that the tools can be utilized best, and in as many steps as possible. In this section, we briefly present some best practices.

15.8.2.1 Reuse of Processes

Quality Assurance (QA) in aerospace is especially critical due to the relatively small production quantities and potentially large impact of failures on safety of operation. Accordingly, well-defined processes and many best-practice approaches exist. The development process for safety-relevant software development in the automotive sector can be derived from the time-tested development process for safety-relevant software development in aerospace. Since cost-effectiveness is a driving force behind innovation in the automobile industry, the efficient reuse of existing components is seen as one of the most effective factors in cost reduction. The savings of development and quality-assurance costs, as well as the robustness that results from time-tested and available components, contribute significantly to the realization of savings potential.

Due to the similar intentions of the above-mentioned standards, the development processes that are used for aerospace, automotive or industrial software development can be quite similar, too. It makes perfect sense to design a series of individual processes for all areas of business in an identical way.

- The processes for formal reviews and for change request management can be carried out with the help of tools that are uniform across the entire company in all areas of software development, and are carried out according to the same rules.

- The use of a common build framework for all lifecycle documents of software development makes it possible to simplify configuration management and to get an overview in the formal domain.

- A proven automatic certified test framework can be used to carry out several unit- and system-level tests in a particularly economical and exactly checkable manner.

15.8.2.2 Extending Checklists

QA stretches through the development, production, and operation and maintenance phases of an aircraft. During the development phase, a "verification and validation plan" has to be created in order to comply with any standard mentioned above. The plan contains checklists in addition to detailed descriptions of the checking procedures, test environments, test tools, documentation and result validation. These checklists are applied in the creation process of the corresponding lifecycle documents, as well as during their formal reviews.

The basis for project-specific checklists are the checklists from the standard software development process which usually exists in any company that develops safety-critical software. They are extended with the checklist points from the respective standard.

15.8.2.3 Use of COTS Products

When developing safety-related systems, testing indisputably causes the biggest overhead compared to development of conventional, non safety-related systems. However, the biggest savings potential also lies within this testing phase. A big proportion of testing time and cost can be saved when using commercial-off-the-shelf (COTS) components that are already safety-certified. Such components can either be complete units, or just sensors, or software modules like software drivers or protocol stacks. If these components or modules have already been tested by the supplier to the necessary degree required for the respective safety level, only the application layer and the interfaces to the COTS components need to be tested. The number of test cases for the application can be therefore significantly reduced.

Important prerequisites for the usage of COTS software modules are:

- A certificate indicating the safety integrity level of the component and the component failure rates that are needed for calculating the overall system failure rate.

- The availability of a safety manual that provides clear guidance on how to use the component in a safety-critical system.

The effort for the remaining required tests can be reduced by making use of appropriate tools for requirements management, configuration management, test execution, checking of coding standards, etc. However, all these tools also have to be qualified for use in safety-critical development.

15.8.2.4   Modular Certification

Modular certification according to DO-297 [270] is a rather new approach based on the need for certification of integrated modular avionics (IMA) and the corresponding system architectures [23]. The standard uses an architectural approach which enables the certification of small, reusable modules and applications. The needed functionality is established by connecting the single parts of the distributed application with a communication system. The standard breaks down the whole system into the following levels to map to the modular approach:

- **Module Acceptance:** A module is a component or a collection of components which may be software, hardware or a combination of both, which provides resources to the application and/or the system platform.

- **Application Acceptance:** An application is based on modules and performs a function.

- **System-level Acceptance:** The system level consists of one or several platforms which provide a computing environment, managing resources for at least one application. Furthermore, it establishes support services and platform-related capabilities like health monitoring and fault management.

- **Aircraft-level Acceptance:** The aircraft level considers the integration of the system into the aircraft and its systems.

Using such an architectural approach forces the reuse of legacy systems and provides the possibility of using modular platforms [175]. Therefore, the certification activities have to consider the certification of modules and especially their integration into the platform. An interesting approach considered for the future is to use formal methods to verify the integration.

The certification of single modules in this approach is fairly similar to the certification effort needed for reusable software components, i.e., for developing COTS products. Therefore, the reduction of certification effort applies here, too. In addition, the communication system which connects the modules needs to be fully approved.

15.8.2.5   Requirements Management

The requirements and design phases at the beginning are the most important parts of the software lifecycle process. The requirements define the expected output, and therefore need to be clear and easy to understand. The design is derived from the requirements and describes how they should be implemented. Every fault or obscurity in this phase has much impact later on. Requirements are the building blocks of the system. Therefore, the quality of the system depends on the quality of every single requirement.

Usually, the outcome of the design phase is stated in requirements, too. They are called low-level requirements, as opposed to high-level (software) requirements or system requirements, which are processed in the requirements phase. It is highly cost-efficient to apply the same processes for review and traceability checking on all requirements, rather than developing new processes for the design. Following this idea, the design thus consists of detailed low-level requirements, which mostly have to be traceable to the high-level requirements, and design components that describe complex algorithms and data structures to support the understanding.

Another major point is the traceability from the requirements to the design and further on to the test cases, down to the source code. This ensures that nothing is missing and everything has a reason for its existence. To ensure a constant quality level for the requirements and guarantee the traceability throughout the process, some basic points have to be recognized.

Tool Support

A database-centric requirements management tool provides a lot of advantages to the development and certification process. Firstly, several process steps are already included in the tool, hence the formal handling is simplified. Secondly, the waterfall-based top-down lifecycle process may be split up, which allows moving forward from requirements to implementation and verification without the need for consideration of other parts of the system. Furthermore, such a tool checks that all relevant traceability information is available. Additionally, some of these tools provide the possibility of creating evidence media which contain all necessary lifecycle and traceability information in an easy-to-review form. According to this efficient way to deal with the process, the effort for these steps may be reduced by about 20% with respect to the process necessary without tooling support.

Standardized Requirements Definitions

There should be standards for requirements definition, which provide guidelines for the development in order to facilitate their understanding. Furthermore, each requirement has to be self-contained because this supports the verification of each requirement.

Design Components

If a requirement describes a complex functionality, the developer should add definitions, figures and additional information which support the understanding. This encourages the demand for self-contained low-level requirements and helps to comprehend the whole system.

Testability

Each requirement has to be checked for testability. This has to be done by the requirements developer and especially by the reviewer. The easiest way to handle this is to write functional test cases in parallel to the requirements to find testability problems at an early stage of the requirements process. If this is not possible, the developer should at least give some advice or hints, regarding what to test, to the verification staff for efficient verification.

15.8.2.6 Test Vectors

In addition to the above, requirements-based test vectors (test cases and the input to automatic test procedures) can be automatically generated for each software product via a tool that is independent of the one used to generate the product code. These test vectors can cover nominal, MC/DC and robustness testing at the software module level. As with the code, these test vectors may have the requirements under test automatically inserted into them for better readability and traceability. All test vectors can then be parsed to create a complete test-vectors-to-requirements traceability matrix that is automatically inserted into the requirements management tool.

With different tools being used to generate test vectors and code, independence can be maintained, and therefore the test vector tool can be qualified as a verification tool as defined in DO-178B. With this qualification, peer reviews of the module (i.e., low-level diagram) tests are not required, resulting in a very large reduction of costs.

15.8.2.7 Test Suite

Another major concern regarding the verification process is the use of a test suite. The advantage of such a test suite is the possibility to automatically verify test cases and structural coverage. If the tool qualification package, which has to be provided to the authority, is already available for the chosen test suite tool, the verification effort may be optimized by about 10% compared to a process implementation without test suite tooling.

15.8.3 Verification Tooling Approach

The (automatic) generation of code and configuration items can be viewed as a step in the build process, similar to compilation. In a typical time-triggered communication system, these items can be grouped into three main blocks:

1. Communication configuration (i.e., MEDL) verification

2. Node-specific COM-layer verification

3. Application (control code) verification

This view eases the discussion about which processes shall be applied, and which measures and quality assurance metrics are applicable to source code generators and configuration generators. This view also implies, especially when applying DO-178B, that such generators are classified as development tools: It has to be shown that the output of said generators is correct with respect to the stated requirements, and that there is no code or configuration that is not covered by requirements. Tool verification is seen as less strict in the other mentioned standards; however, the considerations necessary for DO-178B form a valuable basis [63].

15.8.3.1 Output Correctness

To show evidence for output correctness, basically two different approaches are possible, and both are accepted and described in the standards. The first approach is to develop and test the generating tool in such a manner that for every possible input, the output is correct and adheres to the requirements stated in the input. Although such a development and certification of a code or configuration generator is costly, it removes the requirement to perform verification — often conducted by means of peer reviews — of the code or configuration itself. The tool is considered trustworthy. Thus, the one-time cost of certification is far less than the continual cost of performing verification of code and configuration.

The second approach is to develop the generating tool without respecting any processes. The tool might be non-deterministic, based on unreliable libraries or other components or even produce false output in some cases. It actually may "guess" the output. Obviously, the tool itself and thus its output cannot be considered trustworthy. But such freedom to choose any strategy to get to a possible solution allows for much more advanced algorithms and a higher chance to find a solution for a particular problem. In a subsequent, additional step — the verification — it has to be shown that for the *particular* given input, the output is correct with respect to the requirements stated in that input. It should be noted that if the output of the tool is verified, the tool can be used without qualification according to the standards. Such is the case for nearly all code generators and schedulers.

15.8.3.2 Manual vs. Automated Verification

Verification of the output can be done manually or automated. For manual verification, usually peer reviews are conducted, and checklists and a detailed process description for the reviewers exist. Manual verification can be cost-effective if done only once or only a few times. But the result of manual verification may depend on the assigned reviewers and their experience and expertise, and the result may not be exactly reproducible.

Automated verification pays off if the configuration data is expected to change several times during development. This is definitely the case when iterative development processes are used. It may also pay off if potential changes during the maintenance phase are considered, too. Verification can be done much faster if a verification tool exists. But also with other development processes, automated verification may be advantageous: The expertise of all involved persons gets cumulated in the verification tool, and is utilized in all subsequent versions of the tool. In addition, the result provided by the verification tool is exactly reproducible.

The largest portion of today's software costs is driven by the generation of the test cases and verification data. This is especially true for the development of verification tools. Verification data is required for each possible aspect of a configuration item. Verification data extend the test cases with input vectors and output vectors. The generation of verification data may also be automated, and the same requirements regarding tool qualification apply as for verification tools.

DO-178B classifies tools used during the development phase into two categories:

- **Development tools:** Tools whose output forms part of the airborne software and thus can introduce errors in the source code base (e.g., code generators).

- **Verification tools:** Tools that cannot introduce errors but may fail to detect them. For example, a static analyzer that automates a software verification process activity should be qualified if the function that it performs is not verified by another activity. Type checkers, analysis tools and test tools are other examples.

The use of verification tools is an interesting aspect of DO-178B. It provides the possibility of getting complex algorithms, like schedulers, easily certified. The verification tools have to verify the results of these algorithms, to prove their safe and deterministic behavior. Furthermore, a tool qualification package is needed for the verification tool, which provides confidence regarding the tool. The verification tool and its tool qualification package are mostly less expensive, if the verification for correctness has to be done several times, than to certify the development tool — containing the constructive algorithm — itself. Moreover, it is possible to hide intellectual property in the development tools, as their interior need not be assessed. Only the verification tool is assessed.

15.8.3.3 Qualification of Verification Tools

Tool qualification of verification tools is easier and thus more cost-effective than certification of development tools due to several aspects. Basically, it must only be shown that the tool does not accept any invalid, incomplete, incorrect or malicious code or configuration. However, the tool may (although not favorable) mark correct configurations as incorrect. In such a case, manual verification is necessary. Usually, such an incident results in an updated version of the verification tool, which is able to also handle this case correctly, as the intention of tool-based verification is to have no need for manual revision.

Any configuration that contains at least one element not having a matching requirement, or whose matching requirement implies another value, must be considered incorrect. Quite often, several requirements have an impact on the value of a certain output element. The verification tool does not need to tell the correct value of an output element — it is sufficient if it marks the element (or set of elements) as incorrect. The fact that the verification tool need not be constructive contributes to the cost-effectiveness of verification tools.

Another big advantage in the qualification of verification tools is the possibility to view the tool as black box. Internals need not be assessed. Consequently, there may be unused or even dead code inside the verification tool. It is not necessary to provide a detailed design and low-level requirements. Low-level test cases are not necessary, either. Only high-level requirements and the corresponding test cases are necessary. The total number of test cases and test vectors is thus significantly smaller than for the certification of a development tool. It is also possible to qualify a third-party tool, of which no internals are known. And it is further possible to qualify a tool just for a particular use case.

With the automation of requirements testing (i.e., the verification of the output generated by development tools, with respect to the requirements stated in the input to these tools), and MC/DC testing at the module level, the majority of (manual) testing emphasis can be directed at the system level, toward hardware-software integration and robustness testing. This results in a higher quality product, with reduced testing costs. At the system level there is limited automation because the testing requires system-level knowledge not captured in the software requirements. As such, these tests still need to be created and mostly also executed by hand. Consequently, an ideal process removes much of the manual work required to create safety-critical software, leaving the system and software design engineers to work at the system integration and test level, resulting in an overall product quality improvement.

The verification of MEDLs using ^{TTP}Verify will be discussed next, followed by a discussion of the verification of the configuration of a certain COM layer, the ^{TTP}TD-COM Layer.

15.8.3.4 ^{TTP}Verify

^{TTP}Verify is a comprehensive tool for the verification of TTP cluster designs, based on MEDLs. A TTP cluster contains a number of hosts exchanging messages in a statically defined temporal pattern. Any TTP controller in the cluster has stored this temporal pattern in its MEDL. This MEDL defines the whole transmission behavior on the bus and the local CNI interface behavior to the host controller. ^{TTP}Verify reads the MEDL files and verifies their integrity as well as their conformance to the TTP protocol. It is verified that the MEDLs belong to the same cluster and do not contradict each other. Some aspects of fault tolerance of the whole cluster are also checked.

The output of ^{TTP}Verify is a file that is divided into chapters for better readability. To allow a condensed view of the verification results, the user can customize the report to his needs. But the user cannot influence the verification algorithms to avoid conditions where the tool may fail due to bad user input. The command file structure and the output file structure are especially designed to support automatization (e.g., for extracting specific data), since the purpose of ^{TTP}Verify is to support and improve the verification process for TTP-based systems.

^{TTP}Verify automates the verification of the TTP schedule and the MEDLs where this schedule table is stored inside the TTP communication controllers. The correctness of this schedule is analyzed by ^{TTP}Verify and the resulting report has to be checked by additional tools or manually. Therefore, it is necessary to allow for easy extraction of information by tools as well as to provide a human readable representation of this data. ^{TTP}Verify is designed to specifically support safety-critical application software. Based on the Time-Triggered Architecture (TTA) and the TTP communication system, ^{TTP}Verify supports distributed fault-tolerant hard real-time application software.

^{TTP}Verify not only verifies the correctness of MEDLs, it also provides information about a MEDL or the cluster schedule. ^{TTP}Verify provides a summary for any verified controller, including scalar data (e.g., macrotick length, membership position) as well as different tables summarizing specific properties of a MEDL. This includes properties of all round-slots in any cluster mode which is provided for any controller type. Additional tables will be provided for specific controller-dependent properties. Different controller types will provide different types of properties that are reported. This controller data is not only informative for the user. It can also be used to manually verify issues that are beyond the scope of ^{TTP}Verify (e.g., order of slots). Furthermore, if ^{TTP}Verify detects a problem in a MEDL, the controller summaries may also be of help in finding the root cause behind the reported fault. These controller summaries are written in the report in the respective chapter of the MEDL.

^{TTP}Verify also provides a complete dump of the MEDL content in a human-readable form. This is necessary for verification activities that go beyond the scope of ^{TTP}Verify, and allows a significant gain of productivity for these purposes.

15.8.3.5 ^{TTP}TD-COM-Verify

The TTP Table-Driven Communication Layer (^{TTP}TD-COM Layer) is a static table-driven communication layer between the application and the TTP controller. It is designed for multiple TTP networks that are attached to one single CPU, and includes optimization for redundant messages. The ^{TTP}TD-COM Layer is a static embedded library written in C, which is certified according to DO-178B.

As the name suggests, it is driven by configuration tables. These tables are usually generated by ^{TTP}Build in C source code format, then compiled and linked into the embedded application, and will then reside in the ROM of the embedded target. Since this data influences the correct behavior of the embedded ^{TTP}TD-COM code, the used configuration data needs to be verified. This is the main application of ^{TTP}TD-COM-Verify.

What ^{TTP}TD-COM-Verify is:

- A tool to verify the correctness of the provided configuration data, which is used by the embedded source code of the ^{TTP}TD-COM Layer.

- A tool that checks the configuration data in binary form (as an S19 file, a Motorola-specific ASCII text encoding for binary data). This guarantees an end-to-end verification and no further need to verify a compilation or another transformation step.

- A tool that verifies the configuration data for integrity and consistency.

- A tool that verifies the configuration data for internal and global consistency against all participating hosts' configurations.

- A tool that verifies the correctness of scheduled user-interrupts.

What ^{TTP}TD-COM-Verify is not:

- A verification tool which verifies the correctness of the embedded code.

- A verification tool which verifies the correctness of the code of the configuration generation tool.

- A WCET measurement tool for the given configuration data.

- A blackbox test of the embedded ^{TTP}TD-COM code.

- A verification tool to check the C source code in any form (coding guidelines, correctness, etc.).

^{TTP}TD-COM-Verify reads a tool configuration file (in XML format), which on one hand contains switches for the tool behavior, and on the other hand the input file names of all other involved files. The latter include the requirement specification as an Interface Control Document (ICD), as well as the configuration tables and MEDLs to be verified. The configuration tables and MEDLs are read as S19 images together with unified map-files. In addition, ^{TTP}TD-COM-Verify uses the MHL partition header files. Optionally, the worst-case execution times (WCETs) can be supplied to ^{TTP}TD-COM-Verify to check the timing requirements.

Data Flow

Figure 15.43 shows the interaction between the development tools and the verification tools. On the left side, the standard TTP toolchain with ^{TTP}Plan, ^{TTP}Build and ^{TTP}Load is shown, which finally results in several MEDLs and ^{TTP}TD-COM Layer configurations, one for every host. These source files are compiled by a C-compiler chosen by the customer. They might be linked with the user application and the ^{TTP}TD-COM embedded library. Finally, the linker has to provide an S19 file which serves as the verification input for ^{TTP}TD-COM-Verify.

Additionally, the compiler provides a map-file mapping symbols to addresses inside the S19 image. Since every compiler has its own map-file format, ^{TTP}TD-COM-Verify will only accept a unified XML-based map-file. In this map-file the MHB allocations — which are given in the `tt_tdc_application_data_mhb_alloc_*.h` files — are included as well. This map-file handling is shown in Figure 15.43 between the C-compiler and the `map.xml` file(s). It includes the process of converting a compiler-specific map-file and the MHB message allocations into a unified XML-format map-file `map.xml`. Several requirements ensure that this conversion is done correctly. For checking those requirements, an additional small verifying tool is provided.

The following arrows show activities which have to be done by the user:

- The arrows from the customer database requirement specification files (command file for ^{TTP}Verify, tool configuration file for ^{TTP}TD-COM-Verify and the ICD) show the responsibility of the user to define application requirements inside those files independently of input data to the tool chain.

- The dashed line between ^{TTP}Verify and ^{TTP}TD-COM-Verify illustrates the responsibility of the user to check if all MEDL requirements that are needed for ^{TTP}TD-COM-Verify passed the tests correctly. Before ^{TTP}TD-COM-Verify is allowed to be operated, the MEDLs need to be checked for internal and global consistency by ^{TTP}Verify. To this end, ^{TTP}Verify uses a special command file as input for a cross-check with application requirements. This command file is usually provided directly by the user.

Besides the MEDLs, ^{TTP}TD-COM-Verify needs some further input. Similar to

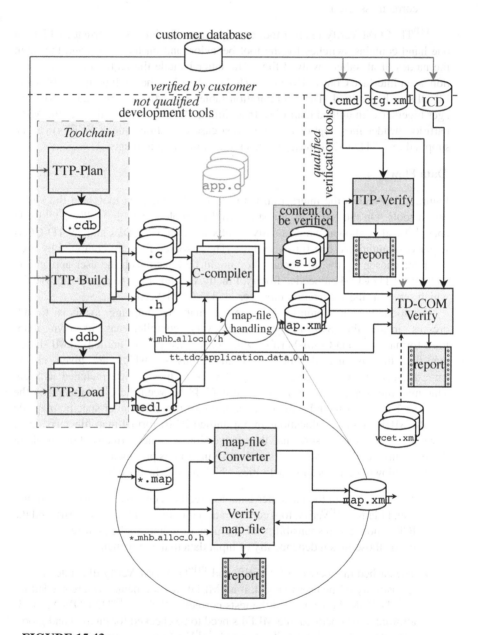

FIGURE 15.43
Interaction of the Development Tools with the Verification Tools

^{TTP}Verify's command file, the application requirements needed for ^{TTP}TD-COM-Verify must be provided as ICD. While ^{TTP}Verify supports verification for different cluster modes, the ^{TTP}TD-COM Layer does not support cluster mode switches and has only one active cluster mode during the whole runtime. This active cluster mode needs to be provided to ^{TTP}TD-COM-Verify through the ICD. The worst-case execution times (WCETs) of the frame copy tasks can be supplied to ^{TTP}TD-COM-Verify in wcet.xml files. Every host needs a separate file. If these files are not present, ^{TTP}TD-COM-Verify will just skip the timing requirements analysis.

Hence, a correct verification process would look like this:

1. Run ^{TTP}TD-COM-Verify without WCET files to guarantee correctness of the binary table data.

2. If the tables are correct, use those tables to measure the WCETs of every frame copy task, and enter these times into the WCET files.

3. Rerun ^{TTP}TD-COM-Verify with the newly created WCET files to check the scheduling timing requirements of the ^{TTP}TD-COM Layer.

Certification Aspects

The host applications contain a number of high-level requirements for operation and interface to the ^{TTP}TD-COM Layer. The ^{TTP}TD-COM Layer has specific requirements for the proper delivery and retrieval of messages to/from the CNI. These requirements are composed of requirements derived from the ^{TTP}TD-COM embedded code and the configuration tables. However, this procedure is very time consuming and expensive for large systems, and might slow down the development cycle dramatically. Therefore, a tool-based approach is considered. In such a tool-based approach, the interface requirements and the high-level requirements are provided as input to ^{TTP}Build, which produces the code containing the C data structures used by the ^{TTP}TD-COM Layer. ^{TTP}Build and the C compiler suite are considered development tools according to the guidelines of DO-178B section 12.2, whereas ^{TTP}TD-COM-Verify is considered a verification tool. ^{TTP}TD-COM-Verify must examine the S19 images containing data from the configuration tables. Additionally, ^{TTP}TD-COM-Verify has to validate them for correctness according to the application high-level requirements, the controller requirements and the ^{TTP}TD-COM high-level and low-level requirements. By qualifying ^{TTP}TD-COM-Verify in accordance with DO-178B, ^{TTP}Build and the C compiler suite do not need to be qualified.

Bibliography

[1] A. Ademaj. Slightly-off-specification failures in the time-triggered architecture. In *Proc. of the 7th IEEE International High-Level Design Validation and Test Workshop*, page 7, Washington, DC, USA, IEEE Computer Society, 2002.

[2] A. Ademaj, H. Sivencrona, G. Bauer, and J. Torin. Evaluation of fault handling of the time-triggered architecture with bus and star topology. In *Proc. of the International Conference on Dependable Systems and Networks (DSN)*, pages 123–132, 22–25 2003.

[3] T. Amnell, G. Behrmann, J. Bengtsson, P.R. D'Argenio, A. David, A. Fehnker, T. Hune, B. Jeannet, K.G. Larsen, M.O. Möller, P. Pettersson, C. Weise, and W. Yi. UPPAAL - Now, Next, and Future. In F. Cassez, C. Jard, B. Rozoy, and M. Ryan, editors, *Modelling and Verification of Parallel Processes*, number 2067 in Lecture Notes in Computer Science Tutorial, pages 100–125. Springer–Verlag, 2001.

[4] E. Anceaume and I. Puaut. A taxonomy of clock synchronization algorithms. Research Report 1103, Institut National de Recherche en Informatique et Systèmes Aléatoires (IRISA), Rennes, France, July 1997.

[5] E. Anceaume and I. Puaut. Performance evaluation of clock synchronization algorithms. Research Report 3526, Institut National de Recherche en Informatique et Systèmes Aléatoires (IRISA), Rennes, France, October 1998.

[6] C. Scheidler, P. Puschner, S. Boutin, E. Fuchs, G. Gruensteidl, and Y. Papadopoulos. Systems engineering of time-triggered architectures—the Setta Approach. In *Proceedings of the 16th IFAC Workshop on Distributed Computer Control Systems*, 2000.

[7] ARINC. *ARINC Specification 629: Multi-Transmitter Data Bus – Part 1: Technical Description*. Aeronautical Radio, Inc., Annapolis, MD, USA, November 1991.

[8] ARINC. *Backplane Data Bus. ARINC Specification 659*. Aeronautical Radio, Inc., 2551 Riva Road, Annapolis, MD 21401, December 1993.

[9] ARINC. Multi-transmitter data bus: Part 1 technical description. arinc specification 629p1-5. Technical report, Aeronautical Radio Inc., Annapolis, MD, USA, March 31st 1999.

[10] ARINC. Arinc specification 429. digital information transfer system (DITS) parts 1,2,3. Standard ARINC 429, Aeronautical Radio Inc., 2001.

[11] ARTEMIS. The ARTEMIS strategic research agenda. http://www.artemisia-association.org/sra, 2006. [Online; accessed 25-August-2010].

[12] K. Arvind. Probabilistic clock synchronization in distributed systems. *IEEE Transactions on Parallel and Distributed Systems*, 5(5):474–487, May 1994.

[13] Association for Standardisation of Automation and Measuring Systems (ASAM). *ASAM MCD-2 NET, Data Model for ECU Network Systems (Field-Bus Data Exchange Format), Version 3.1.0*, 2009.

[14] Atmel Corporation. *AVR 308: Software LIN Slave*, May 2002. Application note available at http://www.atmel.com.

[15] Audi AG, BMW AG, DaimlerChrysler AG, Motorola Inc. Volcano Communication Technologies AB, Volkswagen AG, and Volvo Car Corporation. LIN specification and LIN press announcement. SAE World Congress Detroit, http://www.lin-subbus.org, 1999.

[16] Audi AG, BMW AG, DaimlerChrysler AG, Motorola Inc. Volcano Communication Technologies AB, Volkswagen AG, and Volvo Car Corporation. LIN specification v2.0, 2003.

[17] N.C. Audsley, I.J. Bate, and A. Grigg. The role of timing analysis in the certification of IMA systems. *IEEE Certification of Ground/Air Systems Seminar (Ref. No. 1998/255)*, Dept. of Comput. Sci., York Univ., London, UK, February 1998.

[18] Autosar. *AUTOSAR – Technical Overview V3.0*, 2006.

[19] Autosar. *General Requirements on Basic Software Modules, Release 3.1, Document Version 2.2.2*, 2008.

[20] Autosar. *List of Basic Software Modules, Release 3.1, Document Version 1.3.0*, 2009.

[21] A. Avizienis, J.C. Laprie, and B. Randell. Fundamental concepts of dependability. Research Report 01-145, LAAS-CNRS, Toulouse, France, April 2001.

[22] O. Babaoglue and R. Drummond. (Almost) no cost clock synchronization. In *Proceedings of the 7th International Symposium on Fault-Tolerant Computing*, pages 42–47, Pittsburgh, PA, USA, IEEE Computer Society Press, July 1987.

[23] A. Bahrami. Complex integrated avionic systems and system safety. In *Online Proceedings of the The Europe-US International Aviation Safety Conference*, 2005.

[24] M.B. Barron and W.F. Powers. The role of electronic controls for future automotive mechatronic systems. *IEEE/ASME Transactions on Mechatronics*, 1(1):80 –88, March 1996.

[25] G. Bauer and H. Kopetz. Transparent redundancy in the time-triggered architecture. In *Proc. of the Int. Conference on Dependable Systems and Networks (DSN 2000)*, New York, pages 5–13, June 2000.

[26] G. Bauer, H. Kopetz, and W. Steiner. The central guardian approach to enforce fault isolation in a time-triggered system. In *Proc. of the 6th Int. Symposium on Autonomous Decentralized Systems (ISADS 2003)*, pages 37–44, Pisa, Italy, April 2003.

[27] G. Bauer and M. Paulitsch. An investigation of membership and clique avoidance in TTP/C. In *Proc. of the 19th IEEE Symposium on Reliable Distributed Systems*, pages 118–124, 2000.

[28] G. Behrmann, A. David, K.G. Larsen, O. Müller, P. Pettersson, and W. Yi. UPPAAL - present and future. In *Proc. of 40th IEEE Conference on Decision and Control*. IEEE Computer Society Press, 2001.

[29] B. Beizer. *Software Testing Techniques (2nd ed.)*. Van Nostrand Reinhold Co., New York, NY, USA, 1990.

[30] R. Benesch. TCP für die Time-Triggered Architecture. Master's thesis, Technische Universität Wien, Institut für Technische Informatik, Treitlstr. 3/3/182-1, 1040 Vienna, Austria, June 2004. ARTEMIS

[31] C. Bergenhem and J. Karlsson. A process group membership service for active safety systems using tt/et communication scheduling. In *Dependable Computing, 2007. PRDC 2007. 13th Pacific Rim International Symposium on*, pages 282 –289, December 2007.

[32] M. Bertoluzzo. Experimental activities on ttcan protocol. In *Intelligent Data Acquisition and Advanced Computing Systems: Technology and Applications, 2005. IDAACS 2005. IEEE*, pages 22 –27, 5-7 2005.

[33] P. Binns. A robust high-performance time partitioning algorithm. The Digital Engine Operating System Approach. In *Digital Avionics Systems Conference*. AIAA/IEEE, IEEE, 2001.

[34] P. Bishop. A methodology for safety case development. Technical report, Adelard, London, UK, 1998.

[35] P. Bjorn-Jorgensen and J. Madsen. Critical path driven cosynthesis for heterogeneous target architectures. In *Proceedings of the 5th International Workshop on Hardware/Software Co-Design*, pages 15–19. IEEE Computer Society, 1997.

[36] G. Bloor, G. Karsai, R. Reuter, S. Gulati, and S. Hutchings. The integration of anomaly, prognostics, and diagnostics reasoners to optimize overall vehicle health management goals. In *Proc. of the IEEE Aerospace Conference*, page 469, vol.2, 1999.

[37] B.W. Boehm, R. Madachy, and B. Steece. *Software Cost Estimation with Cocomo II with Cdrom*. Prentice Hall PTR Upper Saddle River, NJ, USA, 2000.

[38] M. Borovicka. Design of a gateway for the interconnection of real-time communication hierarchies. Master's thesis, Technische Universität Wien, Institut für Technische Informatik, Treitlstr. 3/3/182-1, 1040 Vienna, Austria, 2003.

[39] BOSCH. CAN specification - version 2.0. available at http://www.bosch.de.

[40] J.D. Boskovic and R.K. Mehra. Multi-mode switching in flight control. In *Proc. of the 19th Digital Avionics Systems Conferences (DASC)*, pages 6F2/1 –6F2/8, vol.2, 2000.

[41] D. Bosnacki and D. Dams. Discrete-time promela and spin. In *Formal Techniques in Real-Time and Fault-Tolerant Systems*, volume 1486 of *Lecture Notes in Computer Science*. Springer Berlin / Heidelberg, 1998.

[42] D. Bosnacki and D. Dams. Integrating real time into spin: A prototype implementation. In *FORTE XI / PSTV XVIII '98: Proceedings of the FIP TC6 WG6.1 Joint International Conference on Formal Description Techniques for Distributed Systems and Communication Protocols (FORTE XI) and Protocol Specification, Testing and Verification (PSTV XVIII)*, pages 423–438, Deventer, The Netherlands, Kluwer, B.V., 1998.

[43] A. Bouajjani and A. Merceron. Parametric verification of a group membership algorithm. In *Proc. of the Symposium on Formal Techniques in Real-Time and Fault Tolerant System (FTRTFT), LNCS Vol. 2469*, pages pp. 83–105, Oldenburg, Germany, Springer-Verlag, September 2002.

[44] I. Broster, A. Burns, and G. Rodriguez-Navas. Comparing real-time communication under electromagnetic interference. 2004.

[45] T. Bultan and T. Yavuz-Kahveci. Action language verifier. In *Proc. of the 16th Annual International Conference on Automated Software Engineering (ASE 2001)*, pages 382 – 386, 26-29 2001.

[46] E.K. Burke and G. Kendall. *Search Methodologies: Introductory Tutorials in Optimization and Decision Support Techniques*. Springer Verlag, 2005.

[47] D. Butler, T. Schmidt, and T. Waclawczyk. LIN protocol implementation using picmicro mcus. available at www.microchip.com, 2000. Microchip AN729.

[48] R.W. Butler, J.L. Caldwell, and B.L.Di Vito. Design strategy for a formally verified reliable computing platform. In *Proc. of the 6th Annual Conference on Computer Assurance (COMPASS) Systems*, pages 125–133, Gaithersburg, MD, USA, NASA Langley Res. Center, June 1991.

[49] G.C. Buttazzo. *Hard Real-Time Computing Systems: Predictable Scheduling Algorithms and Applications*. Springer-Verlag New York Inc, 2005.

[50] I. Cardei, R. Jha, M. Cardei, and A. Pavan. Hierarchical architecture for real-time adaptive resource management. In *Proc. of the IFIP/ACM International Conference on Distributed Systems Platforms (Middleware '00)*, pages 415–434, Secaucus, NJ, USA, Springer-Verlag New York, Inc., 2000.

[51] T. Carpenter, K. Driscoll, K. Hoyme, and J. Carciofini. ARINC 659 scheduling: problem definition. *Real-Time Systems Symposium, 1994., Proceedings*, pages 165–169, December 1994.

[52] W.C. Carter. A time for reflection. In *Proc. of the 8th IEEE Int. Symposium on Fault Tolerant Computing (FTCS-8)*, page 41, Santa Monica, June 1982.

[53] CAST, Inc., IP Provider. LIN bus controller core, 2010. Available at www.cast-inc.com/ip-cores/interfaces/lin/index.html.

[54] CENELEC. *EUROPEAN STANDARD 50128: Railway applications - Communications, signalling and processing systems - Software for railway control and protection systems*, March 2001.

[55] CENELEC. *EUROPEAN STANDARD 50159-1: Railway applications - Communication, signalling and processing systems; Part 1: Safety-related Communication in closed transmission systems*, March 2001.

[56] CENELEC. *EUROPEAN STANDARD 50159-2: Railway applications - Communication, signalling and processing systems; Part 2: Safety-related Communication in open transmission systems*, March 2001.

[57] CENELEC. *EUROPEAN STANDARD 50128: Railway applications - Communications, signalling and processing systems - Safety related electronic systems for signalling*, February 2003.

[58] P. Cholasta. LIN 2.0 mirror unit slave based on the MC68HC908EY16 MCU and the LIN 2.0 communication protocol. Application Note AN2885, Rev. 0, 11/2004, Freescale Semiconductor, 2004.

[59] G. Ciardo and C. Lindemann. Comments on "analysis of self-stabilizing clock synchronization by means of stochastic Petri nets." *IEEE Transactions on Computers*, 43(12):1453–1456, 1994.

[60] V. Claesson, H. Lönn, and N. Suri. An efficient TDMA start-up and restart synchronization approach for distributed embedded systems. *IEEE Transactions on Parallel and Distributed Systems*, 15(7), July 2004.

[61] D.D. Cofer and M. Rangarajan. Event-triggered environments for verification of real-time systems. In *Simulation Conference, 2003. Proceedings of the 2003 Winter*, pages 915 – 922, vol.1, 7-10 2003.

[62] E.G. Coffman and R.L. Graham. Optimal scheduling for two-processor systems. *Acta Informatica*, 1(3):200–213, 1972.

[63] M. Conrad, P. Munier, and F. Rauch. Qualifying software tools according to ISO 26262. In *Tagungsband Dagstuhl-Workshop MBEES: Modellbasierte Entwicklung eingebetteter Systeme VI*, 2010.

[64] FlexRay Consortium. FlexRay protocol specification ver. 2.1, 2005.

[65] C. Constantinescu. Trends and challenges in VLSI circuit reliability. *IEEE micro*, 23(4):14–19, 2003.

[66] G. Coulouris, J. Dollimore, and T. Kindberg. *Distributed Systems: Concepts and Design*. Int. Computer Science Series, Addison-Wesley, second edition, 1994.

[67] F. Cristian. Probabilistic clock synchronization. *Distributed Computing*, 3:146–158, 1989.

[68] F. Cristian. Reaching agreement on processor-group membership in synchronous distributed systems. *Distributed Computing*, 4:175–187, 1991.

[69] F. Cristian. Understanding fault-tolerant distributed systems. *Communications of the ACM*, 34(2):56–78, 1991.

[70] F. Cristian, H. Aghili, and R. Strong. Clock synchronization in the presence of omission and performance failures, and processor joins. *In Proc. of 16th Int. Symp. on Fault-Tolerant Computing Systems*, July 1996.

[71] F. Cristian and C. Fetzer. Fault-tolerant external clock synchronization. In *Proceedings of the 15th International Conference on Distributed Computing Systems*, pages 70–77, Los Alamitos, CA, USA, IEEE, May 30–June 2 1995.

[72] P.H. Dana. Global Positioning System (GPS) time dissemination for real-time applications. *Real-Time Systems*, 12(1):9–40, January 1997.

[73] C.T. Davies. *Computing Systems Reliability*, Data Processing Integrity, pages 288–354. Cambridge University Press, 1979.

[74] L. de Moura, S. Owre, H. Rue, J. Rushby, N. Shankar, M. Sorea, and A. Tiwari. SAL 2. In *Computer Aided Verification*, volume 3114 of *Lecture Notes in Computer Science*, pages 251–254. Springer Berlin / Heidelberg, 2004.

[75] J.A. Debardelaben, V.K. Madisetti, and A.J. Gadient. Incorporating cost modeling in embedded-system design. *IEEE Design & Test of Computers*, 14(3):24–35, 1997.

[76] S. Dolev. Possible and impossible self-stabilizing digital clock synchronization in general graphs. *Real-Time Systems*, 12(1):95–107, January 1997.

[77] S. Dolev and J.L. Welch. Self-stabilizing clock synchronization with Byzantine faults. In *Proceedings of the 14th ACM Symposium on Principles of Distributed Computing*, page 256. ACM Press, 1995.

[78] K. Driscoll, B. Hall, M. Paulitsch, P. Zumsteg, and H. Sivencrona. The real Byzantine generals. In *Proc. 23rd Digital Avionics Systems Conf.*, volume 6.D.4, pages 61–11, October 2004.

[79] K. Driscoll and K. Hoyme. The airplane information management system: An integrated real-time flight-deck control system. *Real-Time Systems Symposium*, pages 267–270, December 1992.

[80] K. Driscoll, G.M. Papadoupoulos, S. Nelson, G.L. Hartmann, and G. Ramohalli. Multi-processor flight control system. Technical Report AFWAL-TR-84-3076, Honeywell Systems and Research Center, September 1984.

[81] K.R. Driscoll. Apparatus and method for fault detection on redundant signal lines via encryption. Patent U.S. 5307409, Honeywell, April 26th 1994.

[82] K.R. Driscoll. Apparatus and method for transmitting information between dual redundant components utilizing four signal paths. Patent U.S. 5386424, Honeywell, January 31st 1995.

[83] B. Dutertre and M. Sorea. Modeling and Verification of a Fault-Tolerant Real-time Startup Protocol using Calendar Automata. In *Proc. of the Joint Conference Formal Modelling and Analysis of Timed Systems (FORMATS), Formal Techniques in Real-Time and Fault-Tolerant Systems (FTRTFT)*, Lecture Notes in Computer Science. Springer-Verlag, September 2004.

[84] C. Ebert and C. Jones. Embedded software: Facts, figures, and future. *Computer*, 42(4):42–52, 2009.

[85] S.A. Edwards. *Languages for Digital Embedded Systems*. Springer Netherlands, 2000.

[86] P. Eles, A. Doboli, P. Pop, and Z. Peng. Scheduling with bus access optimization for distributed embedded systems. *IEEE Transactions on Very Large Scale Integration(VLSI) Systems*, 8(5):472–491, 2000.

[87] W. Elmenreich. Time-triggered smart transducer networks. *IEEE Transactions on Industrial Informatics*, 2(3):192–199, 2006.

[88] W. Elmenreich and M. Delvai. Time-triggered communication with UARTs. In *Proceedings of the 4th IEEE International Workshop on Factory Communication Systems*, pages 97–104, 2002.

[89] W. Elmenreich, W. Haidinger, P. Peti, and L. Schneider. New node integration for master-slave field-bus networks. In *Proceedings of the 20th IASTED International Conference on Applied Informatics (AI 2002)*, pages 173–178, February 2002.

[90] W. Elmenreich and S. V. Krywult. A comparison of field-bus protocols: LIN 1.3, LIN 2.0, and TTP/A. In *Proceedings of the 10th IEEE International Conference on Emerging Technologies and Factory Automation*, pages 747–753, 2005.

[91] C. Elmore. Electronic controls. *OEM Off-Highway*, November 2008.

[92] C. Engel, E. Jenn, P.H. Schmitt, R. Coutinho, and T. Schoofs. Enhanced dispatchability of aircrafts using multi-static configurations. In *Proc. of the Embedded Real Time Software and Systems*, Toulouse, France, May 2010.

[93] J. Erjavcc and R. Scharff. *Automotive Technology: A Systems Approach*. Delmar Cengage Learning, 5th edition, 2009.

[94] J.A. Estefan. Survey of model-based systems engineering (MBSE) methodologies. *Incose MBSE Focus Group*, 25, 2007.

[95] FAA. Aviation databus assurance. Advisory Circular 20-156, Federal Aviation Administration, August 4th 2006.

[96] Federal Aviation Administration (FAA). Airworthiness directives; dassault model Falcon 2000ex and 900ex series airplanes. Airworthiness Directive Federal Register: (Volume 70, Number 39, Page 9853-9856), FAA, Docket No. FAA-2005-20425; Directorate Identifier 2005-NM-014-AD; Amendment 39-13987; AD 2005-04-15, March 1st 2005.

[97] Federal Aviation Administration (FAA). *Advisory Circular AC 20-115B*, 1993.

[98] C. Ferdinand and R. Heckmann. aiT: Worst-case execution time prediction by static program analysis. *Building the Information Society*, pages 377–383, 2004.

[99] M. Fernström and D. Ungerdahl. TTCAN Reference Application - An investigation on time-triggered network performance. Master's thesis, Chalmers University of Technology, 2006.

[100] C. Fetzer and F. Cristian. Lower bounds for function based clock synchronization. *In Proc. of 14th Int. Symp. on Principles of Distributed Computing*, August 1985.

[101] C. Fetzer and F. Cristian. An optimal internal clock synchronization algorithm. In *Proceedings of the 10th Conference on Computer Assurance*, pages 187–196, Gaithersburg, MD, USA, IEEE, June 1995.

[102] C. Fetzer and F. Cristian. Integrating external and internal clock synchronization. *Real-Time Systems*, 12:123–171, March 1997.

[103] M. Fletcher. Progression of an open architecture: from Orion to Altair and ISS. Companion to report (Presentation) S65-5000-20-0, Honeywell, May 2009. FaultTolerant Spaceborne Computing Employing New Technologies 2009 Conference.

[104] FlexRay Consortium. FlexRay communications system – preliminary central bus guardian specification version 2.0.9. Technical report, BMW AG., DaimlerChrysler AG., Robert Bosch GmbH, and General Motors/Opel AG, 2002.

[105] FlexRay Consortium. BMW AG, DaimlerChrysler AG, General Motors Corporation, Freescale GmbH, Philips GmbH, Robert Bosch GmbH, and Volkswagen AG. *FlexRay Communications System Protocol Specification Version 2.1*, May 2005.

[106] FlexRay Consortium. BMW AG, DaimlerChrysler AG, General Motors Corporation, Freescale GmbH, Philips GmbH, Robert Bosch GmbH, and Volkswagen AG. *Node-Local Bus Guardian Specification Version 2.0.9*, December 2005.

[107] A. Galleni and D. Powell. Consensus and membership in synchronous and asynchronous distributed systems. Technical report, 1995.

[108] GAMA. *ASCB: Avionics Standard Communications Bus Version C*. General Aviation Manufacturers Association (GAMA), Washington, DC, April 15 1996.

[109] M.-C. Gaudel, V. Issarny, C. Jones, H. Kopetz, E. Marsden, N. Moffat, M. Paulitsch, D. Powell, B. Randell, A. Romanovsky, R. Stroud, and F. Taiani. Final version of the DSoS conceptual model. *DSoS Project (IST-1999-11585) Deliverable CSDA1*, December 2002. Available as Research Report 54/2002 at http://www.vmars.tuwien.ac.at.

[110] M. Ghetie, H. Noura, and M. Saif. Fault diagnosis using balance equations methods and the algorithmic redundancy approach. In *Proc. of the 37th IEEE Conference on Decision and Control*, pages 586–591, vol.1, 1998.

[111] M. Ghiassi and K. I. S. Woldman. Dual programming approach to software testing. *Software Quality Journal*, 3(1):45–59, 1994.

[112] Robert Bosch GmbH. E-Ray FlexRay IP-module users manual revision 1.2.7, 2009.

[113] S. Godavarty, S. Broyles, and M. Parten. Interfacing to the on-board diagnostic system. In *Proc. of the 52nd IEEE Vehicular Technology Conference*, pages 2000 –2004, vol.4, 2000.

[114] S. Goldwasser, S. Micali, and R.L. Rivest. A digital signature scheme secure against adaptive chosen-message attacks. *SIAM Journal of Computing*, pages 281–308, April 1988.

[115] R. Gusella and S. Zatti. An election algorithm for a distributed clock synchronization program. In *Proc. of 6th Int. Conf. on Distributed Computing Systems*, pages 364–373, 1986.

[116] R. Gusella and S. Zatti. The accuracy of the clock synchronization achieved by tempo in Berkeley UNIX 4.3BSD. *IEEE Trans. on Software Engineering*, 15(7):847–853, July 1989.

[117] J. C. Palencia Gutiérrez and M. González Harbour. Schedulability analysis for tasks with static and dynamic offsets. In *Proceedings of the 19th IEEE Real Time Systems Symposium*, pages 26–37, December 1998.

[118] J. C. Palencia Gutiérrez and M. González Harbour. Exploiting precedence relations in the schedulability analysis of distributed real-time systems. In *Proceedings of the 20th IEEE Real-Time Systems Symposium*, page 328. IEEE Computer Society, 1999.

[119] B. Heppner and H. Brauner. Assessment of whole vehicle behaviour by means of simulation. Technical report, Daimler AG, 2008.

[120] W. Haidinger and R. Huber. Generation and analysis of the codes for TTP/A fireworks bytes. Research Report 5/2000, Technische Universität Wien, Institut für Technische Informatik, Vienna, Austria, 2000.

[121] B. Hall and K. Driscoll. A new aerospace network family. Presentation to INCOSE, Honeywell, October 2009.

[122] B. Hall, K.R. Driscoll, M. Paulitsch, and S. Dajani-Brown. Ringing out fault tolerance. A new ring network for superior low-cost dependability. *Dependable Systems and Networks, International Conference on*, 0:298–307, 2005.

[123] B. Hall, M. Paulitsch, and K.R. Driscoll. FlexRay BRAIN fusion: A FlexRay-based braided ring availability integrity network. *SAE World Congress, Paper No 2007-01-1492*, 2007.

[124] J.Y. Halpern, B. Simons, R. Strong, and D. Dolev. Fault-tolerant clock synchronization. In *Proceedings of the 3rd ACM Symposium on Principles of Distributed Computing*, pages 89–102, 1984.

[125] F. Hartwich. *TTCAN IP Module - User's Manual*. Bosch, 1.6 edition, 11 2002.

[126] F. Hartwich, B. Müller, T. Führer, and R. Hugel. Timing in the TTCAN Network. Technical report, Robert Bosch GmbH, 2003.

[127] C. Haubelt, J. Teich, K. Richter, and R. Ernst. System design for flexibility. In *Proceedings of the Conference on Design, Automation and Test in Europe*, pages 854–861. IEEE Computer Society, 2002.

[128] K. Hayhurst, C. Dorsey, J. Knight, N. Leveson, and G. McCormick. Stream-lining software aspects of certification: Report on the SSAC survey. Technical report, NASA Technical Memorandum 1999-209519, August 1999.

[129] Health and Safety Executive (HSA). *Reducing Risks, Protecting People – HSEs Decision-Making Process*, 2001.

[130] M. Hecht, D. Tang, and H. Hecht. Quantitative reliability and availability assessment for critical systems including software. In *Proc. of the 12th Annual Conference on Computer Assurance*, Gaithersburg, MD, USA, June 1997.

[131] G. Heiner and T. Thurner. Time-triggered architecture for safety-related distributed real-time systems in transportation systems. In *Proc. of the Twenty-Eighth Annual Int. Symposium on Fault-Tolerant Computing*, pages 402–407, June 1998.

[132] R. Hexel. FITS: a fault injection architecture for time-triggered systems. In *Proc. of the 26th Australasian Computer Science Conference (ACSC '03)*, pages 333–338, Darlinghurst, Australia, Australian Computer Society, Inc., 2003.

[133] D. Höchtl and U. Schmid. Long-term evaluation of GPS timing receiver failures. In *Proceedings of the 29th Precise Time and Time Interval Systems and Applications Meeting*, Long Beach, USA, December 1997.

[134] G.J. Holzmann. The model checker Spin. *Software Engineering, IEEE Transactions on*, 23(5):279 –295, May 1997.

[135] Honeywell. http://www.honeywell.com. accessed August 2010.

[136] K. Hoyme and K. Driscoll. SAFEbus. *IEEE Aerospace and Electronic Systems Magazine*, pages 34–39, March 1993.

[137] I. Hwang, S. Kim, Y. Kim, and C.E. Seah. A survey of fault detection, isolation, and reconfiguration methods. *IEEE Transactions on Control Systems Technology*, 18(3):636 –653, May 2010.

[138] IEC: Int. Electrotechnical Commission. *IEC 61508-7: Functional Safety of Electrical/Electronic/Programmable Electronic Safety-Related Systems – Part 7: Overview of Techniques and Measures*, 1999.

[139] IEEE. *Standard IEEE 802.4 – Information Processing Systems– Local Area Networks—Part 4: Token-Passing Bus Access Method and Physical Layer Specifications*, 1990.

[140] IEEE. IEEE standard 802.3 – carrier sense multiple access with collision detect (CSMA/CD) access method and physical layer. Technical report, IEEE, 2000.

[141] IEEE. *Draft Standard for a Precision Clock Synchronization Protocol for Networked Measurement and Control Systems (V0.19.13)*. IEEE Press, New York, NY, USA, May 2002. IEEE Standard No. P1588; Product No. DS5905-TBR.

[142] IEEE. *IEEE Standard for a Precision Clock Synchronization Protocol for Networked Measurement and Control Systems*. IEEE Press, New York, NY, USA, IEEE Standard No. 1588, March 2008.

[143] Aeronautical Radio Inc. *Avionics Application Software Standard Interface Part 1 – Required Services*, ARINC specification 653P1-2 edition, December 2005.

[144] Aeronautical Radio Inc. *Avionics Application Software Standard Interface Part 3 – Conformity Test Specification*, ARINC specification 653P-3 edition, October 2006.

[145] Aeronautical Radio Inc. *Avionics Application Software Standard Interface Part 2 – Extended Services*, ARINC specification 653P2-1 edition, 12 2009.

[146] National Instruments. FlexRay automotive communication bus overview. Technical report, August 2009.

[147] International Electrotechnical Commission (IEC). *IEC 61508: International Standard Functional Safety of Electrical / Electronic / Programmable Electronic Safety-Related Systems*, 1998.

[148] International Standardization Organisation (ISO). *Road Vehicles – Controller Area Network (CAN) – Part 4: Time-Triggered Communication, ISO 11898-4*, 1993.

[149] International Standardization Organisation (ISO). *Road Vehicles – Interchange of Digital Information – Controller Area Network (CAN) for High-Speed Communication, ISO 11898*, 1993.

[150] International Standardization Organisation (ISO). *Road Vehicles - Controller Area Network (CAN) – Part 1: Data Link Layer and Physical Signalling, ISO 11898-1*, 1993.

[151] International Standardization Organisation (ISO). *Road Vehicles - Controller Area Network (CAN) – Part 2: High-Speed Medium Access Unit, ISO 11898-2*, 1993.

[152] International Standardization Organisation (ISO). *ISO/IEC 15765-3:2004 - Road Vehicles – Diagnostics on Controller Area Networks (CAN) – Part 3: Implementation of Unified Diagnostic Services (UDS on CAN)*, 2004.

[153] International Standardization Organisation (ISO). *ISO/DIS 26262: International Standard Road Vehicles – Functional Safety*, 2009.

[154] C. Jeffrey, N. Dumas, Z. Xu, F. Mailly, F. Azas, P. Nouet, R.J.T. Bunyan, D.O. King, H. Mathias, J.P. Gilles, and A.M.D. Richardson. Sensor testing through bias superposition. *Sensors and Actuators A: Physical*, 136(1):441–455, 2007. 25th Anniversary of *Sensors and Actuators A: Physical*.

[155] S.C. Johnson and R.W. Butler. Design for validation. *IEEE Aerospace and Electronic Systems Magazine*, 7(1):38–43, January 1992.

[156] H. Kantz and N. König. Tas control platform: A vital computer platform for railway applications. *Alcatel Telecommunications Review*, 2nd Quarter 2004.

[157] H. Kantz and C. Koza. The ELEKTRA railway signalling system: Field experience with an actively replicated system with diversity. In *Proc. of the 25th International Symposium on Fault-Tolerant Computing (FTCS)*, pages 453 – 458, 27–30 1995.

[158] R. Kapeller. Design and implementation of a TTP/A master and gateway controller on a 32-bit microcontroller. Master's thesis, Technische Universität Wien, Institut für Technische Informatik, Vienna, Austria, 2001.

[159] J. Karlsson, J. Arlat, and G. Leber. Application of three physical fault injection techniques to the experimental assessment of the MARS architecture. In *Proc. of the 5th Annual IEEE International Working Conference on Dependable Computing for Critical Applications*, pages 150–161. IEEE Computer Society Press, 1995.

[160] J. Karlsson, P. Folkesson, J. Arlat, Y. Crouzet, and G. Leber. Integration and comparison of three physical fault injection techniques. In B. Randell, J. Laprie, H. Kopetz, and B. Littlewood, editors, *Predictably Dependable Computing Systems*, pages 309–327. Springer Verlag, Heidelberg edition, 1995.

[161] S. Katz, P. Lincoln, and J. M. Rushby. Low-overhead time-triggered group membership. In *WDAG*, pages 155–169, 1997.

[162] B. Keinhuis, K. Vissers Deprettere, and P. van der Wolf. An Approach for Quantitative Analysis of Application-Specific Dataflow Architectures. In *Proceedings of the 8th IEEE International Conference on Application-Specific Systems, Architectures and Processors*, pages 338–350, 1997.

[163] K. Keutzer, S. Malik, A.R. Newton, J.M. Rabaey, and A. Sangiovanni-Vincentelli. System-level design: Orthogonalization of concerns and platform-based design. *IEEE Transactions on Computer-Aided Design of Integrated Circuits and Systems*, 19(12):1523, 2000.

[164] M.S. Khan. Political and economic dimensions of Global Navigation Satellite System (GNSS). In *IEEE Proceedings of the Aerospace Conference*, volume 3, pages 3/1271 – 3/1276. IEEE, 2001.

[165] H. Kopetz. Event triggered versus time triggered. In *Proc. International Workshop on Operating Systems of the 90s and Beyond*, volume 563 of *Lecture Notes in Computer Science*, pages 87–101. Springer Verlag, 1992.

[166] H. Kopetz. Sparse time versus dense time in distributed real-time systems. In *Proc. of 12th Int. Conference on Distributed Computing Systems*, Japan, June 1992.

[167] H. Kopetz. TTP/A – A time-triggered protocol for body electronics using standard uarts. In *International Congress and Exposition*, Detroit, MI, USA, The Engineering Society for Advancing Mobility Land Sea Air and Space, SAE International, February-March 1995.

[168] H. Kopetz. Why time-triggered architectures will succeed in large hard real-time systems. In *Proc. of the 5th IEEE Computer Society Workshop on Future Trends of Distributed Computing Systems*, Cheju Island, Korea, August 1995.

[169] H. Kopetz. *Real-Time Systems, Design Principles for Distributed Embedded Applications*. Kluwer Academic Publishers, Boston, 1997.

[170] H. Kopetz. Elementary versus composite interfaces in distributed real-time systems. In *Proc. of the Int. Symposium on Autonomous Decentralized Systems*, Tokyo, Japan, March 1999.

[171] H. Kopetz. *TTP/C Protocol – Version 1.0*. TTTech Computertechnik AG, Vienna, Austria, July 2002. Available at http://www.ttpforum.org.

[172] H. Kopetz. Fault containment and error detection in the time-triggered architecture. In *Proc. of the Sixth Int. Symposium on Autonomous Decentralized Systems*, April 2003.

[173] H. Kopetz. Time-triggered real-time computing. *Annual Reviews in Control*, 27(1):3–13, 2003.

[174] H. Kopetz. The fault-hypothesis of the time-triggered architecture. In *Proc. of the 18th Edition of the IFIP World Computer Congress*, August 2004.

[175] H. Kopetz. From a federated to an integrated architecture for dependable real-time embedded systems. In *Proceedings of the Eighth Annual High Performance Embedded Computing (HPEC) Workshop*, 2004.

[176] H. Kopetz. On the fault hypothesis for a safety-critical real-time system. In *Keynote Speech at the Automotive Software Workshop San Diego (ASWSD 2004)*, San Diego, CA, USA, January 10–12, 2004.

[177] H. Kopetz and G. Bauer. The time-triggered architecture. *IEEE Special Issue on Modeling and Design of Embedded Software*, January 2003.

[178] H. Kopetz, G. Bauer, and S. Poledna. Tolerating arbitrary node failures in the time-triggered architecture. In *Proc. of the SAE 2001 World Congress*, Detroit, MI, USA, March 2001.

[179] H. Kopetz et al. The Time-Triggered Ethernet (TTE) design. In *Proc. of 8th IEEE Int. Symposium on Object-Oriented Real-Time Distributed Computing (ISORC)*, May 2005.

[180] H. Kopetz and G. Grunsteidl. TTP-A protocol for fault-tolerant real-time systems. *Computer*, 27(1):14–23, 1994.

[181] H. Kopetz, M. Holzmann, and W. Elmenreich. A universal smart transducer interface: TTP/A. *International Journal of Computer System Science & Engineering*, 16(2):71–77, March 2001.

[182] H. Kopetz and R. Nossal. Temporal firewalls in large distributed realtime systems. In *Proc. of IEEE Workshop on Future Trends in Distributed Computing*, Tunis, Tunisia, IEEE Press, 1997.

[183] H. Kopetz and W. Ochsenreiter. Clock synchronization in distributed real-time systems. *IEEE Transactions on Computers*, 36(8):933–940, 1987.

[184] H. Kopetz and J. Reisinger. The non-blocking write protocol NBW: A solution to a real-time synchronisation problem. In *Proc. of the 14th Real-Time Systems Symposium*, 1993.

[185] J.M. Krause, M.J. Englehart, and D.A Shaner. Achievable performance of fault tolerant avionics clocks. In *AIAA Computing in Aerospace Conference, 8th*, Technical Papers. Vol. 2 (A92-17576 05-61), pages p. 608–622, Baltimore, MD, American Institute of Aeronautics and Astronautics, Oct. 21-24 1991.

[186] A. Krüger. *Interface Design for Time-Triggered Real-Time System Architectures*. PhD thesis, Technische Universität Wien, Institut für Technische Informatik, Treitlstr. 3/3/182-1, 1040 Vienna, Austria, 1997.

[187] J.H. Lala and R.E. Harper. Architectural principles for safety-critical real-time applications. *Proc. of the IEEE*, 82:25–40, January 1994.

[188] L. Lamport and P. M. Melliar-Smith. Synchronizing clocks in the presence of faults. *Journal of the ACM*, 32(1):52–78, January 1985.

[189] L. Lamport and P.M. Melliar-Smith. Byzantine clock synchronization. In *Proceedings of the 3rd ACM Symposium on Principles of Distributed Computing*, pages 68–74, 1984.

[190] L. Lamport, R. Shostak, and M. Pease. The Byzantine generals problem. *ACM Transactions on Programming Languages and Systems (TOPLAS)*, 4(3):382–401, 1982.

[191] L. Lavagno and C. Passerone. *Embedded Systems Handbook*, chapter 3, pages 3–1–3–22. CRC Press, 2006.

[192] M. Lebedev. GLONASS as instrument for precise UTC transfer. In *Proceedings of the 12th European Frequency and Time Forum*, Warsaw, Poland, March 1998.

[193] E.A. Lee. Cyber physical systems: Design challenges. In *Proc. of the 11th IEEE International Symposium on Object Oriented Real-Time Distributed Computing (ISORC)*, pages 363–369, 2008.

[194] P.A. Lee and T. Anderson. *Fault Tolerance Principles and Practice*, volume 3 of *Dependable Computing and Fault-Tolerant Systems*. Springer Verlag, 1990.

[195] G. Leen and D. Heffernan. Modeling and verification of a time-triggered networking protocol. In *Networking, International Conference on Systems and International Conference on Mobile Communications and Learning Technologies, 2006. ICN/ICONS/MCL 2006*, pages 178–178, 23-29 2006.

[196] J. P. Lehoczky. Fixed priority scheduling of periodic task sets with arbitrary deadlines. In *Proceedings of 11th IEEE Real-Time Symposium*, pages 201–209, 1990.

[197] W. Lewandowski, J. Azoubib, and W.J. Klepczynski. GPS: Primary tool for time transfer. *Proceedings of the IEEE*, 87(1):163–172, January 1999.

[198] C. Li and S. Dey. Software-based self-testing methodology for processor cores. *IEEE Transactions on Computer-Aided Design of Integrated Circuits and Systems*, 20(3):369 –380, March 2001.

[199] R. Lichtenecker. Terrestrial time signal dissemination. *Real-Time Systems*, 12(1):41–61, January 1997.

[200] LIN Consortium. LIN specification package revision 2.1, 2006.

[201] B. Liskov. Practical use of synchronized clocks in distributed systems. In *Proceedings of 10th ACM Symposium on the Principles of Distributed Computing*, pages 1–9. ACM Press, 1991.

[202] C.L. Liu and J.W. Layland. Scheduling algorithms for multiprogramming in a hard real-time environment. *Journal of the ACM*, 20(1):46–61, 1973.

[203] C.D. Locke. Software architecture for hard real-time applications: cyclic executives vs. fixed priority executives. *Real-Time Systems*, 4(1):37–53, 1992.

[204] H. Lönn. Initial synchronization of TDMA communication in distributed real-time systems. In *19th IEEE Int. Conf. on Distributed Computing Systems*, pages 370–379, Gothenburg, Sweden, 1999.

[205] H. Lönn and J. Axelsson. A comparison of fixed-priority and static cyclic scheduling for distributed automotive control applications. In *Proceedings of the 11th Euromicro Conference on Real-time Systems*, pages 142–149. IEEE Computer Society Press, June 1999.

[206] H. Lönn and P. Pettersson. Formal verification of a TDMA protocol start-up mechanism. In *Pacific Rim International Symposium on Fault-Tolerant Systems (PRFTS '97)*, pages 235–242, Taipei, Taiwan, IEEE, December 1997.

[207] T. Losert. *Extending CORBA for Hard Real-Time Systems*. PhD thesis, Vienna University of Technology, Institute of Computer Engineering, 2005.

[208] M. Lu, D. Zhang, and T. Murata. Analysis of self-stabilizing clock synchronization by means of stochastic Petri nets. *IEEE Transactions on Computers*, 39(5):597–604, 1990.

[209] J. Lundelius and N. Lynch. A new fault-tolerant algorithm for clock synchronization. In *ACM Symp. on Principles of Distributed Computing*, pages 75–88, 1984.

[210] J. Lundelius and N. Lynch. An upper and lower bound for clock synchronization. *Information and Control*, 62:190–204, 1984.

[211] J. Luo, K.R. Pattipati, L. Qiao, and S. Chigusa. Agent-based real-time fault diagnosis. In *Aerospace Conference, 2005 IEEE*, pages 3632–3640, 5-12 2005.

[212] S.R. Mahaney and F.B. Schneider. Inexact agreement: accuracy, precision, and graceful degradation. In *Proceedings of the 4th ACM Symposium on Principles of Distributed Computing*, pages 237–249. ACM Press, 1985.

[213] S.M. Mahmud and A. Arora. Performance Analysis of Fault Tolerant TTCAN System. 2005.

[214] R. Maier, G. Bauer, G. Stoger, and S. Poledna. Time-triggered architecture: a consistent computing platform. *IEEE Micro*, 22(4):36–45, July/August 2002.

[215] S. Martello and P. Toth. *Knapsack Problems: Algorithms and Computer Implementations*. Wiley, New York, 1990.

[216] G. Martin, F. Schirrmeister, and C.D.S. Inc. A design chain for embedded systems. *Computer*, 35(3):100–103, 2002.

[217] K. Marzullo and S. Owicki. Maintaining the time in a distributed system. In *Proceedings of the 2nd ACM Symposium on Principles of Distributed Computing*, pages 295–305, 1983.

[218] K.A. Marzullo. *Maintaining the Time in a Distributed System: An Example of a Loosely Coupled Distributed Service*. PhD thesis, Department of Electrical Engineering, Stanford University, Stanford, CA, USA, February 1984.

[219] M. McCabe, C. Baggerman, and D. Verma. Avionics architecture interface considerations between constellation vehicles. In *Proc. of the 28th Digital Avionics Systems Conference (DASC)*, pages 1.E.2–1 – 1.E.2–10. IEEE/AIAA, October 2009.

[220] M.D. Mesarovic and Y. Takahara. *Abstract Systems Theory*, chapter 3. Springer-Verlag, 1989.

[221] B. Meyer. *Object-Oriented Software Construction*. Prentice Hall, 1997.

[222] D. Michaud. *Maintenance Avionique - ATA 100 34 Test Automatique Bus Avionique Langage C*. Institut de Maintenance Aronautique, Universit Bordeaux I, 2006.

[223] V. Mikolasek, A. Ademaj, and S. Racek. Segmentation of Standard Ethernet Messages in the Time-Triggered Ethernet. Technical Report 22/2008, Technische Universität Wien, Institut für Technische Informatik, Treitlstr. 1-3/182-1, 1040 Vienna, Austria, 2008.

[224] D.L. Mills. Internet time synchronization: the network time protocol. *IEEE Transactions on Communications*, 39(10):1482–1493, October 1991.

[225] P.S. Miner. Verification of fault-tolerant clock synchronization systems. Technical Report NASA Technical Paper 3349, NASA Langley Research Center, November 1993.

[226] R. Mores, G. Hay, R. Belschner, J. Berwanger, C. Ebner, S. Fluhrer, E. Fuchs, B. Hedenetz, W. Kuffner, A. Krüger, P. Lohrmann, D. Millinger, M. Peller, J. Ruh, A. Schedl, and M. Sprachmann. FlexRay – the communication system for advanced automotive control systems. In *Society of Automotive Engineers World Congress*, Detroit, MI, USA, SAE International. Document No 2001-01-0676, March 2001.

[227] M. Morgan. *The Avionics Handbook*, chapter Boeing B-777. CRC Press, Boca Raton, FL, USA, 2001.

[228] J. Morris, G. Lee, K. Parker, G.A. Bundell, and P.L. Chiou. Software component certification. *Computer*, 34(9):30–36, September 2001.

[229] MOST Cooperation, Karlsruhe, Germany. *MOST Specification Version 2.2*, November 2002.

[230] Motor Industry Software Reliability Research Association (MISRA). *Development Guidelines for Vehicle Based Software*, 1994.

[231] B. Müller, T. Führer, F. Hartwich, R. Hugel, and H. Weiler. Fault tolerant TTCAN networks. Technical report, Robert Bosch GmbH, 2002.

[232] C.J. Murray. Time-triggered protocol gains aerospace mileage. *EE Times*, September 2002.

[233] NXP Semiconductor. Fault-tolerant CAN/LIN fail-safe system basis chip. product data sheet, 2010. Available at `www.nxp.com/documents/data_sheet/UJA1061.pdf`.

[234] R. Obermaisser. CAN Emulation in a Time-Triggered Environment. In *Proc. of the 2002 IEEE Int. Symposium on Industrial Electronics (ISIE)*, volume 1, pages 270–275, 2002.

[235] R. Obermaisser. Message reordering for the reuse of CAN-based legacy applications in a time-triggered architecture. In *Proc. of the 12th IEEE Real-Time and Embedded Technology and Applications Symposium*, pages 301–310, April 2006.

[236] R. Obermaisser and A. Kanitsar. Application of TTP/A for the Otto Bock Axon bus. Technical Report 27/2000, Technische Universität Wien, Institut für Technische Informatik, Vienna, Austria, July 2000.

[237] R. Obermaisser and P. Peti. A fault hypothesis for integrated architectures. In *Proc. of the 4th Int. Workshop on Intelligent Solutions in Embedded Systems*, June 2006.

[238] Object Management Group. *The Common Object Request Broker: Architecture and Specification*, July 2002.

[239] Object Management Group (OMG). *Smart Transducers Interface V1.0*, January 2003. Specification available at http://doc.omg.org/formal/2003-01-01 as document ptc/2002-10-02.

[240] A. Olson and K. Shin. Fault-tolerant clock synchronization in large multicomputer systems. *IEEE Trans. on Parallel and Distributed Systems*, 5(9):912–923, 1994.

[241] OMG. Smart Transducers Interface V1.0. Available Specification document number formal/2003-01-01, Object Management Group, Needham, MA, U.S.A., January 2003. available at http://doc.omg.org/formal/2003-01-01.

[242] OSEK/VDX. *OIL: OSEK Implementation Language, Version 2.5*, 2004.

[243] S. Owre, J. Rushby, N. Shankar, and F. von Henke. Formal verification for fault-tolerant architectures: Prolegomena to the design of PVS. *IEEE Transactions on Software Engineering*, 21(2):107–125, February 1995.

[244] J. C. Palencia, J. J. Gutiérrez Garcia, and M. González Harbour. On the schedulability analysis for distributed hard real-time systems. In *Proceedings of the Euromicro Conference on Real Time Systems*, pages 136–143, 1997.

[245] J.C. Palencia and M.G. Harbour. Schedulability analysis for tasks with static and dynamic offsets. In *Proceedings of the 19th IEEE Real-Time Systems Symposium*, pages 26–37. IEEE Computer Society, 1998.

[246] M. Papatriantafilou and P. Tsigas. Self-stabilizing wait-free clock synchronization. In *Proceedings of the 4th Scandinavian Workshop on Algorithm Theory*, volume 824 of *Lecture Notes in Computer Science*, pages 267–277. Springer-Verlag Berlin Heidelberg, Germany, July 1994.

[247] R.J. Patton. Fault detection and diagnosis in aerospace systems using analytical redundancy. In *IEEE Colloquium on Condition Monitoring and Fault Tolerance*, pages 1/1–120, 6 1990.

[248] M. Paulitsch and B. Hall. Insights into the sensitivity of the BRAIN (braided ring availability integrity network)–on platform robustness in extended operation. *Dependable Systems and Networks, International Conference on*, 0:154–163, 2007.

[249] M. Paulitsch and B. Hall. Starting and resolving a partitioned BRAIN. *Object-Oriented Real-Time Distributed Computing, IEEE International Symposium on*, 0:415–421, 2008.

[250] M. Paulitsch, J. Morris, B. Hall, K.R. Driscoll, E. Latronico, and P. Koopman. Coverage and the use of cyclic redundancy codes in ultra-dependable systems. *Dependable Systems and Networks, International Conference on*, 0:346–355, 2005.

[251] P. Pedreiras and L. Almeida. Combining event-triggered and time-triggered traffic in FTT-CAN: Analysis of the asynchronous messaging system. In *Proc. of 3rd IEEE Int. Workshop on Factory Communication Systems*, September 2000.

[252] P. Peti, R. Obermaisser, and H. Kopetz. Out-of-norm assertions. In *Proc. of the 11th IEEE Real-Time and Embedded Technology and Applications Symposium (RTAS'05)*, pages 280–291, San Francisco, CA, USA, March 2005.

[253] P. Peti, R. Obermaisser, and H. Paulitsch. Investigating connector faults in the time-triggered architecture. In *Proc. of the IEEE Conference on Emerging Technologies and Factory Automation (ETFA'06)*, pages 887 –896, 20-22 2006.

[254] P. Peti and L. Schneider. Implementation of the TTP/A slave protocol on the Atmel ATmega103 MCU. Technical Report 28/2000, Technische Universität Wien, Institut für Technische Informatik, Vienna, Austria, August 2000.

[255] H. Pfeifer. Formal verification of the TTP group membership algorithm. In *Proc. of Formal Methods for Distributed System Development (FORTE XIII / PSTV XX 2000)*, pages 3–18. Kluwer Academic Publishers, 2000.

[256] H. Pfeifer, D. Schwier, and F.W. von Henke. Formal verification for time-triggered clock synchronization. In *Proc. of the 7th IFIP InternationalWorking Conference on Dependable Computing for Critical Applications (DCCA-7)*, pages 207–226, November 1999.

[257] M. Pfluegl and D. Blough. A new and improved algorithm for fault-tolerant clock synchronization. *Journal of Parallel and Distributed Computing*, 27:1–14, 1995.

[258] S. Poledna. Replica determinism in distributed real-time systems: A brief survey. *Real-Time Systems*, 6:289–316, 1994.

[259] S. Poledna. *Fault Tolerant Real-Time Systems: The Problem of Replica Determinism*. Kluwer Academic Publishers, Boston, 1996.

[260] P. Pop, P. Eles, and Z. Peng. Scheduling with optimized communication for time-triggered embedded systems. In *Proceedings of the Seventh International Workshop on Hardware/Software Codesign*, pages 178–182. ACM, 1999.

[261] P. Pop, P. Eles, and Z. Peng. *Analysis and Synthesis of Distributed Real-Time Embedded Systems*. Kluwer Academic Pub, 2004.

[262] P. Pop, P. Eles, and Z. Peng. Schedulability-driven communication synthesis for time triggered embedded systems. *Real-Time Systems*, 26(3):297–325, 2004.

[263] P. Pop, P. Eles, and Z. Peng. Schedulability-driven frame packing for multi-cluster distributed embedded systems. *ACM Transactions on Embedded Computing Systems (TECS)*, 4(1):140, 2005.

[264] P. Pop, V. Izosimov, P. Eles, and Z. Peng. Design optimization of time- and cost-constrained fault-tolerant embedded systems with checkpointing and replication. *IEEE Trans. on Very Large Scale Integrated (VLSI) Systems Volume*, 17(3):389–402, 2009.

[265] T. Pop, P. Eles, and Z. Peng. Schedulability analysis for distributed heterogeneous time/event triggered real-time systems. In *15th Euromicro Conference on Real-Time Systems, 2003. Proceedings*, pages 257–266, 2003.

[266] T. Pop, P. Pop, P. Eles, and Z. Peng. Optimization of hierarchically scheduled heterogeneous embedded systems. In *Proceedings of 11th IEEE International Conference on Embedded and Real-Time Computing Systems and Applications*, pages 67–71, 2005.

[267] T. Pop, P. Pop, P. Eles, Z. Peng, and A. Andrei. Timing analysis of the FlexRay communication protocol. *Real-Time Systems*, 39(1):205–235, 2008.

[268] D. Powell. Failure mode assumptions and assumption coverage. In *Proc. of the 22nd IEEE Annual Int. Symposium on Fault-Tolerant Computing (FTCS-22)*, pages 386–395, Boston, USA, July 1992.

[269] Radio Technical Commission for Aeronautics, Inc. (RTCA). *DO-178B: Software Considerations in Airborne Systems and Equipment Certification*, 1992.

[270] Radio Technical Commission for Aeronautics, Inc. (RTCA). *DO-297: Integrated Modular Avionics (IMA) Development Guidance and Certification Considerations*, 2005.

[271] D. Ragan, P. Sandborn, and P. Stoaks. A detailed cost model for concurrent use with hardware/software co-design. In *Proceedings of the 39th annual Design Automation Conference*, pages 269–274. ACM, 2002.

[272] P. Ramanathan, K.G. Shin, and R.W. Butler. Fault-tolerant clock synchronization in distributed systems. *IEEE Computer*, 23(10):33–42, October 1990.

[273] J.C. Ramirez and A.S. Piqueras. Learning Bayesian networks for systems diagnosis. In *Proc. of the Electronics, Robotics and Automotive Mechanics Conference*, volume 2, pages 125 –130, September 2006.

[274] Mathias Rausch. *FlexRay Grundlagen, Funktionsweise, Anwendung*. HANSER, 2008.

[275] C. R. Reeves. *Modern Heuristic Techniques for Combinatorial Problems*. Blackwell Scientific Publications, 1993.

[276] FAST Report. Study of worldwide trends and r&d programmes in embedded systems. Technical report, 2005.

[277] RTCA. Software considerations in airborne systems and equipment certification. Standard DO-178B, RTCA, Inc., 1828 L Street, NW, Suite 805, Washington, DC 20036-5133, USA, December 1, 1992.

[278] RTCA. Design assurance guidance for airborne electronic hardware. Standard DO-254, RTCA, Inc., 1828 L Street, NW, Suite 805, Washington, DC 20036-5133, USA, April 19, 2004.

[279] RTCA. Environmental conditions and test procedures for airborne equipment. Standard DO-160E, RTCA, Inc., 1828 L Street, NW, Suite 805, Washington, DC 20036-5133, USA, December 9, 2004.

[280] B. Rumpler and W. Elmenreich. Considerations on the complexity of embedded real-time system design tasks. In *Proceedings of the IEEE International Conference on Computational Cybernetics 2006 (ICCC'06)*, pages 55–60, 2006.

[281] J. Rushby. Partitioning for avionics architectures: Requirements, mechanisms, and assurance. NASA Contractor Report CR-1999-209347, NASA Langley Research Center, June 1999.

[282] J. Rushby. Systematic formal verification for fault-tolerant time-triggered algorithms. *IEEE Transactions on Software Engineering*, 25(5):651–660, September 1999.

[283] J. Rushby. Formal verification of transmission window timing for the time-triggered architecture. Technical report, Computer Science Laboratory, SRI International, Menlo Park, CA 94025 USA, March 2001.

[284] J. Rushby. Modular certification. Technical report, Computer Science Laboratory SRI International, 333 Ravenswood Avenue, Menlo Park, CA 94025, USA, September 2001.

[285] J. Rushby. An overview of formal verification for the time-triggered architecture. In *Proc. of the Symposium on Formal Techniques in Real-Time and Fault Tolerant System (FTRTFT), LNCS Vol. 2469*, pages 83–105, Springer-Verlag, Oldenburg, Germany, September 2002.

[286] J. Rushby and F. von Henke. Formal verification of the interactive convergence clock synchronization algorithm. Technical Report CSL-89-3R, Computer Science Laboratory, SRI International, CA, Menlo Park, USA, February 1989.

[287] SAE. ARP 5107 (aerospace recommended practice). guidelines for time-limited-dispatch analysis for electronic engine control systems. Technical Report Rev. B, Society of Automotive Engineers, November 2006.

[288] I. Saha and S. Roy. A finite state analysis of time-triggered CAN (ttcan) protocol using Spin. In *Computing: Theory and Applications, 2007. ICCTA '07. International Conference on*, pages 77 –81, 5-7 2007.

[289] I. Saha, S. Roy, and K. Chakraborty. Modeling and verification of TTCAN startup protocol using synchronous calendar. In *Software Engineering and Formal Methods, 2007. SEFM 2007. Fifth IEEE International Conference on*, pages 69 –79, 10-14 2007.

[290] J.H. Saltzer, D.P. Reed, and D.D. Clark. End-to-end arguments in system design. *ACM Transactions on Computer Systems (TOCS)*, 2, 1984.

[291] A. Sangiovanni-Vincentelli. Electronic-system design in the automobile industry. *IEEE Micro*, 23(3):8–18, 2003.

[292] A. Schedl. *Design and Simulation of Clock Synchronization in Distributed Systems*. Doctoral thesis, Institut für Technische Informatik, Technische Universität Wien, Treitlstr. 1-3/3/182-1, Vienna, Austria, April 1996.

[293] F. Scheler and W. Schröder-Preikschat. Time-triggered vs. event-triggered: A matter of configuration? In *Proc. of the Workshop on Model-Based Testing*, Nürnberg, Germany, 2006.

[294] U. Schmid. Orthogonal accuracy clock synchronization. *Chicago Journal of Technical Computer Science*, 2000(3):3–77, August 2000.

[295] U. Schmid and K. Schossmaier. Interval-based clock synchronization. *Real-Time Systems*, 12:173–228, March 1997.

[296] F.B. Schneider. A paradigm for reliable clock synchronization. Technical Report TR86-735, Computer Science Department, Cornell University, February 1986.

[297] F.B. Schneider. Understanding protocols for Byzantine clock synchronization. Research Report 87-859, Department of Computer Science, Cornell University, Ithaca, NY, USA, August 1987.

[298] W. Schwabl. *Der Einfluss zufälliger und systematischer Fehler auf die Uhrensynchronisation in verteilten Echtzeitsystemen.* Doctoral thesis, Institut für Technische Informatik, Technische Universität Wien, Treitlstr. 1-3/3/182-1, Vienna, Austria, October 1988.

[299] K.G. Shin and R. Ramanathan. Clock synchronization of large multiprocessor systems in the presence of malicious faults. *IEEE Transactions on Computers*, 36(1):2–12, 1987.

[300] O. Sinnen. *Task Scheduling for Parallel Systems.* Wiley-Blackwell, 2007.

[301] H. Sivencrona, P. Johannessen, M. Persson, and J. Torin. Heavy-ion fault injections in the time-triggered communication protocol. In *Dependable Computing, Lecture Notes in Computer Science*, volume 2847/2003, pages 69–80. Springer Berlin/Heidelberg, 2003.

[302] Society of Automotive Engineers (SAE). *ARP 4754: (Aerospace Recommended Practice) - Certification Considerations for Highly Integrated or Complex Aircraft Systems*, 1996.

[303] Society of Automotive Engineers (SAE). *ARP 4761: (Aerospace Recommended Practice) - Guidelines and Methods for Conducting the Safety Assessment Process on Civil Airborne Systems and Equipment*, 1996.

[304] T.K. Srikanth and S. Toueg. Optimal clock synchronization. *Journal of the ACM*, 34(3):626–645, 1987.

[305] W. Steiner. *Startup and Recovery of Fault-Tolerant Time-Triggered Communication.* PhD thesis, Technische Universität Wien, Institut für Technische Informatik, Treitlstr. 3/3/182-1, 1040 Vienna, Austria, 2004.

[306] W. Steiner. TTEthernet Executable Formal Specification. Research report, 2009. Available at http://www.ttagroup.org/.

[307] W. Steiner. An Evaluation of SMT-based Schedule Synthesis For Time-Triggered Multi-Hop Networks. In *RTSS'10: Proceedings of the 31st IEEE Real-Time Systems Symposium.* IEEE, 2010.

[308] W. Steiner. Synthesis of Static Communication Schedules for Mixed-Criticality Systems. In *AMICS 2011: Proceedings of the 1st International Workshop on Architectures and Applications for Mixed-Criticality Systems.* IEEE, 2011.

[309] W. Steiner, G. Bauer, B. Hall, M. Paulitsch, and S. Varadarajan. TTEthernet dataflow concept. In *NCA*, pages 319–322, 2009.

[310] W. Steiner and B. Dutertre. SMT-Based formal verification of a TTEthernet synchronization function. In *FMICS*, pages 148–163, 2010.

[311] W. Steiner and W. Elmenreich. Automatic recovery of the TTP/A sensor/actuator network. In W. Elmenreich, editor, *Proceedings of the First Workshop on Intelligent Solutions in Embedded Systems*, pages 25–37, 2003.

[312] W. Steiner and H. Kopetz. The startup problem in fault-tolerant time-triggered communication. *International Conference on Dependable Systems and Networks (DSN 2006)*, June 2006.

[313] W. Steiner and M. Paulitsch. The transition from asynchronous to synchronous system operation: An approach for distributed fault-tolerant systems. In *Proc. of the International Conference on Distributed Computing Systems*, pages 329–336, 2002.

[314] W. Steiner, M. Paulitsch, and H. Kopetz. The TTA's approach to resilience after transient upsets. *Real-Time Syst.*, 32(3):213–233, 2006.

[315] K. Steinhammer. *Design of an FPGA-Based Time-Triggered Ethernet System*. PhD thesis, Technische Universität Wien, Institut für Technische Informatik, Treitlstr. 3/3/182-1, 1040 Vienna, Austria, 2006.

[316] K. Steinhammer, P. Grillinger, A. Ademaj, and H. Kopetz. A Time-Triggered Ethernet (TTE) switch. In *Proc. of Design, Automation and Test in Europe*, Munich. Germany, March 2006.

[317] J. Stelzer. LIN bus emerging standard for body control apps. *EE Times Asia*, September 2004.

[318] S. Subbiah and S. Nagaraj. Issues with object orientation in verifying safety-critical systems. In *Object-Oriented Real-Time Distributed Computing, 2003. Sixth IEEE International Symposium on*, pages 99 – 104, 14-16 2003.

[319] Sunplus Technology Co., Ltd. LIN bus master note application using UART module. available at mcu.sunplus.com, 2006. V1.3.

[320] J. Swingler, J.W. McBride, and C. Maul. Degradation of road tested automotive connectors. *IEEE Transactions on Components and Packaging Technologies*, 23(1):157–164, March 2000.

[321] Systems Integration Requirements Task Group, Society of Automotive Engineers. *ARP 4754: Certification Considerations in for Highly-Integrated or Complex Aircraft Systems*, April 1996.

[322] Systems Integration Requirements Task Group, Society of Automotive Engineers. *ARP 4761 (Aerospace Recommended Practice) - Guidelines and Methods for Conducting the Safety Assessment Process on Civil Airborne Systems and Equipment*, December 1996.

[323] B. Tabbara, A. Tabbara, and A. Sangiovanni-Vincentelli. *Function/Architecture Optimization and Co-Design of Embedded Systems.* Springer Netherlands, 2000.

[324] C. Tanzer. TTPos - the time-triggered and fault-tolerant RTOS. In *Real-Time Magazine* 99-4, 1999.

[325] Time-Triggered Technology TTTech Computertechnik AG, Schönbrunner Strasse 7, A-1040 Vienna, Austria. *TTP-Load: The Download Tool for the Time-Triggered Protocol – Version 6.1.6*, 2004.

[326] Time-Triggered Technology TTTech Computertechnik AG, Schönbrunner Strasse 7, A-1040 Vienna, Austria. *TTP Bootloader: User Manual*, November 2005.

[327] K. Tindell and H. Hansson. Babbling idiots, the dual-priority protocol, and smart can controllers. In *Proceedings of the 1st Int. CAN Conference*, 1994.

[328] K. W. Tindell. Adding time-offsets to schedulability analysis. Technical Report YCS 221, Department of Computer Science, University of York, January 1994.

[329] K. W. Tindell, A. Burns, and A. J. Wellings. Allocating hard real-time tasks: an np-hard problem made easy. *Real-Time Systems*, 4(2):145–165, 1992.

[330] K. W. Tindell and J. Clark. Holistic schedulability analysis for distributed real-time systems. *Euromicro Journal on Microprocessing and Microprogramming (Special Issue on Parallel Embedded Real-Time Systems)*, 40:117–134, 1994.

[331] F. Tisato and F. DePaoli. On the duality between event-driven and time-drivern models. In *Proc. of the 13th IFAC DCCS*, Toulouse, France, 1995.

[332] Aviation Today. Parker selects TTTech for fly-by-wire system. Press release, July 2010.

[333] G. Torrisi, J. Notaro, G. Burlak, and M. Mirowski. Evolution and trends in automotive electrical distribution systems. In *Proc. of the IEEE Conference on Vehicle Power and Propulsion*, page 7, 7-9 2005.

[334] W. Townsley, A. Valencia, A. Rubens, G. Pall, G. Zorn, and B. Palter. Layer two tunneling protocol "L2TP." RFC 2661, Internet Engineering Task Force, August 1999.

[335] C. Trödhandl. Architectural requirements for TTP/A nodes. Master's thesis, Technische Universität Wien, Institut für Technische Informatik, Vienna, Austria, 2002.

[336] C.H. Tsai and C.W. Wu. Processor-programmable memory bist for bus-connected embedded memories. In *Proceedings of the Asia and South Pacific Design Automation Conference*, pages 325 –330, 2001.

[337] TTChip. *TTP/C Controller C2: Controller Schedule (MEDL) Structure – Document Protocol Version 2.1*. Schönbrunner Strasse 7, A-1040 Vienna, Austria, September 2002.

[338] TTChip Entwicklungsges.m.b.H. *TTP/C Controller C2 Controller–Host Interface Description Document, Protocol Version 2.1*, November 2002.

[339] TTTech Computertechnik AG, Schönbrunner Strasse 7, A-1040 Vienna, Austria. *TTPPlan The Cluster Design Tool for the Time-Triggered Protocol TTP/C*, April 2002.

[340] TTTech Computertechnik AG. *Time-Triggered Protocol TTP/C, High-Level Specification Document, Document Number D-032-S-10-028, Protocol Version 1.1*, 2003.

[341] TTTech Computertechnik AG. *TTX-AUTOSAR FlexRay Stack User Manual, Document Number D-110-G-70-006, Document Edition 4.3.1*, 2009.

[342] TTTech Computertechnik AG. *Interface Control Document HS-COM Layer, Document Number D-115-G-10-005, Version 0.1.1*, 2010.

[343] TTTech Computertechnik AG. *TTP-Build User Manual, Document Number D-001-G-01-002, Manual Edition 8.1.4*, 2010.

[344] TTTech Computertechnik AG. *TTP-Plan User Manual, Document Number D-001-G-01-003, Manual Edition 8.1.2*, 2010.

[345] Honeywell Tuscon. Design, implementation, and verification of fault-tolerant modular aerospace controls, Honeywell ncc-1-377. http://shemesh.larc.nasa.gov/fm/talks/Honeywell–TTTech.ppt, accessed August 2010, April 2003. Aviation Safety Program Single Aircraft Accident Prevention. Coop. Agreement NCC-1-377.

[346] Vector Informatik GmbH. Product catalog ECU software, page 80-81: CAN embedded LIN communication. available at www.vector.com, 2010.

[347] P. Veríssimo, L. Rodrigues, and A. Casimiro. CesiumSpray: A precise and accurate global time service for large-scale systems. *Real-Time Systems*, 12(3):243–294, May 1997.

[348] D.D. Davidson and V.Y. Chiu. Fail-operational global time reference in a redundant synchronous data bus system. Patent Application US 2005/0102586 A1, Honeywell, May 12, 2005.

[349] C.J. Walter, M.M. Hugue, and N. Suri. *Advances in Ultra-Dependable Distributed Systems*. IEEE Computer Society, 10662 Los Vaqueros Circle, Los Alamitos, CA 90720, January 1995.

[350] H.F. Wedde and W. Freund. Harmonious internal clock synchronization. In *12th Euromicro Conference on Real-Time Systems*, pages 175–182, Informatik III, Dortmund University, Dortmund, Germany, June 2000. IEEE Press.

[351] N. Weininger and D.D. Cofer. Modeling the ASCB-D synchronization algorithm with SPIN: A case study. In *Proceedings of the 7th International SPIN Workshop on SPIN Model Checking and Software Verification*, pages 93–112, Springer-Verlag, London, UK, 2000.

[352] J. Welch and L. Lynch. A new fault-tolerant algorithm for clock synchronization. *Information and Computation (*formerly *Information and Control)*, 77(1):1–36, 1988.

[353] J. Widder. Booting clock synchronization in partially synchronous systems. In *DISC*, pages 121–135, 2003.

[354] A.T. Winfree. *The Geometry of Biological Time*. Springer Verlag, New York, 2001.

[355] B. Witwer. Developing the 777 airplane information management system (AIMS): a view from program start to one year of service. *Aerospace and Electronic Systems, IEEE Transactions on*, 33(2):637 –641, April 1997.

[356] www.softing.com. CAN, CANOpen, DeviceNet. Website, August 2010.

[357] J. Zhang. Improved on-line process fault diagnosis using stacked neural networks. In *Proc. of the International Conference on Control Applications*, pages 689 – 694, vol.2, 2002.

[358] W. Zheng, J. Chong, C. Pinello, S. Kanajan, and A. Sangiovanni-Vincentelli. Extensible and scalable time triggered scheduling. In *Fifth International Conference on Application of Concurrency to System Design, 2005. ACSD 2005*, pages 132–141, 2005.

[359] W. Steiner and G. Bauer. TTEthernet: Time-triggered services for Ethernet networks, 28th Digital Avionics Systems Conference, IEEE, 2009.

Index